MANAGEMENT SCIENCE – THEORY AND APPLICATIONS

SALES DYNAMICS

THINKING OUTSIDE THE BOX

MANAGEMENT SCIENCE – THEORY AND APPLICATIONS

Additional books in this series can be found on Nova's website under the Series tab.

Additional E-books in this series can be found on Nova's website under the E-books tab.

SALES DYNAMICS

THINKING OUTSIDE THE BOX

RAJAGOPAL

Nova Science Publishers, Inc.
New York

LIBRARY OF CONGRESS CATALOGING-IN-PUBLICATION DATA
Rajagopal, 1957-
Sales dynamics : thinking outside the box / Rajagopal.
p. cm.
Includes index.
ISBN 978-1-61728-776-3 (hardcover)
1. Selling. 2. Sales personnel. I. Title.
HF5438.25.R35 2010
658.8'1--dc22
2010025966

Published by Nova Science Publishers, Inc. † New York

CONTENTS

FOREWORD

I have just had the honor of being among the first to read Dr. Rajagopal's book "Sales Dynamics, Thinking Outside the Box". It was immediately clear to me that Dr. Rajagopal has brought his years of academic and practical experience in sales to each of the topics and pages of his book. Throughout the three hundred pages he has integrated his knowledge with the research of others to provide not only a thorough but insightful reading. This makes the book a must read not only for those in academic institutions studying sales but also for those the corporate world who are out there each day facing the challenges of selling.

Sales and the company's perception of the sales force have experienced significant change. Following the Second World War companies of the industrial nations of the word (and particularly the unbombed United States) were run by engineers who lived in a rather happy demand driven economy. Make things and consumers come was a way of life. It was really not until the 60s with the advent of the Japanese, armed with market research and insights into consumer needs and wants, that this ideal world was shaken. With that change came a shift in the role of the sales force. Once just viewed as just the last link in the distribution system they were now evolving into a group that was challenged to sell things to people who had choices as the Japanese came to the market with both knowledge of products and knowledge of consumers. Over the decades since what has been largely missing has been the role that sales can play in interfacing between consumers and firms as direct sources of product and environmental information. Written in a clear and highly organized manner the book offers dynamic insights through text, research and best practices studies into how that role can assist firms to organize people and information for successful operations.

Sales and sales texts are often envisioned in a rather direct how to organize and how to sell context. This book is true to its title and presents a "thinking outside the box" perspective to the reader. It has been fifteen years since Waters and Peterman made the flexible organization popular in academic classes, consulting circles and business discussions. Unfortunately, in truth the idea were not conceptualized to an extent that they became very useful at the more operational levels of business. What Dr. Rajagopal has done is to provide a road map to take the rather abstract idea of thinking outside the box and present it in a manner that can be put to practice to improve not only sales but companywide performance. What has been accomplished is that the sales function is viewed, as Dr. Rajagopal puts it, as, "a complex phenomenon that encompasses tasks of managing buyers, buying institutions, market competition and performance." What this recognizes and is stressed through the ten chapters is that sales is not just a means of placing products and services in consumer's hands but it is a vital source of competitive information, knowledge of consumer needs and wants and feedback on product modification and pricing. This approach is clearly missing in other writings on the sales function. The importance of this cannot be overstated in today's international and highly competitive environment. His material is as practical to those in the new nations entering the world's markets as it is to those countries that have learned sales lessons the hard way over the past decades.

Kip Becker, Ph.D.
Chairman,
Administrative Sciences Department,
Boston University

Dr. Becker is the Chairman of Administrative Department, Boston University and Editor of *Journal of Transnational Management*. He also serves on the editorial board of the Journal of *Advances in Competitiveness Research, Scientific Journal of Administrative Strategic Outsourcing, The Journal of Teaching in International Business, Journal of Business and Information Technology* He is a member of the board of the International Management Development Association and principal of his own international management consulting firm. He is a prolific researcher, and published numerous books and research papers in international journals of repute. Dr. Becker teaches courses in electronic commerce, international business, marketing, and strategy.

PREFACE

Selling is an art largely associated with the behavioral skills of the sale personnel of a sales organization. In a competitive marketplace selling is performed using scientific methods of product presentation, advertising and various approaches drawn to take the customer into confidence. A firm begins to sell its products in a competitive marketplace and thrives continuously on acquiring new customers, launches new product lines or services in order to gain competitive advantage, retain the existing customers, enhance customer value, and gain competitive lead in the market. Increasing market competition has revisited the Darwinian concept of struggle for existence, making firms of all types and at the brink of competition.The principal constituents of sales dynamicscan be illustrated in reference to SCOPE that includes strategy, customer, opportunity, people and energy. People engaged in selling products are the firm's vanguard which encompasses sales representatives at the bottom of the sales pyramid.To compete in a dynamic and interactive marketplace environment, firms must transform their focus from just selling the products and services to value added sales management in order to maximize customer lifetime value and encourage repeat sales. Hence firms should ensure that products and services offered by salespeople must be made subservient to customer relationships. The new generation sales management strategies need to be focused not only on enhancing the volume of sales but also serve customers for generating long-term customer loyalty. The customer loyalty can be generated by adopting new performance metrics, and bringing the corporate image under the umbrella of customer centric vision, including quality of customer service rather than concentrating on augmenting the volume of sales by somehow pushing products [1]. Thus in the contemporary times sales signifies more than achieving volume of turnover of product and services, to a firm that is aiming towards sustainable growth in the competitive marketplace.

As a marketing tool, Sales has always had an important and legitimate role in creating consumer excitement and in clearing surplus or obsolete stock. However, in recent years sales has become ubiquitous and pervasive for business growthin competitive marketplace. Sales effectiveness is developed through cost-control and customer value augmentation process. It has been observed that selling process has changed over time and consequent upon increase in market competition due to fast penetration of global firms, most firms hasadopted customer centric selling process. Changing market conditions have tempted regional sales firms also to identify consumer preferences towards sales and services. However, some personal selling methods are successfully marketed on a worldwide basis with very little local adaptation [2]. Researchers and management educators have begun integrating the consumer behaviorand selling process with the management and executive sales programs.

Sales management is not as simple as it appears by definition. Selling is a complex phenomenon that encompasses tasks of managing buyers, buying institutions, market competition and performance. Hence, it is necessary for the sales oriented firms to develop effective selling strategies and put them into action by proper implementation. However, prior to developing sales strategies it is a must for managers to know what sales strategyis and how it affects performance. The discussions in this book attempt to provide answers to many such overlooked questions by illustrating the sale management process outside the box and provide new insights. Developing a comprehensive conceptualization of sales strategy, and examining its impact on sales force and firm performance have been addressed in this book in reference to sales force design, managing sale territories, industrial selling, account management, sales force automation, recruitment and compensation. Sales strategy can be made operational as a multidimensional construct in reference to customer segmentation, targeting, customer prioritization, framing relationship objectives, developing selling models, and selling through new routes to market [3]. As the knowledge on marketing-sales interface expands, there is a greater need to understand the specific aspects of marketing-sales configurations in reference to direct and business-to-business selling. Using in-depth perceptional models of salespeople and interface with clients, this book presents a dynamic, evolutionary spectrum of contemporary sales configurations for competitive advantage. These configurations are described in detail in terms of sales force structure, decision patterns, information sharing, sales partnering and behavioral and outcome performance. Sales organizations must be valued as a crucial source of critical market intelligenceand sales managers must demonstrate to salespeople how their market feedback contributes to the firm's strategic activities. Firms must leverage the market intelligence that salespeople possess and drive sales force activities towards developing a string of buyer-seller relationship [4].

Customer-centric sales strategyhas emerged as an effective tool for companies to augment their performance in global competition. In this process, businesses firms are learning more about their customers, preferences of industrial clients, strengthening consumer loyalty, and acquiring new customers who are defecting from less user-friendly competitors. There are many cognitive dimensions involved in the sales function that spans from analyzing consumer behaviorto the economic gains on goods and services. Consumer personality traits are determined by multi-dimensional factors such as the individual's behavior, appearance, attitude, and beliefs that affect the selling process. Sales territory management not only envelopes the geographic dimensions that influence the performance of salespeople, but also demographic attributes that influence consumers to develop conviction in buying [5]. Effectively managing the sales function is a prerequisite for success in business. Hence, practitioners are increasingly interested in improving the performance of their sales function. Fortunately, advances in the field of selling and sales management hold the promise of enhancing the scope of in-depth learning for managers and within marketing academia [6].

Reviewing extensively the previous researches on various issues ofselling and sales management, discussions in this book are focused around cognitive, operational and performance related dimensions that support developing new insights on improving the overall growth of sales firms. Drawing on a wide range of best practices, author has offered several managerial strategies that will help firms improve sales performance. Efficiency in delivering sales and serviceshas been driven by globalization as well as by growth of technology over time but fundamentals of sales has not radically changed. Conventional wisdom among salespeople still overrides the sales automation process in some cases.

However, the principal task in sales activity calls forthe ways tomaximize their relationship benefits in reference to acquiring customers, settling price, offering convenience, anddelivering post-service. Though virtual shopping is observed to be a fast growing channel, the Internet has just added another layer of convenience to customers being presented in an innovative manner [7].

This book examines how salespeople can improve their skills in emerging markets by understanding the market, consumer culture and consumption that help in developing values based sales process in the competitive marketplace. The discussions in the book are woven around typology of sales, client portfolio management, selling process mapping, recruiting, training and deploying sales people to drive effective results, territory management, and compensation planning.There is an increasing interest among the students and managers to pursue regular status/non-status courses related to sales management in reference to emerging markets, at the level of undergraduate and graduate programs offered by the universities in North and South America, the Caribbean, and Europe. Studies on sales management in reference to emerging markets have become a continuous learning process. There is not enough work done on strategic and tactical sales in Latin America from the perspective of management learning. This book reviews categorically the previous contributions on the subject and discusses new concepts related to efficiency and effectiveness of selling strategies in business-to-consumers and business-to-business environment.

Unlike text books on sales management, this book is a research work developed upon extensive review of literature and models based on empirical research motivations. There are over seven hundred references cited in various chapters of this books which would not only provide sources for future research but also guide mapping the progress of thoughts in selling process and sales management. Discussions in the book are divided into ten chapters encompassing topics related to various issues on sales management and are exemplified with contemporary company strategies and market research done in reference to different countries. This book significantly contributes to the existing literature and serves as a think tank for students, researchers and business managers. This book also serves as principal text to the under-graduate and graduate students who are pursuing courses on sales management.Hence, undergraduate and graduate students of major business and economics schools in American continent and Europe and Asia would be the potential audience of this book.

The book has a novel feature of building sales models in reference to the managerial experience of successful companies and strategic thinking evidenced in extensive survey of literature. The mini casesof multinational companies have been illustrated throughout the book to support the sales models and effective managerial strategies. All concepts are presented in a simple and easily accessible graphic format. The book provides critical analysis of the broad spectrum of important topics on sales process and management that are not covered in standard text books for the students. Throughout the book, the focus is laid on providing implementable concepts and metrics for salespeople and managers that helps in improving professional skills.

This book has emerged as an outgrowth of my teaching course on advanced selling system in the graduate program at EGADE Business School of Monterrey Institute of Technology and Higher Education (ITESM). Initially, I had worked out a research agenda on sales management and based on empirical studies a dozen of working papers were electronically published. The research work so posted on Internet attracted discussions among peers and some of the working papers were revised and published in refereed international journals. Such refined work has been thought of exhibiting in this book endorsed with applied illustrations and updated research on sales management. I am thankful to various anonymous referees of my research works on sales management issues who helped me to look deeper into the conceptual gaps and improve the quality considering their valuable comments. I sincerely thank my colleagues at Monterrey institute of Technology and Higher Education, Mexico City Campus and students who extended their cooperation in conducting empirical research studies on selling process and performance management topics. Finally, I express my deep gratitude to my wife Arati Rajagopal who copy edited the manuscript and stayed in touch till the final proofs were cross checked. She has been the light of the spirit in carrying this comprehensive work.

Rajagopal
April, 2010

REFERENCES

[1]   Rust, R. T., Moorman, C., andBhalla, G. (2010), Rethinking Marketing, *Harvard Business Review*, 88 (1), 94-101.

[2]   Macquin, A., Rouziès, D.,and Prime, N. (2001),The Influence of Culture on Personal Selling Interactions. *Journal of Euromarketing*, 9(4), 71-88.

[3]   Panagopoulos, N. G., and Avlonitis, G. J. (2010), Performance implications of sales strategy: The moderating effects of leadership and environment, *International Journal of Research in Marketing,* 27(1), 46-57.

[4]   Biemans, W. G., Makovec Brenčič, M., and Malshe, A. (2010), Marketing-sales interface configurations in B2B firms, *Industrial Marketing Management,* 39(2), 183-194.

[5]   Rajagopal (2008), Point of sales promotions and buying stimulation in retail stores, *Journal of Database Marketing and Customer Strategy Management, 15 (4), 249-266.*

[6]   Alzola, L. M., and Robaina, V. P. (2010),The impact of pre-sale and post-sale factors on online purchasing satisfaction: A survey, *International Journal of Quality and Reliability Management,* 27(2), 121-137.

[7]   Maruca, R. F. (1999), Retailing: Confronting the Challenges that Face Bricks-and-Mortar Stores, *Harvard Business Review*, 77 (4), 159-168.

Chapter 1

FRAMEWORK OF SELLING

Selling is an art largely associated with the behavioral skills of the sale personnel of a sales organization. In a competitive marketplace selling is performed using scientific methods of product presentation, advertising and various approaches drawn to take the customer into confidence. A foreign firm begins to sell its products to export markets by switching from domestic market and launches new product lines or services to gain competitive advantage. Sales personnel in international business can be classified in two ways - by performance or by their nationality. According to the task they perform, there are three categories of selling tasks that include sales generation, sales support, and missionary work. Generating sales is the creative task of helping the customer to make a purchase decision. Missionary work is undertaken by a salesperson to stimulate demand to help the distributors.

Globalization has revisited the Darwinian concept of struggle for existence, making firms of all types to stand at the brink of competition.With worldwide sales topping over 100 billion, the scale of selling exhibits greater understanding of its principal constituents wider SCOPE that includes Strategy, Customer, Opportunity, People, and Energy. People engaged in selling products are the firm's vanguard which encompasses sales representatives at the bottom of the sales pyramid. The sales representatives are instrumental to their firms' success and growth and stay dynamic in the sales force. The determinants for the success in sales largely constitute developing appropriate strategies for right products and right customers, selecting the right individuals, maintaining their motivation, developing the appropriate skills, and providing high perceived value and supply [1]. The job of salespeople may be described in reference to:

- Selling function
- Working with others
- Servicing the product
- Managing information
- Servicing the account
- Attending conferences and meetings
- Training and recruiting
- Entertaining

- Traveling
- Distribution

As the race of companies in global marketplace is getting tougher, many are choosing to develop customer centric selling strategies against the market centric approaches. Customer centric selling aims at developing pro-customer strategies to improve the value propositions on buying and deliver complete experience of products or services to the customers. Learning about customers and experimentation with different segmentations, value propositions, and effective delivery of services associate customer in business and help frontline employees acquire and retain customers with increasing satisfaction in sales and services of the firm [2]. The companies engaged in sales and services of high value-high technology goods like hybrid automobiles need to explore new modes of cooperation among customers, retailers and manufacturers resulting from co-design which leads to a customer-centric business strategy. Co-design activities are performed at dedicated interfaces and allow for the joint development of products and solutions between individual customers and manufacturers [3]. The shifts in the paradigm of selling practices and various attributes that affected the corporate philosophy of selling during post-globalization period are exhibited in Figure 1.1.

Organizations seeking to adopt a more customer-focused strategy will learn from the approach DuPont. It took in grappling with this challenge, based on an extensive program of qualitative and quantitative research with customers around the globe. The customer touch-point analysis of the organization facilitated alignment of functional groups within the organization (product, sales, customer service, etc.) and equipped them to deliver on newly developed, segment-specific value propositions. This major initiative has enabled DuPont to reprioritize internal efforts and business practices and has been a catalyst for broader organizational changes notably the dissolution of many functional silos that previously had hindered its ability to deliver against its brand promise [4].

Bottled water sector of consumer products has performed comparatively higher that the soft drinks within the beverages industry. In Latin America, although consumption rates and industry maturity between countries vary significantly, bottled water has likewise outperformed all other soft drink categories. Mexico has been the largest market in Latin America. The sales of bottled water took off in the market with high speed for the reasons of poor quality drinking water, which pushed the sales to the highest per capita consumption in the world with an average of 170 liters drunk by each citizen in 2002. In Mexico, the bottled water market is characterized by necessity where consumers also use bottled water for cooking and washing food in contrast to Brazil, where bottled water sales were spurred by the trend towards wellness and the fad among high income groups to drink bottled water through on-trade channels as an alternative to carbonates. The Venezuelan and Colombian markets showed similar characteristics sales led by the fashion trend with consumers opt for bottled water either as a style statement or as a move towards healthy living. However, in these two countries bottled water still remains a comparative luxury, given the low purchasing power [5].

A strong market oriented strategy of the firm alleviates the possibility of using coercive influence strategies by the competitors and offers advantage to the customers over competitive market forces [6]. Market orientation is an organization-wide concept that helps explain sustained competitive advantage. Since many manufacturing firms have linked their marketing strategies with services delivery attributes, the concept of market orientation is expanding as a system in global corporate settings. The process of market orientation contributes to continuous learning and knowledge accumulation by an organization which continuously collects information about customers and competitors and uses it to create superior customer value and competitive advantage [7].

Salesperson plays different roles in the selling process and his role changes over the course of action. He need to possess different abilities in each stage of the selling process, which include identifying prospects, gaining buy-in from potential customers, creating solutions, and closing the deal. Both the salesperson's job behavior and psychological well being can be affected if there are perceptions of role ambiguity or conflict or if these perceptions are inaccurate. There is a good deal of evidence, for example, that high levels of both perceived ambiguity and conflict are directly related to high mental anxiety and tension and low job satisfaction. In addition, the salesperson's feelings of uncertainty and conflict and the actions taken to resolve them can have a strong impact on ultimate job performance [8]. Salespeople in a firm are influenced by sales drivers to reach high outcome performance in a given region and time. Sales drivers include territory design, compensation, scope of assigned task and cultural interaction in the market. Salespeople in the more effective sales units display higher levels of intrinsic and extrinsic motivation, sales orientation, and customer orientation. Both behavior of salespersons and outcome performance were rated higher by managers in the organizations with more effective sales units [9].

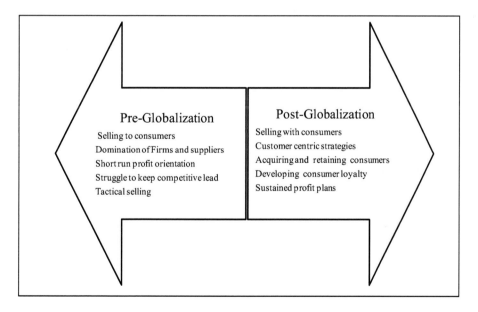

Figure 1.1. Paradigm Shiftin Selling Strategies.

Salespeople should understand that a customer is sensitive to his demand and time. Prospecting customers is an art and value based approach towards target customers would lead towards acquiring customers successfully. On the contrary the eight words for a customer- "let me get back to you on that" may kill the sales as response time to customer was inordinately delayed and customer value for time is not respected by the salesperson. In competitive consumer markets, building and maintaining a good relationship with customers is essential to long-term business survival. In selling process relationship quality, trust, satisfaction and commitment of salespeople are instrumental in prospecting and closing the deals and such attributes help not only in acquiring customer but also retaining him delivering sustainable customer value [10].

PROFILE OF SALESPERSON

A strong salesperson performs pro-customer actions and wastes little time waiting for the customer's attention. He does not take refuge in the notion that some days are good and some are bad and nothing can be done about the bad while weak salespersons lean towards waiting for the customer and shrug off a string of unsuccessful calls and continue calling [11]. However salesperson of either type has to attend to administrative work of his company inevitably. Some tasks of office administration performed by the salespeople include:

- Call reports and files
- Monthly sales reports
- In-side sales, service, shipping
- Forecasting.

Non-selling activities comprise traveling, accompanying superiors, conducting service calls and attending to paperwork. It is observed that about 40 percent time of salespeople is spent on internal activities including management reporting and internal meetings. Firms can evaluate the time-spread of salespeople on sales and non-sales activities by asking the following questions:

- How much time are they spending in meetings with customers addressing new sales opportunities? (These meetings can be telephone or face to face meetings.)
- How many of these meetings do they have a week?
- How much time are they spending preparing for these calls?
- How many of these meetings are with new prospects (within current customers, or with new customers)? How many of these meetings are with current customers?
- How much time are they spending in meeting with current customers on customer service/satisfaction issues? These are important meetings, but too much time could be an indicator of problems.
- How much time are they spending in internal meetings not related to new sales opportunities? What are these meetings accomplishing?
- How much time are they spending doing internal reports? Who is asking for the reports? Are they really necessary? Are they being used?
- How much time are they spending in meetings? What meetings? Are they appropriate?
- How much time are they spending in training?
- How much time are they spending in travel? To/from customers, other travel?

It is observed that finding salespeople with all required attributes of a high performer is an up-hill task for the firms. Sales managers at many companies have depended on top performers to achieve corporate goals though a very small number of salespeople at the bottom of the pyramid to stay competitive. Successful salespeople need to take scientific approach in performing their tasks by developing tactics in response to changing markets and customer preferences. Salespeople should have the zeal to span out to new customers in innovative ways and thrive for increasing productivity. Sales managers who encourage a scientific approach to augment effectiveness of salespeople should drive the bottom line strategies systematically matching the right products with the right customers. Salespeople would also benefit from the applications of the sales process automation and autonomy in making decisions on the sales negotiations. However the performance of salespeople is largely backed by both sales culture in the organization and internal processes. The overall profile of a salesperson should show the capability and competence in increasing the productivity and possibility of exponential growth in volume of sales, contribution to the revenue of firm, lifetime value of customers and number of accounts [12].

Sanofi-Aventis is a pharmaceutical company which encourages salespeople to work in teams and enhance their productivity. The company emphasizes that each sales professional is responsible for establishing key partnerships and for providing accurate and comprehensive information regarding the company and products, thus furthering its commitment to improving health and offering greater access to vital medicines worldwide. The sales force operates in a competitive market environment determining product and customer strong portfolios employing values-based selling techniques to gain competitive advantage. Primary health care sales professional of the company serve as local experts in their marketplace, being extremely knowledgeable about their designated territory, target consumers, and assigned products including competitors to those products. Salespeople attend over eight calls a day each consisting of pre-call planning, driving brand messaging, providing visual aids and/or reprints, and closing the sale. Sales Professionals also execute promotional programs to key customers in set timeframes, and they attend all company, regional and divisional meetings, as well as medical education programs [13].

Salespeople should acquire contemporary knowledge, strategies and action orientation by observing the market. It is important for them to get out on the street and study competitors, visit their stores or the locations where their products are offered, analyze the location, customer volumes, traffic patterns, hours of operation, busy periods, prices, quality of their goods and services, product lines carried, promotional techniques, positioning, product catalogues and other handouts. The salespeople of company should know the customers and update business information. The principal sources of market information that may help salespeople to attune with the prevailing practices include:

- Competitor learning models
- Neighboring business experiments
- Trade suppliersbehavior
- Downtown business associations
- Trade associations
- Trade publications
- Trade directories.

It is important for the company to consider how well competition satisfies the needs of potential customers in the selected markets, as well as to determine the best fit of the firm's strategies in the determined market segment or niche. Should the firm offer a better location, convenience, better price, later hours, better quality, and better service, may be the further interest of the firm to gain the competitive advantage. Most salespeople are employed in various kinds of retail selling and this would facilitate them to acquire information on sales trends existing in the market. These jobs involve selling goods and services to ultimate consumers for their own personal use like door-to-door selling agents, insurance agents, real estate brokers, and retail store clerks do. A much larger volume of sales, however, is accounted for by industrial selling - the sale of goods and services at the wholesale level [14].

SALES EFFECTIVENESS

Selling is a dominant factor in the growth of a business organization. As a marketing tool it has always had an important and legitimate role in creating consumer excitement and in clearing surplus or obsolete stock. However, in recent years the sales has become ubiquitous and pervasive in emerging firms as it plays significant role in contributing to business growth. Sales effectiveness is developed through cost control and customer value augmentation process. The sales effectivenesslargely depends on various management practices, such as accurate demand forecasting, which the effective group uses more and aligned with the corporate objectives of the sale [15]. However, different 'sales' objectives require different marketing management practices. Major factors that affect the sales effectiveness and overall performance of sales include:

- Customer face time
- Productivity of salespeople
- Team communications
- Personal selling skills.

Customer face time may be understood as the span of interaction between the salespeople and customers during the process of negotiation. It is observed that the longer the interface time with customer the higher the confidence in selling process. In a call center agents may hang over the telephone call of a customer for reasonably long till a solution is reached and customer is satisfied. A call center can dramatically improve an organization's ability to serve its customers. It is found that selling skills of employees in call centers show persuasive, sustained and customer driven attributes and lean towards ensuring satisfaction and spending more time in face-to-face interaction [16].

Salespeople in a competitive marketplace are considered as human capital of the organization who directly contributes towards the generation of revenue. Returns on the salespeople should be accounted as their per capita productivity.Companies need to develop a system that allows sales managers to monitor and evaluate the human capital periodically on the basis of cost-productivity metrics both to predict individual performance and to guide organizations' investments in people. The principal drivers for measuring the productivity of salespeople include leadership practices, employee engagement, knowledge accessibility, workforce optimization, and customer relationship management competencies [17]. Consumer sovereignty is a growing concept which has emerged as the core driver of customer value. It is widely assumed that the customer satisfaction is a crucial variable in determining the long-run prosperity of a business. Since the sales force often has the maximum contact with the customer, it seems logical that their efforts toward satisfying the customer would be imperative. Salespeople can practice customer-oriented selling without fear of losing sales [18].Salespeople *learn by doing* over their tenure on the job and it is observed that the more time they spend selling a particular product, the more productive the sales effort. The firms have objective to maximize profits by optimizing the size of all sales force as well as their productivity. Some studies evidence that optimal sales force size increases with both sales productivity and the learning rate, and decreases with sales force costs (e.g., wage per representative), product production costs and consumer price sensitivity [19].

In communicating sales related policies to field salespeople, the environment that would be challenging for a manager is towards understanding and adapting to individual behavior of the salespeople. Managers also need to know the different cultural groups and develop communication with the salespeople accordingly. Japanese people believe in implicit communication with a thumb rule of implied is better than spoken and appreciate interdependence to work in teams. As regards keeping group communication on target and schedule, with effective listening, the American and Mexicans differ in terms of explicit communication style of the former and implicit communication style of the latter. The power axis for American may reflect on speaking independently while Mexicans like Japanese would reveal on interdependence with the other members of the team [20]. The American team members admire to have confidence, trust and commitment to work in a team while Mexican member rely on the facts more revealed from the antecedents and precedents than trust to carry on the team work. However, Mexicans like to achieve confidence once the project rolls-on, than measuring the confidence before beginning the team works [21].

Selling strategies for hygiene and cleanliness products are largely driven by the communication that drives consumer emotions. Getting rid of dirt, or merely its absence, by extension, is a good thing. "A clean bill of health", "a clean record", "clean sweep", or "good clean fun" evoke wholesome flawlessness, renewal or order. The act of washing, whether of a corpse in Jewish culture or of the hands and feet of the happy couple in a Hindi wedding ceremony, carries ritual symbolism in many cultures and faiths. Dirt-hounding has even become a spectator sport, judging from the success of prime-time reality-TV shows, such as Britain's "How clean is your House?" In this series, Kim and Aggie, the queens of clean, descend upon suspect houses armed with mops, scourers and wagging fingers. When swabs from surfaces reveal stratospheric levels of bacteria, up go howls of disapproval and in comes the industrial cleaning equipment. Salespeople describe that Bathrooms have become temples of relaxation. Mosaic tiles and sunken baths nod to Greek and Roman inspiration; organic bath essences promise spa-like regeneration. Americans still buy more bars of soap than Europeans each year. But, if liquid washes are included, Europeans today only just trail Americans in spending per person. The idea gained a safe and steady ground for salespeople. Unilever, for instance, had an advertising campaign for OMO, its laundry-cleaning brand, and entitled "Dirt is good!" The firm's market research showed that mothers were frustrated by the message that dirty clothes are bad. "We wanted to reposition dirt as an expression of freedom," explains Mr. Weed. Let children get dirty, goes the message, safe in the knowledge that OMO will clean up afterwards [22].

Emotional communication plays significant role in acquiring new consumers and retaining those who are associated with the firm. Managers of competing firms should provide training to the sales people on communication skills that generate cognitive influence among consumers. Emotions in selling can be developed through psychological and socio-cultural associations and meanings of products, company, perceived and prescribed use value, and keywords used during the negotiation process. Effective sales communication in a cross-cultural selling perspective is considered as the strongest bridge between salespeople and consumers. As cultural values, sales objectives and desired customer relationship levels influence the consumer value, it is argued that an emotional communication in selling process is imperative for acquiring and retaining consumers and develops sustainable sales strategies for competitive advantage [23].

Most global companies are leaning towards personal selling practices to reduce the overhead costs. Personal selling is carried out by the self-employed individuals; these representatives are instrumental to their firms' success and growth. The success of direct selling practices depends on selecting the right individuals, maintaining their motivation, developing the appropriate skills, and providing high perceived value and supply. With an independent sales force framework, it is imperative for direct selling firms to implement skill development programs towards prospecting, negotiation, and customer relations, which help in finding and keeping strong individual sellers [24]. Persuasion is another important skill that needs to be acquired by the salespeople. The process of persuasion can be effectively administered by establishing credibility of salesperson, framing concepts to find common ground, providing vivid evidence, and connecting emotionally with the prospect. Credibility of salespeople grows out of expertise and relationships. The former is a function of product or process knowledge and the latter a chronological map of listening to and working in the best interest of others. But even if a salesperson's credibility is high, his position must make sense--even more, it must appeal--to the audience. Therefore, a salesperson must frame his position to illuminate its benefits to everyone who will feel its impact. Persuasion then becomes a matter of presenting evidence--but not just ordinary charts and spreadsheets [25].

KNOWLEDGE SHARING

There are two preconditions for increasing sales effectiveness that include acquiring knowledge on products, process, competition, and consumers by salespeople and sharing knowledge of salespeople among consumers. There are many organizations that tend to take full advantage of their own as well as of market knowledge on preferences and value perceptions of consumer. Competing sales oriented firms should develop processes to encourage knowledge sharing among consumers as knowledge is power and salespeople can deliver best ideas among consumer to reap the biggest rewards. The secret of making knowledge sharing as core organizational competency lies in flipping the liberal culture around so that sharing becomes central to salespeople. Over long term sharing knowledge would help in improving selling effectiveness. Hence selling knowledge should constitute:

- Breaking islands of customer, product and competitive information
- Route to market and navigation

- Validity of information.

Often consumers develop perceptional bias against brand, company, shopping outlets, sales people and services offered by the company. A prolonged biasness on a particular or multiple perceptions would cause mind block of the consumers. Such cognitive dimensions result into low access to consumers and poor response on various prospecting stages during sales process. Mind-block and perceptional response are interrelated cognitive determinants. It has been observed that creative salespeople are not common in business organizations. Hence, companies employing smart salespeople within the available employees use patching, a process of mapping and remapping thought process to develop positive consumer perception by sharing contemporary knowledge, which is highly focused on products or services for sale and tightly aligned with consumer preferences. Such sales strategies driven by sharing knowledge to develop positive consumer perceptions can respond to selling dynamics in a competitive marketplace [26]. There are many factors that affect the self-perception on buyer-seller relationships. It is observed that younger consumers have more positive self-perceptions, but concordance with peers is often conflicting and elderly people get the advantage of such attribute by just interacting with the salespeople. Salespeople may observe such difference among the two generations of consumer-younger and older in age and experience. The perceptions of the fairness in selling processes, is often considered an antecedent to mutual trust developed between buyer and seller [27]. Consumer mind-blocs that stand as barrier in prospecting can be diffused by the salespeople with effective leadership styles. Re-generating self-consciousness through effective sales leadership impel the consumer involvement in the selling process turning confidence level high. Supportive, participative, and achievement-oriented leadership styles of salespeople motivate consumers and help in diffuse the predetermined mind –blocks against the products or services of a firm [28].

Many multinational companies have worked towards developing new formulae in skin nourishing creams and anti-agers in Italy with a better impact on wrinkles, and have focused on pushing premium brands. This made it hard for private labels to follow, as private label manufacturers lack the required research budgets. Health and beauty retailers, para-pharmacies, drugstores and beauty specialist retailers continue to gain ground in Italian marketplace at the expense of other distribution formats, such as supermarkets and hypermarkets, which are losing share. Specialist channels are mainly strong in color cosmetics, fragrances, skin care and baby care. Sales of cosmetics through the Internet continue to grow, although this channel is still underdeveloped in Italy. Nevertheless, Italian consumers are gradually gaining trust in on-line purchasing, particularly in terms of payment methods and safety, while delivery times have been reduced over the last few years. The direct selling channel is also growing, confirming Italians' preference for distribution channel which offers assistance in buying [29].

Route to market is the multi-channel retailing strategy that caters to the wide preferences of shopping to the customers at varied price, delivery and services options. A company may set more than one routes for its products and services to facilitate consumer convenience in buying. It is necessary that sales people should know these routes to markets and prospect consumers at their most convenient route. In multi-channel selling strategy salespeople can prospect consumers for offering superior products, typically accompanied by superior service outputs, to be sold at higher prices for premium market segment while low price strategy is followed for mass market retail locations [30]. However, luxury goods are not commonly sold through the catalogue, e-bays or call centers and differentiated products usually need relatively more intermediary support to be delivered satisfactorily to the end customer. However, urban shoppers incur higher search costs when searching for a product across retailing channels and gathering information on prices as the urban shoppers are more guided by the value for money considerations in shopping. It is observed that price-sensitive customers always intend to strike a beneficial deal over the costs they incur during searching for such bargain through various channel options [31]. A route to market is a distinct sales process followed by salespeople towards prospecting customers for a selected product or service through a specific market channel. Agents of call centers, representatives of e-bay, and personal selling representatives engaged in prospecting consumers may be assigned to manage sales in different routes to market. Globalization and innovative selling practices have introduced multiple channel selling strategies to improve customer satisfaction and strengthen customer-retailer dyadic loyalty [32].

Salespeople should know one underlying fact in prospecting consumers that all information that is delivered by the salespeople is validated by the consumers sooner or later. Hence it is argued that whatever information is given by a salesperson to consumers, whether verbal or printed, he delivers a promise beyond just information. Firms should know that potential payoff for using valid evidence is even greater when it comes to managing effective sales to consumers. At the same time, however, indirect information delivered by sales people is much slower and less effective because of tacit knowledge. Thus, all information that is provided to consumers should be evidenced based [33].

PIPELINE VISIBILITY

Winning a consumer by successfully passing through all stages of selling is a complex task as consumer perceptions and convictions are unpredictable. Managing pipeline visibility enables the company to: grow the top line without growing the sales force headcount, achieve a culture of high performance selling, "photocopy" its best reps, and supercharge sales results by optimizing quotas, territories and incentives. Pipeline visibility is usually a function of CRM that helps the salespeople view the opportunities available in the market. Having pipeline visibility will enable the organization to see the opportunities and sales leads more clearly, this will save time in briefing about the different options and activities available to serve these opportunities. It enables the sales force to have a clear vision of the demand and predict the required leads needed to meet annual targets. It gives a clear knowledge about the sales productivity and to organize all the company's functions and business units to provide maximum output. Managing pipeline visibility enables salespeople to see opportunities from beginning to end, enables them to manage closure, see bottlenecks and discover inadequate coaching practices in order to improve these. There are different stages of reaching to prospects and develop social network enhancing the pipeline visibility. The role of salespeople at different converging points of pipeline visibility path is exhibited in Figure 1.2

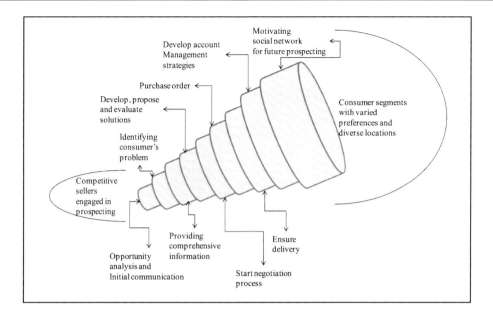

Figure 1.2.Pipeline Visibility in Selling Process.

Although pipeline visibility strategy for salespeople is complex, it might bring with it a hard realization for some consumers who look for quick value for money. Thus salespeople should develop skills to see the contours in the pipeline and find solution to overcome performance gap in the selling process. Salespeople should be trained in conducting gap analysis to identify the weaknesses in the selling process [34]. The pipeline visibility can be managed by salespeople by considering the following factors:

- Need to see opportunity from beginning to end
- Need to manage closure
- Need to see bottlenecks
- Adequate coaching.

Following the pipeline visibility process is the right way for direct salespersons as well as sales force management systems towards prospecting new customers. In any step of the sales process prospects drop out of it, and from the large number of initially interested persons on the narrow end of orders only a fraction of the initially interested people remain and actually place an order. In order to retain the sales of a large account the sales rep must be able to manage this pipeline visibility and be able to respond in every step of the process. To generate traffic the sales force can: develop on site seminars, create community events, give agents incentives to create traffic and leads, make direct contacts, and use mail and phone.

UNDERSTANDING CONSUMERS

Effective selling can be achieved by working closely with consumers. It is necessary for salespeople to understand the buying behavior of consumers to prospect them with confidence. At the onset of the selling process, salespeople should identify the actual decision makers. It is argued that purchasing power does not necessarily establish perfect correlation with the level of decision makers. All buyers act selfishly, but they sometimes miscalculate. However, moving ahead with the selling process salespeople need to collect and apply psychological intelligence. It is necessary for salespeople to make sure that sales calls are highly productive and informative, listen to the sales force, and reward rigorous fact gathering, analysis, and execution to help managers increase sales effectiveness [35].

There are as many as eleven determinants to the consumer behaviorthat influence buying process. These factors include economic, relational and personality led factors which affect the consumer psychology in making buying decisions, getting associated with the product and developing loyalty towards the product, brand or a retail store. The determinants affecting the buying behavior of consumers are exhibited in Figure 1.3.

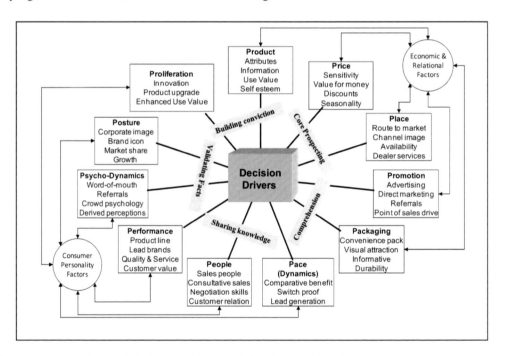

Figure 1.3.Determinants of Buying Decision and Sales Process Integration.

Consumer satisfaction is perceived to be a key driver of long-term relationships between salespeople and consumers, especially when consumers are well acquainted with products and markets, and when industries are highly competitive. Sales efficiency is one of the principal factors which influence consumer satisfaction in a business-to-consumer and consumer-to-consumer context that help laying buyer-seller dyadic relationship. The key sales performance indicators (SPI) include effective communication, cross-functional teams, and seller-supplier integration that are followed to develop long-term relationships. It is widely acknowledged that consumer satisfaction leads to higher performance of sales and contributes to revenues while relationship between consumer satisfaction levels and quality of consumer services influences acquisition of new consumers as well as retaining the existing consumers [36]. The sales of high technology products are positively associated with performance of sellers and suppliers towards enhancing consumer service quality, growth in sales and increase in market share. Sales and services elasticity is widely recognized as a critical component in achieving competitive advantage in the marketplace and improving corporate reputation to augment consumer value without escalating sales costs and time overrun [37].

Effective sales strategies have a direct bearing on possessing the relative brand equity and growth of the business. Sales strategies are the directional statements and that need to be converted into the step-by-step plan of action for effective plan implementation. The strategic sales directions have four options that can be expressed by 4As - arena, advantage, access and activities. The arena may be defined as competitive prospecting of the target consumers or key accounts through an appropriate scale of information flow, advantage appropriation and customer relations. Marketplace arena is challenging ground for the salespeople to show their performance and establish their lead among competitors. For example, traditionally, sales force of multinational companies target consumers in premium segment while a plethora of domestic companies look for prospecting low end consumer segment, often unprofitably [38]. The key selling factors for high effectiveness and result orientation are broadly identified as detailed below:

- Listening skills
- Follow-up skills
- Ability to adapt sales style from situation to situation
- Tenacity– sticking with a task
- Well organized
- Verbal communication skills
- Proficiency in interacting with people at all levels of a customer's organization
- Demonstrated ability to overcome objections
- Closing skills
- Personal planning and time management skills.

Need for the competitively advantageous strategies may further be justified as a large number of firms are increasingly productive in reference to the rapid diffusion of the technologies. The consumers' bargaining power also works out to be an instrument to either broaden or narrow the differences between the competitors. The companies that use intermediaries are often encountered with balancing the power of distribution and delivery of services. In consumer markets the retail trade is forcing major concessions on the multi-national brands. Such strategies hold the access to the retail network through a long chain of channels. Conventionally the choice of appropriate scale in business and scope thereof were guided by the concepts of *the bigger is better* and *umbrella control of activities*. In the current era of globalization the decentralization of activities and production sharing have become more effective tools in marketing. The profit centre approach (PCA), control circles and total quality management practices has endorsed the success of small integrated units operating in a well defined market. In view to promote the PCA concepts and maintain the control circles, the large companies are increasingly creating the autonomous, small and entrepreneurial units to find responsive solutions to the consumer problems in the well defined market niches (Frederick 1989). Corporate structures are changing in order to accommodate the concept of PCA and control circles and are exploring for the long term advantages by way of heavy investment to develop the core competencies.

> BMW, Honda, and Toyota, among other companies, begin with a strong brand that imparts sales momentum to each model. Brands that are weak—because their products have acquired a reputation for shoddy workmanship, their designs are not evocative, or their models bear little relationship to one another—cannot pursue this top-down approach. But a company stands a good chance of selling more cars and, step by step, of rehabilitating the brand if managers take pains to match each model to the consumer segments most likely to be interested in it, identify and overcome the obstacles that keep browsers from becoming purchasers, and emphasize both the functional and the process and relationship benefits of the model in question. BMW Direct is an initiative of BMW (GB) to help selected company car fleet buyers streamline their service for employees. BMW Direct is a web based fully personalized, car configuration and ordering system for the purchase of new BMWs. This highly efficient rules based web application delivers a level of information previously unavailable outside of a showroom. The BMW Direct solution provides users with the ability to view details on all eligible cars online and then go on to configure them against a full menu of accessories. BMW Direct is truly 'CRM' compliant, providing two-way communication via automated alerts and e-mails and incorporating a Contact Centre to ensure immediate access to trained product advisors. Users can track online the status of their individual orders whether by web, phone, fax or email. The call centre functionality includes phone and e-campaign generation, consumer enquiry handling and profiling to customized promotions [39]. Post-sales support is delivered using a thin client solution, (using Citrix) to BMWs contact centre in Croydon and order management centre in Bracknell in UK.

The technological changes are the main impetus behind new market opportunities. The extent of such change may be explained from super technologies to the appropriate and intermediate technologies. The strategic choices have wide ranging ripple effects through the organization that determine the key success factors and growth performance. Some companies would be making right strategic choices by improving the implementation process of competitive advantages. These companies are guided by the shared strategic vision and are driven by the responsive attitude towards the market requirements. They emphasize the continuous strive to satisfy the consumers. A strategic vision in managing markets may be understood as the guiding theme that explains the nature of business and the future projections thereof. These projections or business intentions depend on the collective analysis of the environment that determines the need for new developments or diversifications. The vision should be commissioned on a concrete understanding of the business and the ability to foresee the impact of market forces on the growth of business. The vision will motivate the organization for collaborative business planning and implementation. The powerful visions are also the statements of intent that create an obsession with winning the organization. The business strategy broadly incorporates the following dimensions:

- Consumer needs
- Consumer segments
- Technology and resources

- Activities in the value added chain.

Strategic thrust has a significant magnitude and direction in sailing the business though turbulent situation. The factors associated with the competitive advantage and business investments uphold the strategic thrust to achieve the business objectives though the positive channel efforts. The competitive advantage may be assessed in reference to the superior consumer value and lowest delivered cost. Such combination of the strategies may be termed as competitive superiority that explains cost effective delivery strategy to enhance the consumer value. An overall edge is gained by performing most of the activities at a lower cost than competitors. This would enable the company to optimize its cost of delivery of the new products and simultaneously enhance the consumer value to up-hold the strategic thrust of the company.

Selling is a complex process as it moves though various cognitive stages of decision making among consumers and often remains unpredictable of buying decision. There are four cognitive stages though with consumer is driven by the salespeople, which include:

- Awareness
- Comprehension
- Conviction, and
- Action.

Salespeople invest a reasonably long time with consumers in generating awareness about the products its advantages and the company. Comprehension on the products, services and customer relations is delivered by the salespeople during the process of responding to the questions of prospecting consumers and clarifying their doubts hindering the decision making process. Comprehension in the selling process would generate higher level of confidence among consumers and develop their inclination towards making buying decision. This stage is defined as conviction and is often very delicate because at this juncture consumers borrow time to validate the information given by the salespeople and acquire second opinion from their nears and dears. Consumers also use their self - reference criterion to validate information and review their initial decision. One of the weak points that often strike down the efforts of salespeople is the lack of follow-up to watch the decision movements of consumers at this stage, which may cause escape of prospecting consumers without any notice. In the last stage of action consumer physically acquires the product evidencing the successful selling. Adhering to the core set of values salespeople should ensure the effective use of resources and develop a competitive advantage in selling process to generate action. By blending selling process through four different stages as discussed above, firms may develop lead in selling products and services to consumers in a competitive marketplace [40].

FUNCTIONAL EFFICIENCY OF SALESPEOPLE

Salespeople are considered as creative managers in their field of specialization. They not only sell products or services but also help consumers in offering:

- Solutions

- Information
- Ideas
- Service

Successful salespeople find ways to reach their consumers' problems, offer solutions and motivate them to buy solutions not products. They do this by identifying a thorny issue in the search of consumers for an appropriate solution and develop an original, compelling point of view about it. Smart salespeople pitch this point of view to a carefully chosen line executive in one crucial meeting and then prove its worth with a short diagnostic study. Such provocation linked solution selling leads to significant success in increasing the volume of sales as well as enhancing business opportunities [41]. Salespeople should provide information to the consumers on the products they sell and also generate awareness on alternatives of their firm's product line. Such attitude helps building consumer confidence and also opens them up towards creative thinking. Creative consumers lean to adapt, modify, or transform their decision on buying based on offering of salespeople representing new business opportunities. Central to business is the need to create and capture value, and creative consumers demand a shift in the mindsets and business models of how firms accomplish both [42]. Salespeople play multifaceted roles in business of a firm that includes:

- Solve problems
- Represent the company
- Communicate with customers
- Develop relationships, partnerships, alliances
- Discover needs
- Gather information
- Educate customers
- Catalyze change
- Help people buy
- Serve customers
- Treat people with respect.

Behavior of a salesperson is an important predictor that significantly affects organizational sales results in developed countries. There were many studies conducted in the past, which evidenced that the relationships found in developed countries are also relevant in developing countries. The external and internal environment of a company and personality traits affect of salespeople together influence the buyer-seller relationships. Internal communication and the choice of a control system especially affect ethical decision making. It has been observed that informal internal communication affects the personality traits while the process of monitoring influences the ethical climate of the salespeople. Ethical climate and salespeople's personality traits also affect the ethical decision making during the prospecting, negotiation, closing the deal and post sales services process [43]. There is a significant impact of organizational culture on the values of salespeople which in turn affect the performance of their tasks. The determinants of organizational culture, conflicts and values that drive the behavior of salespeople are exhibited in Figure 1.4.

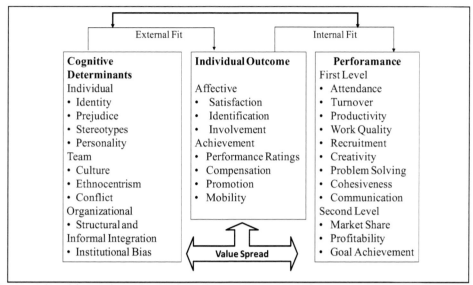

Source: Adapted from Rajagopal and Rajagopal, A. (2008), *Team Performance and Control Process in Sales Organizations, Team Performance Management- An International Journal, 14 (1), 70-85.*

Figure 1.4. Culture, Conflicts and Value Spread among Sales people.

Some management studies have found significant relationships between personal and organizational variables such as job experience, closeness of supervision, performance feedback, influence in determining standards, span of control and the amount of role conflict and ambiguity perceived by salespeople, which affect their performance in the selling process. Other studies relate personal characteristics to variations in motivation by showing that salespeople's desire for different job-related rewards (*e.g.,* pay, promotion) differ with such demographic characteristics as age, education, family size, career stage, or organizational climate [44]. Salespeople in a firm are influenced by sales drivers to reach high outcome performance in a given region and time. Sales drivers include territory design, compensation, scope of assigned task and cultural interaction in the market [45].

TYPES OF SELLING PRACTICES

In the global marketplace, selling practices are largely determined in the context of buyer behavior and competitor strategies. Regarding the task they perform, there are three categories of selling tasks - sales generation, sales support, and missionary work. Generating volume of sales in the competitive marketplace is the creative task of helping the customer to make a purchase decision. It is concerned with after-sale service. Missionary work is undertaken by a salesperson to stimulate demand to help the distributors. There are four categories of sales which include every type of sales position, as listed below:

- Consultative sales
- Technical sales

- Business-to Business sales and
- Direct sales
- Bottom of the Pyramid sales.

Performance of salespeople is also determined by the measures of cost effectiveness adopted by them during the selling process. Sales force consumes a hefty share of the budget of company. If salespeople sell effectively, the company gains both profits and customers. Yet, in many organizations, salespeople cannot give their best on the job because the company mismanages them. One of the most risky propositions in selling faced by the salespeople is when some firms send conflicting messages about job priorities of the salespeople. Firms expect salespeople to be proactive in prospecting customers and at the same time adhere to management control practices that often stall the sales process. Under such circumstances salespeople often fail to satisfy both the firm's goals and customers. Some companies don't coordinate salespeople's efforts with other customer-facing teams, such as store clerks and Web site administrators [46].The efficiency of sales depends up on the type of salesperson and the techniques used in selling goods or services. The consumers are the decision makers in buying the goods and services proposed for sales and hence play key role in the sales process. These selling categories require a specific management approach to deal with the market.

Consultative Sales

Consultative sales may be defined as the selling of specific goods or services to technical organizations. The consultative type of companies keep accounts of their clients and assign the sales force to attend to specific client accounts. Computer systems management, structural planning of factories, human resources management, capital market services and the like may be listed in the category of consultative sales. Consultative sales skills of salespeople are used to solve customer problems and provide value based solutions. A consultative sales approach process permits the salesperson to develop association with the prospective customer or client with an emphasis on a benefit to gain or a loss to avoid value proposition that they may not have considered. Consultative skills techniques focus on enhancing the relationship by leveraging the credibility gained by the supplier through knowledge and performance. The key element in successful execution of the consultative sales approach technique is to focus on cost justifying the solution. Understanding what the prospect or client perceives as value is an important element for salespeople towards knowing how to enhance the value solution so the prospect cannot refuse to accept the offer. Consultative selling provides the salespeople with the ultimate product to sell known as customer profit. As customer profit becomes the product, salespeople who offer 'benefits' or even 'solutions' often quit the prospects [47]. Consultative Selling has proven to be one of the most valuable sales tools. It has provided with a common language for the personal communications with customers, and clearly focuses on their real issues of financial results.

Since performing sales for goods and services of this category demand high skills, they require a low-key, low-pressure approach by the sales personnel. Consultative sales require comprehensive knowledge of the product or services and the user orientation. A successful salesperson in this field needs to have great confidence and an experience of successes achieved periodically. Of all sales types, the consultative salesperson is probably the most professional and demands the highest compensation for his skills and abilities.

Technical Sales

Product knowledge, its application, relevance to contemporary technological development and sales skills are the essential characteristics needed to be present in salespeople engaged in the sale of goods and services of technical products. The dominant industries in this field are electronics, engineering products of all branches, medical equipments and the like. The buying influence is generally drawn from a professional manager. The personal sales in this category are limited. However, organizational sales of technical goods and services are higher compared to the former. The compensation given to the sales force of this category is often linked with performance apart from a low salary. Technical sales process brings variety back into the business system, product by product, and carefully forecasts the resulting impact on sales as well as the cost implications across the value chain. When the analysis shows that the costs rise to overwhelm the added revenues, salespeople should be able to drive innovation fulcrum by selling through technology based prospecting strategies. By deconstructing their companies to a zero-complexity baseline, managers can break through organizational resistance and deeply entrenched ways of thinking to find the right balance between innovation and complexity [48].

Business - to - business Selling

Selling products or services to a business is not complex like selling to consumers as the selling process is followed through an established system. However, selling products to industrial clients often face setback with pricing and volume of order issues. In this process, while some companies simply cave into price pressure, others try to create and capture more value through sales approaches such as enterprise selling, key-account management, or solutions selling. Regardless of the operational problems in business - to - business selling process, buyers and sellers involve building highly collaborative relationships building opportunity for future transactions [49].

Inter-organizational dynamics not only relates to learning, competence development, or adaptation, as suggested by other studies, but also to how such dyadic relationships are governed [50]. Such developments in the business may also escalate switching costs. Consequently, the channel switching costs increase the bargaining power of the manufacturer or service provider in negotiations for the second contract and to preempt improved transactional relationship in a given business environment [51]. Although communication, trust and satisfaction are always important as determinants of elements, their importance is higher when "core" products are considered. When selling to a business, it is vital to know who the key decision maker is within that business. This decision maker may be an individual, or in the case of a company's board, a group of people. The greatest mistake a salesperson can make is to spend all of their time selling themselves and their product to a person within the business who has very little input into the actual decision. When selling to a business, it is vital to know who the key decision maker is within that business. This decision maker may be an individual, or in the case of a company's board, a group of people. The greatest mistake a salesperson can make is to spend all of their time selling themselves and their product to a person within the business who has very little input into the actual decision [52].

In business - to - business selling process firms generally encourage team selling. Product based sales force generally belongs to a company and functions as a team. Examples may be cited of the service for computer industries products in the market. To market such a device, a group of professional sales and service engineers are required to boost-up the market for the principle product as well as for the service packages of the ancillary unit. The major steps for effective selling in any category of the sales organization are listed below:

- *Prospecting and qualifying* - collecting basic information about consumers, references, joining a prospect's organization, initiating communication
- *Negotiation* - about the orders, delivery, pre- and post-sales service, organizational benefits and the like
- *Approach* - the sales package
- *Presentation and Demonstration* - convincing the buyer
- *Problem solving* - resolving conflicts pertaining to all Ps and As
- *Delivering* - closing the deal by satisfying the consumer
- *Follow-up* - post-sales service and maintaining of relationship.

Direct Selling

A large sales force is engaged in the commercial sales of consumer goods and services. These sales are very much individual-consumer oriented. The sales personnel operate as sales representatives of the company or as facilitators of an in-house retail store. The major task of the sales representatives is to collect orders from consumers or organizations. In performing direct sales, the representatives attempt to generate an emotional appeal and this motivates the buyer. It has been observed that the turnover rate in direct sales require a strong persuasive ability and skills to identify a customer's buying motives. Direct salespeople receive compensation in terms of commission on the volume of sales. Direct selling is a cost effective strategy for a firm that generates personalized prospecting and servicing the consumers with innovative ideas. The difference between having good ideas and getting them implemented often comes down to a keen sales mindset. Not only does selling have a logical component, the delivery of objective information to consumers has also a strong affective component, which emerges out of direct selling approaches. After all, salespeople are not selling just to the company but also selling to individuals with specific concerns and motivations [53]. Thus direct selling is the process through with consumer-company bonding is established and such relationship on one hand builds loyalty and makes company sustainable on the other. For a successful direct selling, transparency needs to be maintained on the following issues by the salespeople:

- Tell consumers who they are, why they're approaching him and what products they are selling,
- Promptly start and end a demonstration or presentation at consumer's request,
- Provide a receipt with a clearly stated cooling-off period permitting the consumer to withdraw from a purchase order within a minimum of three days from the date of the purchase transaction and receive a full refund of the purchase price,
- Explain how to return a product or cancel an order,
- Provide consumers with promotional materials that contain the address and telephone number of the direct selling company,
- Provide a written receipt that identifies the company and salesperson, including contact information for either,
- Respect consumer privacy by calling at a time that is convenient,
- Keep confidential and secured the private information of consumers,
- Promptly end a demonstration or presentation at your request,
- Provide accurate and truthful information regarding the price, quality, quantity, performance, and availability of their product or service,
- Offer a written receipt in a language consumer can understand, and
- Offer a complete description of any warranty or guarantee.

Avon Cosmetics is one of the largest direct selling companies in the world. All over the global marketplace, sales and service of Avon's products depend on the entrepreneurial efforts of independent sales representatives. The sales network of the company runs over five million direct sales representatives reaching extensively into the community. The sales representatives of the company also use e-selling through virtual shop on Internet to provide convenience to the consumers. Direct sale of Avon products are carried out to families and friends in neighborhoods, colleges and workplaces. The company insists that prospective sales representative to fit into the direct selling framework should exhibit the behavior that evidences enjoying meeting people in their neighborhood or workplace and hold sufficient time to work. Avon provides adequate knowledge on products and market, selling skills, and promotional support to the direct sales representatives to make their businesses thrive. These attributes of sales representatives are further supported by the company through global advertisement and high brand equity of its products. Sales representatives are also benefitted by the guidance of sales meetings, training sessions, and personal support from their District, Zone or Area Sales Managers. The potential of earnings of direct sales representatives depends on their ability to provide reliable and courteous service to customers. They can broaden their sales skills by training as a Avon Beauty Advisor as in many markets, the company conducts leadership programs that can help sales representatives developing their business and making it bigger. By introducing others to the benefits of selling for Avon, a sales representative can build and mentor his own down line of new representatives [54].

Direct selling has to be followed by the salespeople with strong ethical values to ensure customer value and corporate reputation of the firm. The Direct Selling Association has developed the code of ethics for salespeople engaged in this type of selling process that addresses to both consumer and the seller. It ensures that salespeople will make no statements or promises that might mislead either consumers or fellow salespersons.

Bottom of the Pyramid Sales

Global companies are targeting business in the rural and semi-urban markets consisting of large consumer communities with small consumptions based on personal brand relationships, with local institutions retailers or distributors of global brands in the region. Urban culture and consumerism have emerged when the semi urban marketplaces near rural markets are modernized, experienced the dissolution of the conventional patterns of buying preferences among consumers within the proximity of semi-urban marketplaces. Bottom of the pyramid consumerism, which has emerged as a result of globalization, has brought into force two separate rural or semi-urban behaviors among consumers. Such market taxonomy has influenced the core behavior of consumers [55].

The Bottom of the Pyramid (BoP) market segment which constitutes large number of small consumers has become the principal target of most of the consumer brands emerging from multinational firms. It has been argued that companies penetrating at the bottom of the pyramid of market should provide constancy and agility at the same time. Constancy is required if the brand strategies of the firm is to build awareness and credibility while agility in the brand builds perceived values among consumers while agility is required if the brand is to remain relevant in a free marketplace [56]. Competing firms penetrating in the BoP market segment are largely influenced by the consumption needs, promotions, lifestyle and societal indicators that affect consumer behaviorin relation to purchasing featured brands of up-stream markets. The BoP brands develop acquaintance and familiarity of consumers with the firm and buying behavior of consumers towards the acquainted brands which is referred as brand association. Consumers have only one image of a brand, created by the deployment of the brand assets at their disposal: name, tradition, packaging, advertising, promotion posture, pricing, trade acceptance, sales force discipline, customer satisfaction, repurchases patterns, etc. Clearly, some brand assets are more important to product marketers than service marketers, and vice versa. Some competitive environments put more of a premium on certain assets as well. Quality and price do not exist as isolated concepts in consumers' minds and are interrelated [57]. The Firms intending to penetrate in BoP market segment need to focus on a profit-oriented approach in order to access commercial advantage; inculcate repeat buying behavior among consumers, boost volume of buying by standardizing products, measuring brand trial effects, and should run on low price strategies. Such strategies of brand penetration in the BoP market segment scale up customer loyalty, which ultimately improves the brand association of not only the clients themselves but also helps the brand to lead in the mass market [58].

The convergence of three forces in managing BoP brands consisting of customers, manufacturers and retailers affects mother brands of firms. These forces are put-up with dual responsibility to serve customers of premium and regular market segments as well as cater brand resources to customers in the BoP market segment and provide competitive gains to the firms at the lower market end. Buying decision of customers largely depends on the BoP brand environment wherein customers evaluate the level of satisfaction and generate customer-to-business pull, in case of satisfaction being significantly high [59]. Although the per capita response to sales may be lower in the bottom-line markets, the aggregate buying power of customers is actually quite large, representing a substantial variety of goods and services. Since these markets indicate buying potential, there exists long-run sales advantage for large number of companies. Thus, managers must shift their thinking towards the bottom-line market which holds value of high-volume but high-margin businesses in the long run [60].

TEAM SELLING

'Team' conceptualizes group of people engaged in delivering a common task. In ideal situations the individual and group behavior in a team is integrated towards the common objectives and the task delivery process is shared which leads to set the group dynamics. The basic attributes of a good team include clear identification of goals, clarity of roles, common feeling, motivation, commitment and collaborative attitude. The efficiency of group approach is a function of many behavioral factors which may be expressed as [61]:

$$p = f(m, a, g)$$

Where, p denotes the degree of performance, m represents motivation, a exhibits abilities of the individuals associated with the team and g is expressed as realization of goals. The team may not function effectively if any of the above factors or associated variables thereof are incoherent. The reward and punishment issues in a team emerge as a post-process synergy of all associated variables and are largely governed by the common factors such as feeling, motivation, commitment and collaborative attitude [62]. In selling products or services, making strategic decisions, delivering solutions, or driving innovation, most companies engage team selling today to accomplish results. However, since the early 1990s, teams have evolved from more stable groups-where members were co-located, dedicated to a common mission. As teams have become more dynamic, substantial challenges are carried out by firms to provide traditional advice on team formation, leadership, roles, and process [63].Hence teams are collections of people who must rely upon group collaboration, if each member is to experience the optimum success and goal achievement.

Changing technology and markets have stimulated the team approach in multinational companies for performing the organizational tasks. Moreover, the complexity of the society and human needs prompted team work as a significant tool in managing the corporate tasks [64]. The SMART variables may be considered to administer sales teams which include-strategy orientation, measurability, approach, reality and time frame. The strategy orientation would drive the brainstorming discussion to result orientation and the measurability would count on the success of the deliberations. Teams which need to work within an organization and across functional activities such as sales, marketing, purchasing, personnel and finance find that team working fosters as collaborative tool rather than a competitive approach. It is important that terms of reference of teams must be laid clearly by the firm to make salespeople capable of doing the job [65].

Sales team design in reference to a particular territory is also considered as an important factor, which has received little analytical attention in the traditional literature, but which appears to be an important influence on the effectiveness of the sales operation. It has been argued that organizational effectiveness is determined by outcome performance and behavioral performance of sales team, as well as through systems control approach. Although conventional theory has suggested that behavior performance and outcome performance result from different stimuli, behavior-based control is positively associated with both behavior performance and outcome performance [66]. Mexican sales teams have the promising behavior intending to offer a pleasant, positive and rewarding scenario of the situation under discussion and the unpleasant consequences are kept undisclosed. The members of the sales team avoid any negativity in their conversation as far as possible. Sales team culture in general is embedded with 3-T power-grid comprising a synergy of task (commitment), thrust (driving team) and time (punctuality), which is weak in Mexican sales teams and affects their overall performance [67]. With the emergence of Team Selling, or Selling Centers, in complex business-to-business relationships, their effectiveness becomes a concern to sales managers.

Sales people of the organization strive to improve customer relationship and sales through customer satisfaction and effective project promotion. A strong commitment to innovation, excellence and sound business practices, and a philosophy of long-term investment in human and financial resources, enable the local facility centre developers to consistently attract and retain a fiercely dedicated, capable, performance-oriented workforce. Sales are driven by differences in beliefs about the valid scope and focus of activity, time focus, valid sources of knowledge, differences in perceived status, and the relationship with the business environment. Recent research investigating customer-oriented selling has indicated that greater attention needs to be focused on organizational or personal antecedents influencing customer-oriented selling behaviors [68]. The environment of sales department would include its goals, objectives, and culture, as well as the behaviors, beliefs, and attitudes of top and middle management. Expectations in a sales organization are transmitted to its salespeople through its corporate culture and environment, who then implement the policies and carry out organizational strategies.

Firms engaged in selling services work closely with each client to identify current needs and create innovative solutions to reach short and long-term investment goals and demonstrate value engineering attributes which allows its salespeople to achieve their development goals on or below budget. Changing technology and markets have stimulated the team approach in multinational companies for performing the organizational tasks. Furthermore the complexity of the society and human needs devised to meet augmented needs have endorsed the team spirit as a significant tool in managing the corporate tasks. The team management is employed largely in the organizations where activities are less repetitive and predictable. Such an approach demands effective liaison, appropriate delegation of powers, judicious allocations of roles of team members, sharing of information and accuracy in evaluation of team performance [69].

Team selling strategies can be effectively implemented if the firm organizes its team model considering the following points:

- Create a vision: It is necessary for a firm to ensure that team members should be put through a brainstorming session to understand their role, objectives, and work process. Sales teams should perceive that they have a stake in decision making and accomplishing results.
- Develop the strategies: To make this vision a reality, you need a playbook. Take your strategies and translate them into assigned responsibilities. Instead of pitting people against each other, give them tasks based on their strengths that complement each other. Focus on alignment, interdependence and teamwork, consistent with everybody's talents and aptitudes that they can use to benefit the whole team.

- Develop routines: No business runs on automatic pilot for very long without turning off and crashing. Salespeople should set the routine for prospecting and follow-up actions and divide the necessary duties down into specific habits for following them every day. It is observed that salespeople should develop a planned routine to carry out tasks to increase their productivity. Also working on a systematic routine would reduce competition among salespeople within the organization and allow a fair and ethical sales environment.

- Celebrate success: Salespeople should be rewarded for their accomplishments. Rewarding the sales force for growth would encourage them to look for different ways to prospect customers and get them in the buying process. The firm's approach to rewards forces salespeople to get more pleasure from the success through individual or team efforts. Rewards to salespeople can also be set through the seniority-based system that makes salespeople with the firm for the long haul. By developing such reward based strategy, both the salespeople and firm could reap the benefit [70].

- Focus on culture: Finally, firms should deploy sales people in teams and put them through contemporary roles and processes. The roles and processes of members of sales team should be clearly defined.

The selling skills are determined by the degree of experience of salesperson and effectiveness of training while the adaptive selling techniques are related with increased performance. Firms should be able to apply the time-based philosophy of revenue management to their sales teams. Managers should guide salespeople to manage their time and productivity. Hence, a different type of proposed measure, revenue per available salesperson hour, is proposed to better integrate the value of the salesperson's time as a factor in sales potential and revenue calculation [71]. The sales unit effectiveness is an overall evaluation of outcomes (*e.g.,* sales, profit contribution) of the sales unit for which the field sales manager is responsible [72]. Evaluating effectiveness is important since it provides an indication of the performance of the manager's sales unit.

SALES ENVIRONMENT

There are five essential qualities of aesthetic judgment of consumers during the selling process spread over different stages from prospecting to closing the deal. These attributes, including *interest, subjectivity, exclusivity, thoughtfulness, and internality* need to be nurtured among customers to develop conviction in buying. The quality of aesthetic judgment driven by aura and arousal on products and services, exercised by the customers in association with the salespeople, determines the extent to which product advocated by the salespeople would enhance quality of life [73]. Convergence of sales promotion, customer's perceptions, value for money and product features drive arousal among customers. The nature of customer-retailer relationship functions as the key in selling and buying process in reference to in-store promotions. However, in this process the perceptional problems with customers can greatly devaluate the customer-promoter relationship and brand as a whole [74]. Consumer appreciation of premium-based promotional offers is more positive when the premium is offered through an easy process and in combination of relatively lower quantity of products to purchase. It has also been found that when value of the premium is mentioned and brand perception is positive, compulsive buying tendencies are higher among customer [75].

> Grupo Financiero Banamex Accival SA de CV(BANACCI) provided banking and other financial services in Mexico during 2001-06. The services of the company included investment services to foreign and local companies and securities brokerage services in United States for investment on the Mexican Stock Exchange. The company also provided multiple banking operations including corporate and private banking, financial leasing of fixed assets, credit and debit cards, mortgage, financing, deposits, loans, foreign exchange, execution of trust agreements and insurance services. AEGON-USA was active in the Mexican market through joint ventures with Grupo Financiero Banamex Accival SA de C.V. BANACCI is the major player in the insurance market in Mexico that delivers the insurance products covering life, health, accident, retirement and saving benefits. BANACCI in collaboration with AEGON thrived to maximize the shareholder value in the domestic market by offering competitive sales and services to its customers. The branch offices of the bank promoted the sales of the insurance and associated products through augmenting the portfolio value and specialized financial services to the customers. The Mexican banks found two distinguished advantages in integrating the insurance products into the banking culture. Firstly, the brand name of the bank associated with the financial services developed confidence among the customers to derive longer benefits from the organization. Secondly, the brand name of the bank encourages the staff to quote the referral process through which the product is promoted. The bank paid greater attention to the clients of the key account segment. The premiums were relatively low in the BANACCI as compared to the other competing insurance companies. The life term insurance was offered for 5, 10 and 20 years period and the premium is fixed in monthly, quarterly, half - yearly or annual mode. The banking industry was bringing-in the independent brokers to distribute the insurance packages through their branch network that are automated under the turn-key projects in Mexico and other Latin American countries [76].

Arousal during the prospecting may be seeded among the consumers or clients by the salespeople through multifaceted information and activity that may be performed in various ways and embody different consumer feelings. It is also argued that there is a need to focus more on the influence of retail ambience on shoppers engaged in leisure shopping [77]. The three distinct dimensions of emotions, which include pleasantness, arousal and product attractiveness, have been identified as major drivers for making buying decisions among shoppers. The ambience of shopping malls whether pleasant or unpleasant moderates the arousal effect on satisfaction and in-store buying behaviors. Satisfaction in pleasant retail ambience where music, hands-on experience services, playing areas and recreation are integrated maximizes the consumer arousal. It has been observed that young consumers perceive positive effect on in-store behaviors if shopping arousal is high. Thus, retailers need to pay attention not only to the pleasantness of the store environment, but also to arousal level expectations of shoppers [78].

The impact of ambience developed by the salespeople during the prospecting can be measured in reference to the degree of stimulation and pleasure gained by consumers. Interactive tools on product learning provided in the retail stores significantly affect the level of arousal and pleasure which contribute towards experience, and thereby influence the buying behavior. As higher stimulation or interactive learning provided by the suppliers focuses on gaining initial experience on the product use, consumers tend to engage in higher arousing activities by acquiring the product [79]. However, salespeople at times fail to recognize what influences buyers' satisfaction and consequently do not develop effectively the prospecting ambience to stimulate buying decisions. Hence, they need to vigilantly manage the quality of arousal by developing adequate customer involvement in the buying process [80].

SELECTED READINGS

Harvard Business School (2007), *Strategic Sales Management*, Harvard Business School Press, Boston, MA.

Miller, W. (2009), *Proactive Sales Management: How to Lead, Motivate, and Stay Ahead of the Game*, AMACOM, American Marketing Association, Chicago, IL.

Yip, G. S. and Bink, A.J.M. (2007), *Managing Global Consumers: An Integrated Approach*, Oxford University Press, New York, NY.

REFERENCES

[1]   Crittenden, V.L. and Crittenden, W.F. (2004), Developing the Sales Force, Growing theBusiness: The Direct Selling Experience, *Business Horizons*, 47 (5), 39-44.

[2]   Selden, L. and MacMillan, I. C. (2006), Manage customer-centric innovation systematically, *Harvard Business Review*, 84 (4), 108-116.

[3]   Berger, C., Möslein, K., Piller, F. and Reichwald, R. (2005), Co-designing modes of cooperation at the customer interface: learning from exploratory research, *European Management Review*, 2 (1), 70-87.

[4]   Rajagopal (2009), Effects of Customer Services Efficiency and Market Effectiveness on Dealer Performance, *International Journal of Services and Operations Management*, 5 (5), 575-594.

[5]   MIlenkovic, Z. (2003), *Bottles water sales get Latin fever*, Euromonitor International on line, April 14.

[6]   Chung, J., Jin, B., and Sternquist, B. (2007), The role of market orientation in channel relationships when channel power is imbalanced, *The International Review of Retail, Distribution and Consumer Research*, 17 (2), 159-176.

[7]   Slater, S. and Narver, J. (1995), Market orientation and the learning organization, *Journal of Marketing,* 59 (3), 63–74.

[8]   Singh, J. (1993), Boundary Role Ambiguity: Facets, Determinants and Impacts, Journal of Marketing, 57 (2), 11-31.

[9]   Baldauf, A., Cravens, D.W., and Grant K (2002), Consequences of sales management control in field sales organizations: a cross-national perspective *International Business Review*, 11 (5), 577-609.

[10] Chang, H. (2007), Critical Factors and Benefits in the Implementation of Customer Relationship Management, *Total Quality Management & Business Excellence*, 18(5), 483-508.

[11]   Gellerman, S. W. (1990), Tests of a good salesperson, *Harvard Business Review*, 68 (3), 64-69.

[12]   Ledingham, D., Kovac, M. and Simon, H. L. (2006), New sciences of sales force productivity, *Harvard Business Review*, 84 (9), 124-133.

[13]   For details see Sanofi-Aventis Pharmaceuticals, corporate web site http://www.sanofi-aventis.us; also see Petitt, B. S. (2005), *Sanofi-Synthelabo and Aventis: The Birth of a National Champion (A)*, Discussion Case, Harvard Business School Press, Boston, MA.

[14]   Rajagopal (2007), *International marketing-Global environment, corporate strategies and case studies*, Vikas Publishing House, New Delhi.

[15]   Merrilees, B. and Fam, K. (1999), Effective methods of managing retail 'sales', *International Review of Retail, Distribution and Consumer Research*, 9(1), 81-92.

[16]   Adria, M. and Chowdhury, S. D. (2002), Making Room for the Call Center, *Information Systems Management*, 19(1), 71-80.

[17]   Bassi, L. and McMurrer, D. (2007), Maximizing your return on people, *Harvard Business Review*, 85 (3), 115-123.

[18]   Pettijohn, C. E., Pettijohn, L. S. and Parker, R. (1997), An Exploratory Analysis of the Impact of Salesperson Customer-Orientation on Sales Force Productivity, *Journal of Customer Service in Marketing & Management*, 3(4), 5-24.

[19]   Misra, S., Pinker, E. J. and Shumsky, R. A. (2004), Salesforce design with experience-based learning, *IIE Transactions*, 36(10), 941-952.

[20]   Haru, Y. (1997), *Different Games Different Rules*, Oxford University Press, New Cork, 54-55.

[21]   Harris, P. R. and Moran, R. T. (1999), Managing *Cultural Difference- Leadership Strategies for a New World of Business*, Huston, TX, Gulf Publishing Company, 106-273.

[22]   For details see The Economist (2009), The joy of dirt, *The Economist Print Edition*, December 17.

[23]   Aslam, M. M. (2006), Are You Selling the Right Colour? A Cross-cultural Review of Colour as a Marketing Cue, *Journal of Marketing Communications*, 12(1), 15-30.

[24]   Crittenden, V.L. and Crittenden, W.F. (2004), Developing the Sales Force, Growing theBusiness: The Direct Selling Experience, *Business Horizons*, 47 (5), 39-44.

[25]   Conger, J. (1998), Necessary art of persuasion, *Harvard Business Review*, 76(3), 84-95.

[26]   Eisenhardt, K. M. and Brown, S. L. (1999), Patching: Re-stitching Business Portfolios in Dynamic Markets, *Harvard Business Review*, 77(3), 72-82.

[27]   Hubbell, A. P. and Chory-Assad, R. M. (2005), Motivating factors: perceptions of justice and their relationship with managerial and organizational trust, *Communication Studies*, 56(1), 47-70.

[28]   Williams, D. J. and Noyes, J. M. (2007), How does our perception of risk influence decision-making? Implications for the design of risk information, *Theoretical Issues in Ergonomics Science*, 8(1), 1-35.

[29]   For details see Euromonitor (2009), *Cosmetics and toiletries in Italy*, Market Research Report, Euro Monitor, May.

[30]   Jindal, R. P., Reinartz, W., Krafft, M., and Hoyer, W. D. (2007), Determinants of the variety of routes to market, *International Journal of Research in Marketing*, 24 (1), 17–29.

[31] Rajagopal (2008), Point of sales promotions and buying stimulation in retail stores, *Journal of Database Marketing and Customer Strategy Management, 15 (4), 249-266.*

[32] Wallace, D. W., Giese, J. L. and Johnson, J. L. (2004), Customer retailer loyalty in the context of multiple channel strategies, *Journal of Retailing,* 80 (4), 249-263.

[33] Pfeffer, J. and Sutton, R. I. (2006), Management Half-Truth and Nonsense: How to Practice Evidence-Based Management, *California Management Review,* 48 (3), 77-100.

[34] Nelson, S. J. (2002), Dou you know what's in your pipeline leadership, *Harvard Business Publishing Newsletter,* May 01.

[35] Bonoma, T. V. (2006), Major Sales: Who Really Does the Buying? *Harvard Business Review,* 84 (7/8), 172-181.

[36] Rajagopal (2010), Bridging Sales and Services Quality Functions in Retailing of High Technology Consumer Products, International Journal of Services and Operational Management, 6 (5), in press.

[37] Oberoi, J. S., Khamba, J. S., Sushil, and Kiran, R. (2008), An empirical examination of advanced manufacturing technology and sourcing practices in developing manufacturing flexibilities, International Journal of Services and Operations Management, 4 (6), 652- 671.

[38] Orit, G., Philip, L., and Till, V. (2007), Battle in China's good-enough market, Harvard Business Review, 85 (9), 80-89.

[39] Rajagopal (2003), *Building Consumer Loyalty Through Relationship Networking : A Case of BMW Mexico,* Discussion Case, ITESM, Mexico City Campus, 1-16.

[40] Beverland, M. (2004), Brand Value, C, Flexibility, and New Zealand Wine, *Business Horizons,* 47 (5), 53-61.

[41] Lay, P., Hewlin, T. and Moore, G. (2009), In a downturn, Provoke your customers, *Harvard Business Review,* 87 (3), 43-47.

[42] Berthon, P. R., Pitt, L. F., McCarthy, I., and Kates, S. M. (2007), When Customers Get Clever: Managerial Approaches to Dealing with Creative Consumers, *Business Horizons,* 50 (1), 39-47.

[43] Verbeke. W., Ouwerkerk, C. and Peelen, E. (1996), Exploring the Contextual and Individual Factors on Ethical Decision Making of Salespeople, *Journal of Business Ethics,* 15 (11), 1175-1187.

[44] Chonko. L., Tanner, J. F. and Weeks, W. A.(1992), Selling and Sales Management in Action: Reward Preference of Salespeople, *Journal of Personal Selling and Sales Management,* 13 (4), 65-72.

[45] Rajagopal and Rajagopal, A. (2008), *Team Performance and Control Process in Sales Organizations, Team Performance Management- An International Journal, 14 (1), 70-85.*

[46] Anderson, E. and Onyemah, V. (2006), How right the costumer should be? *Harvard Business Review,* 84 (7), 58-67.

[47] Hanan, M (2003), *Consultative selling: The Hanan formula for high-margin sales at high levels,* AMACOM Publishers, New York.

[48] Gottfredson, M. and Aspinall, K. (2005), Innovation versus Complexity: What Is Too Much of a Good Thing? *Harvard Business Review,* 83 (11), 62-71.

[49] Hancock, M. Q., John, R. H., and Wojcik, P. J. (2005), Better B-to-B selling, *McKinsey Quartely,* June.

[50] Halldórsson, Á. and Skjøtt-Larsen, T. (2006), Dynamics of relationship governance in TPL arrangements – A dyadic perspective, *International Journal of Physical Distribution & Logistics Management*, 36 (7), 490-506.

[51] Mehmet, B. (2000), Switching Costs and Screening Efficiency of Incomplete Contracts, *Canadian Journal of Economics*, 33 (4), November, 1034-1048.

[52] Rajagopal (2009), *Buyer-supplier Relationship and Operational Dynamics, Journal of Operations Research Society, 60 (3), 313-320.*

[53] McFarland, J. (2004), Inside sales job, *Harvard Business Publishing Newsletter*, August 01.

[54] For details see Avon Products Inc., Corporate web site- http://www.avoncompany.com. Also see Tao, Z., Li, D., and Chan, I. (2007), *Future of Avon's China: Direct Sales, Retail Sales or Both*, Discussion Case, Harvard Business School Press, Boston, MA.

[55] Cruickshank, J. A. (2009), A play for rurality: Modernization versus local autonomy, *Journal of Rural Studies*, 25 (1), 98–107.

[56] Blumenthal, D. (2002), Beyond 'form versus content': Simmelian theory as a framework for adaptive brand strategy, *Journal of Brand Management*, 10 (1), 9-18.

[57] Rajagopal (2008), *Measuring Brand Performance through Metrics Application, Measuring Business Excellence, 12 (1), 29-38.*

[58] Akula, V. (2008), Business basics at the base of the pyramid, *Harvard Business Review*, 86 (6), 53-57.

[59] Rajagopal (2010), Conational Drivers Influencing Brand Performance among Consumers, Journal of Transnational Management, 15 (2), 186-211

[60] Rajagopal (2009), Managing Brands in bottom line markets, *Innovative Marketing*, 5 (1), 33-38.

[61] Rajagopal and Rajagopal, A. (2006), *Trust and Cross-Cultural Dissimilarities in Corporate Environment, Team Performance Management-An International Journal, 12 (7-8), 237-252.*

[62] Rajagopal (2006), Innovation and Business Growth through Corporate Venturing in Latin America: Analysis of Strategic Fit, *Management Decision*, 44 (5), 703-718.

[63] Cross, R., Ehrlich, K., Dawson, R., and Helferich, J. (2008), Managing Collaboration: Improving Team Effectiveness through a Network Perspective, California Management Review, 50 (4), 78-99.

[64] Dyer W (1987) *Team Building*, Reading, MA, Addison-Wesley, 20-23.

[65] McGreevy, M. (2006), Team working: How are teams chosen and developed?*Industrial and Commercial Training*, 38 (7), 365-370.

[66] Piercy, N. F. (2006), The Strategic Sales Organization, *The Marketing Review*, 6 (1), 3-28.

[67] Rajagopal and Rajagopal, A. (2006), *Trust and Cross-Cultural Dissimilarities in Corporate Environment, Team Performance Management-An International Journal, 12 (7-8), 237-252.*

[68] Martin, C. A. and Bush, A. J. (2003), The potential influence of organizational and personal variables on customer-oriented selling, *Journal of Business and Industrial Marketing*, 18 (2), 114-132.

[69] Harris, P. R. and Moran, R. T. (1999), *Managing Cultural Difference- Leadership Strategies for a New World of Business*, Huston, TX, Gulf Publishing Company, 106-273.

[70]   Zehnder, E. (2001), Simpler way to pay, *Harvard Business Review*, 79 (4), 53-61.

[71]   Siguaw, J. A., Kimes, S. E. and Gassenheimer, J. B. (2003), B to B Sales Team Productivity: Applications of Revenue Management Strategies to Sales Management, *Industrial Marketing Management*, 32 (7), 539-551.

[72]   Churchill, G. A., Ford, N. M., Walker, O. C., Johnston, M. W. and Tanner, J. E. (2000), *Sales Team Management* (6th ed.), Chicago, IL: Irwin.

[73]   Dobson, J. (2007), Aesthetics as a foundation for business activity, *Journal of Business Ethics*, 72 (1), 41-46.

[74]   Platz, L. A. and Temponi, C. (2007), Defining the most desirable outsourcing contract between customer and vendor, *Management Decision*, 45 (10), 1656-1666.

[75]   d'Astous, A. and Jacob, I. (2002), Understanding consumer reactions to premium-based promotional offers, *European Journal of Marketing*, 36 (11), 1270-1286.

[76]   Rajagopal (2003), Cross promotion strategy for insurance marketing: A case of Banamex-Aegon Mexico, Discussion case, ITESM, Mexico.

[77]   Backstrom, K. (2006), Understanding recreational shopping, *International Review of Retail, Distribution and Consumer Research*, 16 (2), 143-158.

[78]   Wirtz, J., Mattila, A. S. and Tan, R. L. P. (2007), The role of arousal congruency in influencing consumers' satisfaction evaluations and in-store behaviors, *International Journal of Service Industry Management*, 18 (1), 6-24.

[79]   Menon, S. and Kahn, B. (2002), Cross-category effects of induced arousal and pleasure on the internet shopping experience, *Journal of Retailing*, 78 (1), 31-40.

[80]   Miranda, M., Konya, L. and Havira, I. (2005), Shopper's satisfaction levels are not only the key to store loyalty, *Marketing Intelligence and Planning*, 23 (2), 220-232.

Chapter 2

PERSONAL SELLING

Personal selling is one of the customized promotional method in which a salesperson applies preferential skills and techniques for building personal relationships with consumers to drive buying decision. This approach of selling on one hand develops a cognitive ambience for prospecting and results in enhancing value on the other. In most cases performance of the salesperson is realized through the nature of the sale order obtained, financial rewards on the sales achieved, and developing customer relationship for exploring future opportunities while the customer's value is realized from the benefits obtained by consuming the product. However, preparing a customer to purchase a product is not always the objective of personal selling but also is conducted to deliver business information. Personal selling involves nurturing contacts through face-to-face meetings or via telephone conversation, besides newer technologies like text messaging on mobile phones and face book used to manage the customer relationships. Effective sales management should be able to achieve a balance among defining the role of personal selling, deploying the sales force, managing the accounts, and understanding the selling costs. However, as sales depend on the consumer, the overall concern of the firm should be oriented towards managing the customer first, not the sales force [1]. Personal selling is considered as an effective tool because salespeople can:

- Probe customers to learn more about their problems,
- Adjust the marketing offer to fit the special needs of each customer,
- Negotiate terms of sale, and
- Build long-term personal relationships with key decision makers.

Most multinational companies have found that personal selling has stronger impact on consumer value than the selling through a store located in shopping mall. It is observed that through building personal relationships and through research on cognitive determinants of consumers salespeople can develop elaborative files on existing and potential customers on various indicators including education, family, particular interests, and life-style, besides business information. The goal of personal selling is to focus on the individual across the table and such attention to detail requires well-trained, alert salespeople [2]. Changing economic conditions in the global marketplace and increasing sales competition in emerging markets have driven firms to identify customer centric activities such as personal selling, to facilitate individual decision makers. Some personal selling methods are successfully marketed on a worldwide basis with very little local adaptation. Personal selling interactions are largely influenced by both consumer and organizational culture and the relationships between cultural factors and the dynamism of salespeople determine the consumer satisfaction during the selling process [3].

Personal selling approach is also reflected in building consumer confidence by selling solutions to their problems. Firms engaged in manufacturing and marketing high-technology products have also switched to personal selling strategies to highlight the importance and potential contributions of the sales force, especially in business-to-business selling activities. The traditional sales force strategies that are associated with high-technology products are mostly supply driven while in order to enhance the success of high-technology products and services, firms need to be more demand driven in their sales structures such as personal selling. It is imperative that high-technology firms adopt solution selling practices, which can be implemented to enhance high-technology adoption as well as enhance competitiveness of the firm [4].

Acorn International which was established in 1998 is creating infomercial magic in the marketplace and selling products directly to the consumers through virtual hop. The company operates as one of China's largest direct selling companies in terms of revenue and air time buying for television direct sales business. The company telecasts programs for selling hawking wares that include cell phones, consumer electronics, learning devices, and health and wellness items through direct sales. The direct sales advertisements that are telecast by the company are on average five to ten minutes in length on about 50 channels. These advertisements are telecast on local, nationwide China Central Television (CCTV), and satellite networks. In addition, the company has expanded into other forms of direct selling, including catalogs, outbound calls, and an e-commerce site. Demand for direct sales is driven also by the construction needs of companies and governments, and the desire of industrial customers to improve the efficiency of operations. In addition to marketing and selling through its TV direct sales programs and its off-TV nationwide distribution network, Acorn also offers consumer products and services through catalogs, outbound telemarketing center and an ecommerce website. Leveraging its integrated multiple sales and marketing platforms. The business of Acorn is dependent on having access to media time to televise our TV direct sales programs. The direct sales of the company accounted for 54.7%, 70.3% and 68.5% of total net revenues in 2006, 2007 and 2008, respectively, generated through TV direct sales platform. In addition, to TV advertisement, direct sales were also dependent on the participation of nationwide distribution network [5].

In prospecting business-to-business accounts through personal selling method, salespeople should examine carefully to whom to communicate in the prospecting organization. If salespeople communicate to the persons above or below the level of prospect who has the decision making authority, the messaged delivered at this level would not make impact on the decision maker. In order to utilize the time and efforts of salespeople prolifically it is necessary for them to identify the right contact person in the client organization, and his experience and receptivity towards discussing technical issues. Personal selling always needs to be attention driven and to draw the most interest to the product or service offering, salespeople should center their message on the issues that can specifically deliver the solution. In business-to-business personal selling process it is required for salespeople to understand they are selling not only to a company, but also to an individual who specifies, recommends and makes a buying decision on behalf of the firm. At the end, a statement on the guarantee of product or service would enhance the confidence of buyer and also build initial loyalty towards the company that is offering products or services.

There are several developments, emerging in the operating environment of direct sales organizations which affect the strategic roles of the personal selling function. To fully realize the multiple benefits of personal selling, direct sales organizations must be firmly committed to the trust-based relationship selling paradigm. Such a sales approach requires salespeople to move past a short-term transaction orientation to fulfill roles such as counselors, ombudsman, and ambassadors [6].

PERSONAL SELLING SKILLS

Personal selling is a critical marketing tool in both business-to-business and business-to-consumer sectors. In 2006, U.S. businesses spent $800 billion on over 20 million salespeople working full-time in direct-to-customer selling activities of multinational companies like Avon and Amway. This amount almost accounts for three times the amount spent that year on advertising, which evidences the significance and role of personal selling among growing business firms. This selling approach has a stronger effect than advertising on sales per dollar spent. Though this statistics does not stand a yard stick on making budgetary allocations on personal selling, managers may empirically derive quantitative guidance in reference to market sector and consumer preferences [7]. Salespeople engaged in direct-to-customer selling activity need to develop smart strategies and work hard with varied consumer environment. Working hard might be more important for salespeople who perform repetitive, routine tasks, while working smart might be more important for salespeople in highly creative roles. Understanding the relative importance of these behaviors managers should provide useful insights for sales managers considering motivational programs that have best cognitive fit [8]. Accordingly, personal selling skills may be developed among salespeople in the following areas:

- Control questions & tie downs
- Embedded commands
- Agreement framing
- Customer temperaments
- The art of pacing& leading
- Active listening
- Pattern interrupts
- Overcoming any objection
- Advanced closing techniques
- Responsive word tracks
- Closing on the final objection.

It is important to develop personal selling skills in the areas of adaptive selling, customer orientation, listening, and consulting efficiency among the salespeople. It is observed in some research studies that listening behaviors positively influence customer orientation behavior, which, in turn, influences adaptive selling behavior of salespeople. The modified listening behavior further results in a feedback loop of reinforcing behaviors among salespeople towards prospective consumers. Developing selling skills of consulting oriented salesperson needs stronger behavioral modification than a formal sales training [9]. It is common that whenever a company launches a new product or service into a new market, the temptation is to augment sales force capacity immediately to acquire customers within a short span. However, for carrying the sales efficiently, the entire organization needs to learn how to acquire customers and involve them in the product use process. This process may be defined as the sales learning architecture. Customers transfer knowledge and experience back and forth during the selling process which helps driving personal selling effective. Sharing consumer experience is one of the important tools in direct-to-customer selling process as it builds social network among consumer. As customers adopt the product, the firm modifies both the offering and the processes associated with making and selling it [10].

> Amway International Inc. is one of largest multinational company based on the direct-to-customer business model. The Amway business model is based on the Business Owner Compensation Plan which delivers a low-risk and low-cost business opportunity is open to interested entrepreneurs. The sales opportunities are supported by the business rewards for accomplishing predetermined volume of sales by a salesperson as well as for sponsoring others who do the same. As business grows for a salesperson so do his rewards for specific targets. Selling makeup and vitamins to friends and family is a way of life at Amway. One of the world's largest direct-sales businesses, Amway boasts more than 3 million independent business owners (IBOs) who sell as many as 450 personal care, home care, nutrition, and commercial products. Amway also markets products and services of other companies in more than 80 countries. Business revival techniques are used to motivate its distributors (mostly part-timers) to sell products and find new recruits so that the sales efforts of none of the salespeople drain out. With this global success, it expanded further, into the Chinese market. However, the company had to revise its strategy after the Chinese government implements regulations on the direct marketing business model [11].

In the personal selling process customer relationship management (CRM) plays a pivotal role in acquiring and retaining customers by delivering sustainable customer value. The CRM in direct-to-customer selling strategy has emerged as an up-market concept and provide customer director services management through Internet. By making the enterprise accessible to customers, personal selling goals have been enhanced as customers would service themselves better and at a much lower cost. Accordingly personal selling would not only generate sales but also create customer satisfaction through self-managed efforts by the customer under supervision of the salespeople [12]. Direct selling is a vibrant channel in the age of e-commerce and successful tool of generating high sales at low cost. Increasing domination of personal selling practices has emerged as a potential new channel for many traditional companies. As it becomes more difficult to continue the expansion of traditional retail, catalog and online selling channels, many companies are switching their strategies to direct selling as 'guerilla marketing'. Direct-to-customer selling is operated by a large number of independent sales consultants or representatives in a company, who work on absolute commission basis. It is an ultimate pay-for-performance selling paradigm that motivates many salespeople who seek an income with flexible work hours and the "be your own boss" allure [13]. The ideal process of personal selling should follow the steps given below for gaining efficiency in performance:

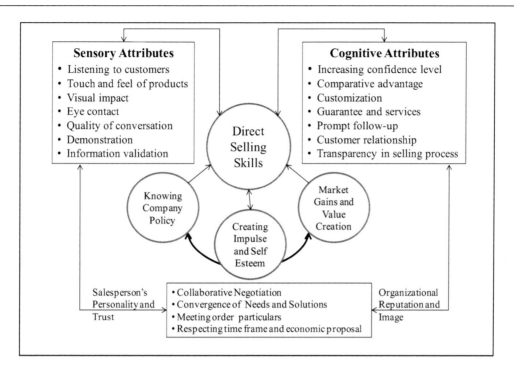

Figure 2.1.Mapping Direct Selling Skills.

- Prospecting
- Classifying leads
- Pre-call planning
- Approach/relating
- Needs discovery
- Presentation
- Handling objections
- Closing
- Follow-up and servicing.

Companies engaged in direct selling should understand that a salesperson passes through four levels of personal selling process that takes him from the initial level of 'approach', through the 'interview' and the 'solution' stages, to the 'action'. The balance between sensory and cognitive skills that are required for effective direct selling are exhibited in Figure 2.1.

The selling skills largely depend on the personality of salespeople and organizational policies, which are influenced by the behavioral factors including personality, image, reputation and trust (PIRT) concerning both salespeople and the organization. The concepts of image and reputation have been increasingly emphasized in the fields of public relations and competitive posture of the company in the market. It is argued that consumer creativity is nurtured with the long association of salespeople, and product-specific emotions and attitudes including PIRT attributes drive the passion of buying among consumers. In this process, knowledge of salespeople is also considered as an important determinant of consumers' response to develop initial buying conviction and get their decisions endorsed with the fellow consumers [14]. Previous studies have established that there is a close relationship between the consumer attributes and the salesperson's personality in driving the emotional values. This relationship in turn influences the consumer's responses towards building trust on salesperson, in the selling process and the company [15]. The quality connection between personality traits of salespeople and consumer association depends on the perceived attractiveness of the product and selling process to a large extent. Salespeople should be trained on several issues in this process, more importantly on managing open-ended questions, demonstrating strong product knowledge, using product features, and closing. While offering significant solutions to clients salespeople need also to deliver practical advice knowing when, how, and what fits into the customers profile [16]. Companies may consider imparting two types of selling skills that reveal learned proficiencies at performing job activities to improve the performance of salespeople:

- Vocational skills related to workplace job profile and specific skills such as technical knowledge and vocabulary related to the firm's product line, the company, and its policies, and
- Sales presentation skills concerning effectively conducting the personal selling process comprising the series of inter-related steps that salespeople commonly use to engage with and influence customers.

Sales process includes prospecting for new customers, arranging to meet with the potential customers identified, uncovering and understanding customer needs through proper questioning, developing a product solution based upon customer needs, presenting the solution back to the customer, handling customer objections and questions regarding the proposed solution, closing the sale, and negotiating the transaction. The training on above subjects is positively associated with salesperson skill in reference to active listening, adaptive selling, handling objections, closing, negotiating, and prospecting consumers [17]. Creative skills for effective direct-to-customer selling process should also include:

- Audience Targeting
    - Attention appeals, Appeals to Needs, Floating Audience Appeal, cross selling
- Product Presentation
    - Feature dramatization, USP, Selling Benefits, Building an Edge, To the point
- Involvement Devices
    - Quizzes and questions, Fascination, Value, Personalization, Play, Win-win Dimensions, Options
- Convenience Factors
    - Realizing commitments, Response at convenience for customers, Ready reckoner, Ease of Payment, Ultimate Interactivity
- Immediacy Incentives
    - Urgency Copy, Command Terminology, Adding Immediacy, Deadline, *Me Too* Feeling
- Speed of Delivery
    - Convergence of Service, Post-purchase Relationship, Creating Referral

- Credibility
  - Hired Selling Agencies, Brand-wagon (Popularity), Endorsers, Corporate Personalization, Testimonial, Strategic Alliance for outside Sales and Service
- Style
  - Professional, Entertaining, Educative, Challenging, Dynamic (Core-R&D-Projection-comparative), Categorical and Segment Specific.

Developing creative ideas might not be an easy task for salespeople and selling them to consumers might be still harder. It has been observed that to get competitive lead among the fellows, a salesperson can often go to great lengths to demonstrate how his new concepts are practical and profitable. However, all creative efforts are risk averse and face rejection by the consumers or clients who do not seem to understand the value of the ideas. This can be explained considering a salesperson as 'pitcher' who demonstrates creative ideas and consumer stands as 'catcher' of the ideas at the receiving end. Often, the pitcher's ability to come up with workable ideas can quickly and permanently overshadow the catcher's feelings [18]. Hence, it is required that salespeople should work continuously with new ideas and measure how they convince consumers to reap the advantage.

The Pampered Chef, a company established in 1980 through Kitchen Shows, has grown as a direct selling organization. The company recruits and trains direct selling representative designated as Pampered Chef Kitchen Consultants. The Pampered Chef is operating with over 70,000 independent Kitchen Consultants in the USA, Canada, the UK and Germany accounting for the sales turnover of over $700 million per annum. The company believes that multipurpose tools are the cornerstone of an efficient kitchen, and is committed in developing timesaving tools, tips and recipes that enhance shared mealtime and suit busy lifestyles. The Pampered Chef's direct-to-customer sales force offer multipurpose kitchen tools, demonstrate easy food preparation techniques and share recipes that can help consumers entertain with style. These Pampered Chef Kitchen Consultants do more than the direct selling representatives in other organizations by demonstrating kitchen techniques and sharing recipes that help the consumers. A staff of home economists, food scientists, chefs, dieticians and product developers edit, test and develop more than 400 recipes every year. The company is continually expanding its recipe and product collection to align with top flavor, ingredient and lifestyle trends. The company respects consumer experience and if negative they will not be repeat customers. The company recognizes that long-term customer loyalty is a critical consideration in its product development [19].

Salespeople engaged in direct-to-customer selling operations should acquire following skills to get through the process successfully:

- The magic of the delivery,
- Flexible follow up procedures,
- Generating strong referrals for future prospecting,
- Professional telephone skills,
- Effective use of the corporate information,
- Developing personalized follow up system, and
- Painless prospecting.

A large sales force in emerging markets is engaged in the personal selling of consumer goods and services. These sales are very much individual consumer oriented. The sales personneloperate as sales representatives of the company or as facilitators of an in-house retail store. The major task of the sales representatives is to collect orders from consumers or organizations. In performing direct sales, the representatives attempt to generate an emotional appeal and this motivates the buyer. It has been observed that the increasing the volume of sales through direct selling approach requires a strong persuasive ability and skills to identify a customer's buying motives among salespeople. Direct salespeople receive compensation in terms of commission on the volume of sales.

STIMULUS RESPONSE MODEL

The major attribute of personal selling is the underlying arousal during the process of prospecting customers. In direct-to-customers selling practice consumers are continuously blitzed with relationship calls and marketing messages including television commercials, e-mail solicitations, and business circulars of the company. However, often persuasion punches on right customers and stimulate the buying need to elicit the desired response on the prospected product or service. It might be very difficult for salespeople to identify what drives consumer behavior, largely because there are so many possible combinations of stimuli. Although innovative selling strategies have always been a creative endeavor, adopting a scientific approach to it could make the selling process easier and more cost effective and enhance the customer value. 'Experimental selling' techniques, which have long been applied by some prominent direct selling companies such as Avon International, Dell Computers and the like, let the salespeople engender stimuli by testing just a few of selling designs on customers. In personal selling process salespeople may lean towards testing combinations of critical attributes including perceived use value, price sensitivity, prolonged guarantee, loyalty benefits and life time service not only to gain confidence of the consumers but also to optimize resources, revenues, and profitability of the firm [20].

There is a strong relationship between stimulus and response in the selling process. Such cognitive interrelationship can be explained as an approach to selling that relies on the salesperson's ability to provoke the prospect by ways of delivering comprehensive knowledge on the product or service (stimulus) in order to obtain a conviction from the buyer (response) towards his possible association. A balance between the extent of stimulus and quality of response needs to be developed in order to measure the performance of the direct selling approach. Stimulus and response are the attributes of an individual, hence the direct selling strategies should not be generalized and applied to a group of consumers, and rather they need to be crafted case by case by optimizing selling efforts. The stimulus-response (SR) model in the context of direct selling may be stated as a cognitive progression of cause and effect relationship. In managing in SR course, after identifying the needs of customers and problems, the salespeople need to deliver a comprehensive presentation so that the thoughts of the prospect will flow naturally from the prospect's mind-set to the salesperson's product. The thinking process of consumer must be guided to experience in the manner which reveals that the salesperson's product as the solution to a problem or need [21]. The four major approaches to designing process of direct selling include:

- Stimulus-response analysis,
- Mind mapping of consumers,
- Problem solution, and
- Need satisfaction.

Many psychological experiments have shown that subjects will respond in a predictable manner when exposed to a specific stimulus. When subjects are rewarded for correct responses, the responses may become automatic. Salespeople using the SR approach concentrate on saying the right thing at the right time to develop a favorable response from the prospect. Knowing how prospects normally respond to certain stimuli helps salespeople build a sequence of favorable responses. The SR approach is most appropriate in simple straightforward selling situations such as selling a vacuum cleaner in a home. There are various intrinsic and extrinsic factors that play role in the SR process that is exhibited in the Figure 2.2 illustrating the SR Paradigm.

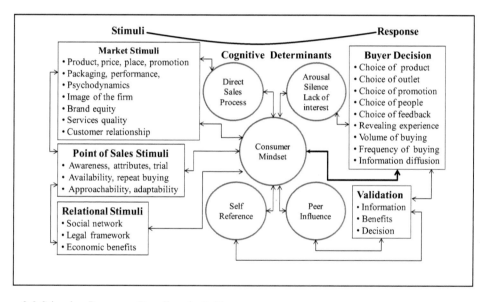

Figure 2.2.Stimulus-Response Paradigm in Selling.

Stimuli are derived from various interrelated factors that affect the cognitive process of a person in a given time and space. Consumer stimuli are derived from market related point of sales offers and relational factors, which steers the analytics in the mindset of consumers as illustrated in Figure 2.2. The opinion of fellow consumers, peers and self-reference criterion also contribute to the information steering process. The factors of market stimuli are generally driven by the salespeople during direct selling interface. Churning of information from all available sources generate arousal, silence or alienation among the consumers which is emerged as response to various buying preferences. Peer influence is generated by the free customers who play a vital role in promoting a vast array of businesses, such as media companies, employment services, and even IT providers. These customers not only influence customers but also generate revenue indirectly for the firm, figuring out the true value of those customers. This phenomenon takes into account not only direct network effects where buyers attract more buyers or sellers more sellers attract but also indirect network effects revealing that buyers attract more sellers or vice versa [22].

During the selling process if salespeople are not able to trigger the thought of creating an alternative platform to persuade the consumer, they can switch to Mind Mapping exercise to chart their thoughts and make sure the prospect absorbs the context and contents of the discussion. The effect of stimulus-response mappings in complex consumers is a difficult process wherein generating stimuli might differ from self-reference stimuli, which consumer acquires during his search for products or services. Therefore, the degree of consistency and variability of information delivery by the salespeople affect the stimulus-response quality and time of consumers. It is observed that consistent SR mapping led to faster and more accurate initial consumer decision making on buying. Consistent SR mapping also helps salespeople in acquiring improved skill on direct selling process [23].

> Discovery Toys Inc. (DTI) is engaged in direct selling operations of creative toys for children online and through educational consultants who deliver products directly to the consumers' home. Educational consultants are trained to drive stimulus among parents, guardians and children on creative toys that help in the mental development of children. Consumers can benefit from asking questions and hearing about the rewards that the products of the company offer to child's specific stage of development. The creative toys of the company are broadly oriented to the community and educational consultants demonstrate the products in community events or in a family gathering. The stimulus is generated through entertaining presentation leading to benefits associated with the formation of child's behavior and mental ability. The company orients educational consultants as hosts and banks on the philosophy of encouraging hosts working with prospects and letting them discover what fun a night in the living room that can be for all with love for learning and giving to children, and the joy of shopping. DTI encourages educational consultants to join their sales activities through Own the Opportunity compensation program, which is designed to give valuable rewards and an incremental income on selling products, recruit team members, and promote them with designation. The company also offers paid family vacations to educational consultants various tourist destinations, who qualify the sales benchmarks [24].

Mind Maps track the natural progression of prospects' brain by allowing a salesperson to connect each new thought to the ones that have come before, whereas linear notes or lists deliberately cut off each idea from the ones proceeding and following it, thereby stimulating the natural thinking process. Mind Maps can also be used by the salespeople in informal meetings with consumers as a sort of visual agenda on Facebook or live chat in any electronic platform towards delivering more lively and participatory learning process [25].

After securing the prospect's agreement that the need exists for the products or services offered, the salespeople should offer a solution to satisfy the need. The major advantage of this approach is that it forces the salesperson to practice the marketing concept. The need satisfaction approach is designed for more experienced and sophisticated salespeople. Salespeople using this approach must understand the psychology of communication and persuasion. They must also spend time and effort to determine the prospect's needs and to demonstrate how their product satisfies those needs. The problem solution approach begins with the salesperson identifying the prospect's need. The salespeople should then help the consumer to list possible alternative solutions and to evaluate the advantages and disadvantages of each alternative. Finally, the salespeople need to work with the consumer to select the best alternative. In using this approach, the salesperson builds a relationship with the customer that resembles a consultant-client relationship [26].

Despite the acknowledged importance of determinants affecting the salesperson-customer interaction, this issue still remains an enigma. It should be understood that there are two main philosophical stances that drive the stimulation among consumers, which include innate behavior of an individual that is different from the acquired behavior and those that emerged during the salesperson-customer interaction. In order to improve current understanding of the salesperson-customer interaction, the competing philosophical stance offered by radical behaviorism can be pondered over by the salespeople to drive buying behavior as a response to stimulus, within the sales interaction context [27].

ENHANCING CUSTOMER VALUE

The customer value is an intangible factor which has significant role influencing the buying decisions. The customer value includes broadly psychometric variables like brand name, loyalty, satisfaction and referral opinions. The customer lifetime value is built over time by the business firms which also contributes to the individual perceptions of the customers and augments their value. The new school of business thoughts and contemporary researchers have emphasized that, towards maximizing the lifetime value of customers, a firm must manage customer relationships for the long term. In a disagreement to this notion a study demonstrates that firm profits in competitive environments are maximized when managers focus on the short term with respect to their customers [28]. When salespeople construct a customer value proposition, they often simply list all the benefits their offering might deliver. But the relative simplicity of this all-benefits approach may have a major drawback of benefit assertion, which may lead the salespeople to claim advantages for those features of their products and services that customers do not care [29]. However, salespeople should look into the following strategies for enhancing the customer value:

- Enhancing customer esteem – *You be the best judge*
- Million Dollar Words- *Satisfaction guarantee or money back*
- Unconditional Guarantee
- Conditional Guarantee
- Time Limit, Warrantee, Usage Conditions, Maintenance only
- Creating Impulse – *Double Guarantee, Dramatize Statement*
- Competitive Guarantee
- Keeper Offer- *Premium value to customers*
- Value Protection Guarantee – *Security against frauds.*

Successful multinational companies organize their customer value and satisfaction (CVS) strategies around various key activities by developing customer focused culture, executive support to salespeople in improving the capability and competence in prospecting customers, and improving customer listening tools. Enhancing customer value is basically an organizational issue where every employee from manager to salesperson plays his role. Hence, managers need to indentify improvement opportunities and plan intensive trainingprograms for the salespeople [30]. It has been observed that there is increasing number of customer goods and services offered in recent years suggesting that product-line extensions have become a favored strategy of product managers. A larger assortment, it is often argued, keeps customers loyal and allows firms to charge higher prices. There also exists a disagreement about the extent to which a longer product line translates into higher profits keeping the customer value higher. The academics, consultants and business people speculated that marketing in the new century would be very different from the time when much of the pioneering work on customer loyalty was undertaken [31].

The key marketing variables such as price, brand name, and product attributes affect customers' judgment processes and derive inference on its quality dimensions leading to customer satisfaction. An experimental study indicates thatcustomers use price and brand name differently to judge thequality dimensions and measure the degree of satisfaction [32]. The value of corporate brand endorsement across different products and product lines, and at lower levels of the brand hierarchy also needs to be assessed as a customer value driver. Use of corporate brand endorsement either as a name identifier or logo identifies the product with the company, and provides reassurance for the customer [33]. Amidst increasing market competition, some companies have outmaneuvered by focusing through value-for-money strategies enabling consumers to economize (manufacture at low cost and make consumers spend less), and become more efficient (manufacturing at the same cost for lesser margin of profit), or become more effective (manufacture more at the low cost but and leas for relatively low profit). This strategy has helped low end firms to enhance customer value competitively and lead the marketplace [34].

Blyth Inc. a direct selling company founded in 1977 is the largest candle maker in the USA. The company prefers to light up the party with one of its wicked products. Blyth sells its scented and unscented candles under the Colonial Candle, PartyLite, Ambria Table Lighting, and Seasons of Cannon Falls names, among others. The product portfolio of the company extends beyond the candle business with Sterno butane products and lighting, Two Sisters Gourmet sauces and condiments, and ViSalus nutritional supplements, as well as a variety of catalog and online businesses that market household goods and gifts. Blyth's products are sold through home parties and retailers worldwide. It also supplies institutional customers, such as restaurants and hotels. Blyth's business strategy is unique in the global home expressions market and as a leading designer and marketer of home decorative and fragrance products, the company seeks to reach consumers across each distribution channel in which they make their purchases, offering a wide variety of products to satisfy multiple needs for consumers to express themselves in their homes. The company entered the direct selling distribution channel in 1990 with its acquisition of Colonial Candle of Cape Cod and, with it, a small division comprising less than $10 million in sales with PartyLite brand. Independent Sales Consultants of PartyLite are engaged in selling premium fragrance candles and a wide range of related decorative accessories to consumers through the home party plan method of direct selling. PartyLite operates in all 50 states in the U.S., in Puerto Rico, the U.S. Virgin Islands, Canada, Mexico, Germany, the U.K., Austria, Switzerland, France, Finland, Australia, Denmark, and in 2009 began its operations in Poland [35]. The company introduced Party Pulse facilitates communication to multiple tiers including corporate-to-consultant and consultant-to-customer through electronic newsletters that build network-wide relationships, maximize the effectiveness of selling events, and drive higher sales performance. *Candle Connection* e-newsletter helped in building high sustainable customer value byconveying images of beauty, warmth, and family living through its design and captivating photos. Each issue contains on average four articles on hostess specials (rewards for hosting PartyLite events), entertainment and decorating tips, guest specials (purchasing incentives), consultant opportunities, and other timely messages [36].

Customer lifetime value (CLV) is a key-metric within customer relationship management. Although, a large number of marketing scientists and practitioners argue in favour of this metric, there are only a few studies that consider the predictive modelling of CLV. Customer lifetime value also represents the net present value of profits, coming from the individual customer, which creates a flow of transactions over time. Firms look at their investments in terms of cost per sale, rate of customer retention, and also conversion of prospects. CLV is, then, used as a convenient yardstick of performance. The concept of the lifetime value of a customer is well established in the theory and practice of database marketing. The lifetime value of a customer, defined to be the expected present value of the net cash flows from the firm's relationship with the customer over his or her lifetime, is often used as an upper limit on spending to acquire the customer [37]. Many firms agree that their efforts should be focused on growing the lifetime value of their customers. However, few have come to terms with the implications of that idea for their marketing management with focus on decision making and accountability of customer values. The customers' lifetime value is constituted by three components- customer's value over time, length of customers association and the services offered to the customer. The satisfaction is the customer's perception of the value received in a transaction or relationship and it helps in making re-patronage decisions on the basis of their predictions concerning the value of a future product. It may be thus stated that the customer value paradigm is contemporary, which includes many elements of the customer satisfaction paradigm and is being more widely adopted and deployed by the firms [38].

Customer value is nurtured through three distinct dimensions of emotions, including pleasantness, arousal and dominance, which have been identified as major drivers for making buying decisions among consumers. Convergence of sales promotion, consumer's perceptions, value for money and product features drive arousal among consumers. Consumer values are created towards the new products through individual perceptions, and organizational and relational competence [39]. Firms need to ascertain a continuous organizational learning process with respect to the value creation chain and measure performance of the new products introduced in the market. The product attractiveness consists of product features including improved attributes, use of advanced technology, innovativeness, extended product applications, brand augmentation, perceived use value, competitive advantages, corporate image, product advertisements, and sales and services policies associated therewith. These features contribute in building sustainable consumer values towards making buying decisions on the new products [40]. The attractiveness of new products is one of the key factors affecting the decision making of consumers and in turn is related to market growth and sales. The higher the positive reactions of the consumers towards the new products in view of their attractiveness, higher the growth in sales.

Consumersin Mexico are largely influenced by the product attractiveness and show higherstore-loyalty, irrespective of higher prices than normal shopping conditions. When choosing food products and store, consumers evaluate both the fixed and variable utilities of shopping; the fixed utility does not vary from trip to trip whereas the variable utility depends on the size and composition of the shopping list [41]. Preferences and perceptions of Mexican consumers on new food products also depend on the social and cultural values. New products introduced in the Mexican market are generally expensive and price is considered as a major factor influencing consumption. Mexican consumers put more emphasis on the country of origin of food products than on brand names. The product-country evaluations of Mexicans seemed to be affected by a strong home country bias [42].

Mass customization is also considered by many firms as a tool for success in cross retailing as a comprehensive strategy that frames current Web initiatives specifically in terms of customer value. The strategy of mass customization will demand an understanding mass customized "e-consumer services" performed by salespeople during direct selling process. Such mass customization products include e-retail services as well as service-related consumer products that are defined and sold via the Internet. The overall success of the broader virtual marketplace can be determined not just by corporate policies but also by the development of rich standards that allow for the highly customized bundling of products. More importantly, salespeople provide Internet interface with customer to comprehensively support customer value driven customization process [43].

LEAD SELLING

Sales leadership can be defined as demonstrating skills, experience, and direction to provide solution to consumers to improve customer value and outperform competitors. However, managerial leadership to motivate salespeople consists primarily of giving actual feedback to reinforce what they do well while suggesting ways and means to sell products and services as well as improves the customer value. This implies to all role players in an organization so that they can improve and it is hoped that everybody wants to get better at what they do [44]. The global corporate leaders such as GE, Wal-Mart, and Toyota have explained to the business community that customer is important and salespeople should develop their strategies around them which enhance customer value. Sales leadership is driven by regionally focused customer centric strategies that can be used in conjunction with local and global initiatives and significantly boost market performance. Some companies adopt the leadership strategies in combined approaches that move the sales and market share of the company in the competitive marketplace. The automobile manufacturing companies have also adopted direct selling strategies as an alternative channel to institute flexibility and creativity in the growth process. A company must decide what constitutes a region, identify customers, and choose the most appropriate strategies to harvest a powerful competitive advantage [45]. Sales leadership should emerge within the organization at fist stance. It is observed in organization where there is the more flexibility; a sustainable leadership can be developed. However, from outside a firm, strategy making usually appears to be a top-down process whereby the CEO and top management make decisions that the rest of the firm must then implement. Sometimes, though, the ideas underlying strategic moves originate with lower level employees [46]. The attributes of lead selling should include:

- Lead flow considerations
  Maintaining lead supplies, supply to order, supply to inventory
- Lead Freshness – Prompt follow-ups

- Lead Reports – Sales and forecasts, Breakthroughs, R&D
  For competitors, For customers and For Salespersonnel
- Lead Customer Relations
  Front end referrals, Back end referrals*, sales gate keepers
- Lead Response Management
- Lead in Developing Sales Ambience
  Cross promotions, portraying positive attitude, creative tactics.

Lead in sales can be generated by applying the most effective leadership style, which includes participative, supportive, directive, and outcome oriented in specific market environment and consumer cultural. Adopting bottom-up leadership styles can provide global sales managers with powerful tools for successfully managing their diverse salespeople and consumers on a global basis. This conceptualization builds current understanding of sales manager leadership by developing leadership style-national culture matrix in different cultural settings [47]. Managerial strategies close to the leadership process include involvement of all role players in business, marketing, and selling decisions;. The leadership attributes also envelop intelligence as a basis for added value, integration of cross-functional contributions to customer value and building appropriate organizational infrastructure that could support the competitive lead role. Broader organizational consequences towards developing sales leadership consists of inspiration and integrity in ethical standards in the operational matters. The sales leadership also emerges from the strong corporate social responsibility initiatives and, an international perspective on managing sales and customers [48].

Hello Direct Inc.(HDI) is the company engaged in direct selling of telephones and accessories including headsets, amplifiers to small and mid-sized businesses and the very telemarketers in USA who contact consumers just when dinner is served. The company sells the merchandise through its catalogs, its Web site, and its own nimble-fingered staff of telemarketers. HDI sells the phones that screen those annoying telemarketer calls. Sales in the company are achieved largely through outbound telemarketing that is directed at corporate accounts. The company is making a significant investment to increase the size of outbound sales force and improve training and marketing support in order to increase its market share. Though selling though catalogues is getting slow, it remains the primary component of the marketing program, other channels provide significant growth and profit opportunities. The company also offers recording, distance-conferencing, mobile navigation, and VoIP products. Historically, HDI has operated its business according to four core principles consisting of fast delivery, easy ordering, total support and full guarantee coverage. In addition to taking input very seriously and translating it into effective product solutions, the company offers a complete package. It sells more than 50 brands including AT&T, GE, Panasonic, and Motorola. HDI keep its client accounts and individual consumers updated through its bi-monthly newsletter Newswire, an e-mail edition featuring product reviews, telephony tutorials, industry news, and advance notice about upcoming special offers. The company has grown from a small handful of visionaries to a well recognized business-to-business leader in telecommunications solutions and was acquired in 2000 by GN Netcom, a Denmark-based maker of hands-free telecom equipment [49].

Team selling is considered as one of the tools for achieving lead in sales against competing firms in the marketplace. The team selling approach is followed by many multinational companies for the complex product and services, which the customer faces as a first-time buy and salespeople need to support such negotiations with comprehensive information needs. The team selling would also be advantageous when an account requires special treatment or a large number of people are involved in the process of buying decision. In addition, team selling is more likely to be employed when the potential sale is large for the representative firm and when the product is new to the product line of salespeople. The sales leadership in an organization can be achieved through the implementation of following strategies effectively [50]:

- Strategic realignment needed for effective organizational change
- Need to capture key account performance as a basis for making strategic choices
- Impact of the leaning of operations and lean enterprises spanning value chains on buyer-seller relationships
- Need to locate sales ethics in an enterprise-wide ethics system that represents a company's key values and its undertakings to stakeholders
- Importance of accommodating sales interface issues in the transformation process and,
- Focusing on the processes of strategizing the traditional sales organization for the new type of role it is expected to play.

Leadership has also three organizational styles which encompass consultative, participative and delegative leadership styles of sales managers. The consultative, participative and delegative leadership processes drive collective decisions with senior managers, while in contrast with younger sales managers who appear happy to take decisions that may not necessarily get the approval of the salespeople in the field. It was interesting to find that hierarchy is directly related to consultative and participative leadership style, but not to directive or delegative leadership. While gender alone does not appear to affect the result in any significant and systematic way, a number of multiple variables, including gender and hierarchy, for example, proved to be useful in explaining the complex styles of managers [51]. Organizations need to develop a culture of ethics to truly make their quality initiatives work. Ethical excellence should be incorporated by design into the continuous improvement process in order to sustain the quality journey over time. Moral leadership is also the most important and the most critical variable for long-term business success [52].

There are two patterns of lead selling that include hardeners and softeners implementing quality of leads. The hardeners approaching to develop quality leads in sales demonstrate transparency in negations with consumers and follow evidenced based disseminations of information to consumers. On the contrary salespeople who are softeners try to avoid long and complex conversations with consumers and play indirectly. Consequently, softener salespeople trigger low level of arousal and are less successful as compared to hardeners. Following are the principal attributes of the salespeople following these two leadership patterns:

- Quality of Leads - Hardeners
  - Transparency in price and value offers,
  - Sales calls, comprehension,

- Customer information,
- Optimize offer that is very relevant to product or service,
- Require a stamp or proof of authorization,
- Charge nominal (do not let anything be free)
• Quality of Leads – Softeners
- Triggering low curiosity
- Less comprehension,
- Mechanical response management,
- Add convenience,
- Give more charge less,
- Less questioning and better value additions.

The quality of lead in sales can be obtained through the sales manager's competence and the extent of management's support that can be solicited during the decision making process in selling. The lead in sales can also be achieved by effective monitoring and feedback analysis of consumers and laying the selling process on an interactive platform to ascertain success in the relationship. The factors that adversely affected the quality performances in sales include conflict among customer-salesperson dyadic relationship, hostile socio-economic environment, harsh market condition, salesperson's ignorance and lack of knowledge, basing negotiation on translucent terms, and aggressive competition. Sales manager's competence and organization support to salespeople contribute significantly in enhancing the quality of lead and sales performance in the competitive marketplace [53].

ATTRIBUTES OF CONSUMER BUYING

There are three types of buying situations among the consumers in the international markets and they need to be managed carefully by the foreign firms. These buying situations include straight re-buy, modified re-buy and new task buy. The least complex buying situation is the straight re-buy buying situation, wherein the customer has considerable experience in using the product and is satisfied with the current purchase arrangements. In this case, the buyer is merely reordering from the current supplier and engaging in routine response behavior. A modified re-buy buying situation exists when the customer has previously purchased and used the product. Although the customer has information and experience with the product, it will usually want to collect additional information and may make a change when purchasing a replacement product. A new task buying situation, in which the organization is purchasing a product for the first time, poses the biggest problems for the buyer. Because the account has little knowledge or experience as a basis for making the purchase decision, it will typically use a lengthy process to collect and evaluate purchase information.

Medifast Inc. is a US based nutritional supplement direct selling company, which is helping people slim down and shape up. The company develops, manufactures, and markets health and diet products under its Medifast brand name. The products, which are manufactured by Medifast subsidiary Jason Pharmaceuticals, include food and beverages (meal replacement shakes, bars), as well as disease management products for diabetics. Medifast sells its wares direct to consumers online and through doctor's offices. Its Take Shape for Life subsidiary distributes the products through a direct marketing and sales network of independent distributors (often people who have had success with Medifast) whom the company calls Health Coaches. Medifast was recently recognized to the Take Shape for Life Health Coach Network; the company operates or franchises 46 Medifast Weight Control Centers, maintains a significant physician network and features one of the largest web-based "direct-to-consumer" Internet marketing programs in the industry. These distinctive programs are designed to meet the unique needs of consumers searching for the appropriate level of support to help them succeed in losing weight and improving their health on the Medifast program, which is clinically proven and has been recommended by more than 20,000 physicians over the past 30 years. Take Shape for Life National Convention also provided forums for Health Coaches to meet and greet, share success with each other and learn best practices from top coaches. At the convention, the Company had offered Health Coaches the opportunity to preview the latest tools for prospecting customers [54].

Traditionally, decision making process either in organizations, community or families, or individuals has rarely been the focus of systematic analysis to show a pattern for salespeople. That may account for the astounding number of recent poor calls in selling products and services, such as direct selling of financial services to invest in and securitize subprime mortgage loans or to hedge risk with credit default swaps. Under such delicate prospecting situations salespeople should adhere to following steps in streamlining their selling process:

- List and prioritize deliverables of information and evidences that could support prospect in making a decision,
- Assess the factors that go into each, such as who plays what role, how dependable is the decision made, and what information is available to support it
- Design the roles, processes, systems, and behaviors organization needs, and
- Institutionalize selling tools and assistance.

Leading salespeople should bring multiple perspectives in stimulating prospects in their decision making [55]. The decision-making process in this type of situation is often called extensive problem solving. The salesperson must first gather information about the prospective customer's operations to determine whether there is a need for the products or services he or she has to offer, and the concerns of various personnel within the customer organization who might influence the final purchase. Then, many low-key sales calls may be necessary to educate the various purchase influencers about the seller's offerings and to establish credibility within the customer's organization. Even after a sale is made, the representative or other members of an account management team-may have a lot of work to do supervising the installation of the product, training the customer's personnel to use it, providing other post-sale services to ensure the customer will be satisfied and to increase the chances for future sales. In managing large accounts and complex decision making units, salespeople should develop skills on:

- Listening to prospects and probing the customer need
- Understanding attributes of decision making units
- Analyzing latent needs and perceptions of customers

- Do not make creeping commitment, and
- Spend lots of time.

An important part of sales planning involves determining the customer value as different employees of a customer-centric organization may be active at different stages of the purchase process,. Salespersons should move to different levels of customer relationship with required supporting information to persuade the deal. In many cases, however, the roles played by various members of the buying center are sufficiently consistent across similar types of firms that a company can establish policies to guide their salespeople. A widely recognized framework identifies seven steps that organizational buyers take in making purchase decisions:

- Anticipation or recognition of a problem or need,
- Determination and description of the characteristics and the quantity of the needed item,
- Search for and qualification of potential suppliers,
- Acquisition and analysis of proposals or bids,
- Evaluation of proposals and selection of suppliers,
- Selection of an order routine, and
- Performance evaluation and feedback.

The customers' buying centers are likely to involve a wider variety of participants when they are considering the purchase of technically complex, expensive products such as a computer system - than when the purchase involves a simpler or less expensive product. Consequently, firms that sell technically complex capital equipment sometimes organize their salespeople into sales teams or utilize multilevel selling, with different salespeople calling on different members of the customer's buying center to reach as many decision participants as possible and to give each participant the kinds of information that person will find most relevant. It is necessary for the foreign firm to review periodically the policies or organizational structures to guide its salespeople in dealing with customers' buying centers.

Behavior of consumers vary for products and the approach followed by the salespeople towards prospecting the consumers. They may differ in their wants, resources, locations, buying attitudes and buying practices. As buyers have unique needs and wants, each buyer is potentially a separate market and salespeople should address his needs distinctly.

For example, in the health supplements market, the salesperson identifies market segments, and develops solution selling strategies to sell health supplements products tailored to the consumer needs and problems. Such buying situation may be addressed by the salespeople in different product categories and buyers in reference to the age, income levels, price and other competitive factors associated thereof. Consumers might have bought such products many times in their life or some might have purchased it at least once in life time, hence there is no single way to segment a market. The mapping of buyer behavior as a guiding tool to develop sales strategies in a customized way is exhibited in Figure 2.3. The most important factors influencing a consumer's involvement level in the buyer-seller

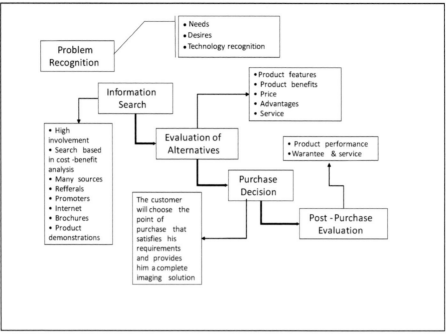

Figure 2.3.Managing Buyer Behavior.

interface are their perceived risks and expected value for money.

The purchase of any product involves a certain amount of risk, which may include:

- Product Failure – risk that the product will not perform as expected. (e.g., will the health supplements make me feel better and prevent me from diseases?)
- Financial – risk that the outcome will harm the consumer financially (e.g., will buying a formal suit cause me financial hardship?),
- Operational – risk that consists of alternative means of performing the operation or meeting the need (e.g., is there any maintenance required for the ball dress?)
- Social – risk friends or acquaintances will decide the purchase (e.g., what will my friends think of me in this dress?)
- Psychological – risk that the product will lower the consumer's self-image (e.g., will I look stupid if I wear this dress?)
- Personal – risk that the product will physically harm the buyer (e.g., is this health tablet meant to do me any good? Will it perform as it claims or will it ruin my health instead?) In a high degree of perceived risk, decisions in this case may require significant financial commitments and involve social or psychological implications.

Salespeople should map the needs, desires and technology preferences for enhanced perceived use value of the products among customers, and provide necessary information and evidences to help them in decision making. Salespeople should comment during the interpersonal communication with the prospecting consumer on the authenticity of information delivered, cost-benefit ratio of the negotiation, referrals views and available point of purchase promotion with the dealers in order to gain confidence. The alternatives need to be evaluated by the salespeople in an informal ways to justify the customer focused decision on the basis of the product attribute analysis and comparative advantages in reference to price, quality and services of the products of the company. Salespeople can stimulate the purchase decisions of buyers by providing the most favored purchase driver like choice of product, store format, payment options, and services requirements to optimize the level of satisfaction in terms of personal attention and solution handling. The post-purchase evaluation needs to be done in reference to the product performance, warranty administration and customer services by the call centre division of the company.

Tupperware Brands Corporation (TBC) practices direct selling as a principal tool in selling products. This strategy is defined by the company as the sale of a consumer product or service through a home party, through product catalogs on a one-to-one basis and has grown to include television and online shopping. The sales force usually consists of independent contractors who market products directly to consumers. The company makes and sells household products and beauty items. Tupperware parties became synonymous with American suburban life in the 1950s, when independent salespeople organized gatherings to peddle their plastic ware. TBC deploys a sales force of about 2.3 million people in about 100 countries. Tupperware's methods of conducting parties have evolved along with its product line. The company now offers the functionality to allow direct sellers to host virtual, online Tupperware parties. Hosts can organize an online party which is up and running for two weeks. Hosts can then do all the traditional things done to invite guests call on the phone, talk about it in face-to-face interaction with friends and acquaintances and write notes. In addition, the online party format is well suited to e-mail announcements. The company also sells its products online. Brands include Armand Dupree, Avroy Shlain, BeautiControl, Fuller, NaturCare, Nutrimetics, Nuvo, Swissgarde, and, of course, Tupperware. Its BeautiControl unit sells beauty and skin care products and fragrances in North America, Latin America, and the Asia/Pacific region. At TBC salespeople strive to listen to consumers to offer superior products and simplify everyday living. To that end, the company has launched a program that allows consumers to share product ideas for improving existing products or new product ideas [56].

In view of the rapid escalation of selling costs the multinational companies are using new technologies to reduce the selling cost and increase sales efficiency. The automation of sales force is one of new technologies that are being used by the multinational companies. Such technologically advanced sales strategies help the companies in reallocating sales effort to customer retention, minimizing the non-selling tasks increase the efficiency of the sale persons. Most salespeople are employed in various kinds of retail selling. These jobs involve selling goods and services to ultimate consumers for their own personal use, like door-to-door selling agents, insurance agents, real estate brokers, and retail store clerks do. A much larger volume of sales, however, is accounted for by industrial selling - the sale of goods and services at the wholesale level. Industrial selling activity includes sales to resellers, sales to business users (such as when a manufacturer sells materials or parts to another manufacturer, who uses them to produce another product or when a Xerox salesperson sells a law firm a copier to be used in conducting the firm's business) and sales to institutions. The activities involved in both retail and industrial selling, and managing sales forces for these types of sales activities, are very similar. Success in either type of selling requires interpersonal and communications skills, solid knowledge of the products being sold, an ability to discover the customer's needs and problems, and the creativity necessary to show the customer how a particular product or service can help satisfy those needs and offer solutions to the problems. While globalization makes the organization of the sales force more complicated, firms tend to resolve organizational issues in international markets in largely the same way as they do in the developed countries. The situational and strategic factors that influence firms' organizational decisions appear to be similar in both types of markets, and those factors seem to affect organizational choices in similar ways both at home and abroad. However a good coordination of sales and marketing activities is important for a firm to service its customers and compete effectively, observations indicate that many organizations do not assign responsibility for all their marketing activities to a single unit.

DIRECT SELLING THROUGH OUTSOURCING

Relationship advancement through sales promoters in the firm has a positive significant impact on the growth of sales and developing customer value. Outsourced salespeople identify appropriate customer from different markets, facilitate the dialogue with the firm and bring customers together with the firm to process the sales. Sales promoters support interactive learning processes and bridge the gap between the customers and firm [57]. The bargaining power of firms increases with outsourced salespeople who stimulate the demand for products and contribute to the enhanced sales at retail outlets. It has been observed that pull effect for the brands supported by the sales promoters increases at the retail stores as customers gather the pre-buying information from sales promoters [58]. Firms, which introduce new products frequently or emerge as less familiar brands in the marketplace, engage in pushing sales by working with customers, and outsourced salespeople. A broad set of process standards makes it easy to determine the selling strategies for the outsourced teams in a given market. Such standards help firms evaluate the costs vs. the benefits of outsourcing. Eventually these costs and benefits will be visible to buyers in terms of satisfaction and develop demand pull effects [59]. Firms must consider and alter four factors over time: the differing roles that internal salespeople and external selling partners should play, the size of the sales force, its degree of specialization, and the way salespeople apportion their efforts among different customers, products, and activities. These variables are critical because they determine how quickly sales forces respond to market opportunities, influence salespersons' performance, and affect companies' revenues, costs, and profitability [60].

A new strategy among the multinational companies engaged in competition of identical products like carbonated drinks and purified bottled water has emerged which can be explained by two-dimensional framework in reference to distributors' capability levels (low, medium, and high) and outsourcing sales force to enhance market coverage. A match between distributor capability levels and outsourcing sales force needs is posited to be the key to a sustainable relationship between suppliers and their customers [61]. However, some researchers have found that relationship of customers with the outsourced salespeople may not be sustainable in the long run. There is a common set of key contractual elements that exist between most outsourcing contracts. The nature of customer-promoter relationship functions as the key in the selling process for new products. In this process the perceptional problems with customers can greatly devalue the customer-promoter relationship and brand as a whole [62].

Driving consumer arousal and merriment as a major influencing factor in making buying decision is a recent strategy of retailers and an innovative concern as these factors reveal personalized enjoyment during shopping. Arousal during shopping may be seeded through multifaceted activity that may be performed in various ways and embody different consumer feelings. It is also argued that outsourced salespeople need to focus more on the consumers engaged in leisure shopping [63]. It has been observed in some studies that consumers who intend to do shopping in short notice, generally lean towards impulsive or compulsive buying behavior driven by arousal effect in the retail stores. Gender, age, leaning towards unplanned purchases, and tendency to buy products not on shopping lists, serve to predict compulsive tendencies [64]. However, sales promoters at times fail to recognize that what influences buyers' satisfaction is not the same as what engenders store loyalty, and consequently do not effectively develop the cognitive drive to stimulate buying decisions. Hence, they need to vigilantly manage the quality of arousal by developing adequate customer involvement in the buying process [65].

The customer value is created towards the new products through individual perceptions, and organizational and relational competence. The firms need to ascertain a continuous organizational learning process with respect to the value creation chain and measure performance of the new products introduced in the market. In the growing competitive markets the large and reputed firms are developing strategies to move into the provision of innovative combinations of products and services as 'high-value integrated solutions' tailored to each customer's needs than simply 'moving downstream' into services [66]. The product attractiveness may comprise the product features including improved attributes, use of advance technology, innovativeness, extended product applications, brand augmentation, perceived use value, competitive advantages, corporate image, product advertisements, and sales and services policies associated therewith which contribute in building sustainable customer values towards making buying decisions on the new products [67]. The attractiveness of new products is one of the key factors affecting the decision making of customers and in turn is related to market growth and sales. The higher the positive reactions of the customers towards the new products in view of their attractiveness, higher the growth in sales.

It has been observed in some studies that consumers who intend to do shopping in short notice, generally lean towards impulsive or compulsive buying behavior. Consumer's compulsive buying is an important area of inquiry in consumer behavior research [68]. Practically consumers react favorably to leisure sales campaigns of goods and services. However, customer value plays a decisive role in the shopping process. It may be argued that individual consumer behavior to a buyingoption is a functionof preference for theoption, whether it isa considered option or any choice constraintis personally directed. The consumerresponse is also affected bysearch associated with making adecision on leisure shopping in reference to the innovative products and campaign through the sales promoters creating the shopping arousal [69].

There is an increasing importance of building behavioral thrust among the potential buyers through effective communication and sales induction at customers' convenience. This process is largely managed by the outsourced sales people with focus on eight underlying factors of varying character important to customer satisfaction. These are awareness, selection, persuasion, convenience, trial, attention, location, promotional activities and merchandising policy [70]. Sales promoters instill emotions among customers in terms of merchandise choice, visual merchandising, store environment, sales personnelattitude, pricing policies and promotional activities during the pre-purchase stage. These factors are the very foundations of consumer satisfaction and decision drivers towards buying products [71].

Consumption among consumers has often been dichotomized in terms of its arousal-hedonic nature and has been closely associated with the level of satisfaction leading to determine the customer value [72]. As the new and exciting products are introduced, firms prospect the consumers through inter-personal negotiations managed by the sales promoters and inculcate high arousal among customers towards buying these products. The Do-It-Yourself (DIY) and computer aided simulations act as stimuli to consumers who intend to elicit apositiveresponse. Further, itis predicted that themagnitude of such positive response will be proportionalto the value ofan option to make buying decision at the available price. Visual effects associated with products often stimulate the buying decisions among young consumers. Point of sales brochures, catalogues and posters build assumption on perceived use value and motivational relevance of buying decisions of product. Emotional visuals exhibited on contextual factors such as proximity or stimulus size, drive perception and subjective reactions on utility and expected satisfaction of the products [73]. A pleasant negotiation ambience with sales promoters where music, hands-on experience facilities, and recreation are integrated helps in maximizing the consumer arousal towards buying. It has been observed that consumers perceive positive effect during interaction with sales promoters if arousal is high [74].

The impact of initial interaction of customers with the sales promoters can be measured in reference to degree of stimulation and pleasure gained by customers. Interactive tools on product learning provided by the sales promoters significantly affect the level of arousal and pleasure which contribute towards experience, and thereby influence the buying behavior. As higher stimulation or interactive learning provided by the sales promoters focuses on gaining initial experience on the product use, consumers tend to engage in higher arousing activities by acquiring the product [75]. The magnitude of consumerresponse to clearance sales isweighed in two ways-evaluative and behavioral. Firstly, consumer satisfactionwith the decision process leading to the expected level of *satisfaction* is measured, which may be expressed as oneof a number ofcognitive and affective responses. The satisfaction is the customer's perception of the value received in a transaction or relationship and it helps in making re-patronage decisions on the basis of their predictions concerning the value of a future product. Hence, many retailers develop innovative approaches to prospect new customers for new products by strengthening customer relationship and value management strategies [76].

Systematically explored concepts in the field of customer value and market driven approach towards new products would be beneficial for a company to derive long term profit optimization strategy over the period. On a tactical level, managers need to consider the optimum spread of customers on a matrix of product attractiveness and market coverage. A Company may also need to consider emphasizing an integrated promotion strategy for new brands in reference to attributes, awareness, trial, availability and repeat (AATAR) principle. One of the challenges for the manager of a retail store is to enhance the in-store ambience to influence the young consumers for prolonged stay in the store for shopping and explore the zone of experience of new products. An augmented and sustainable customer value builds loyalty towards the product and the brand. Systematically explored customer preferences and arousal driven retailing approach towards new products would be beneficial for a company to derive long term profit optimization strategy over the period. This needs careful attention and the application of managerial judgment and experience to generate consumer arousal and develop appropriate point of sales strategies for stimulating the buying decision [77].

As customer satisfaction has become one of the principal measures of retailing performance and leisure shopping behavior towards buying innovative and fashion products, the retail promotions need to be oriented towards augmenting the customer values and in view of the various factors affecting shopping arousal of customers. A strategy paradigm for augmenting such effect is exhibited in Figure 2.4. Appropriate promotional strategies considering the economic and relational variables discussed in the study may be developed by the managers upon measuring the intensity of leisure shopping and the scope of expanding the tenure of leisure shopping in view of maximizing consumer satisfaction and increase the volume of sales. Sales organizations should consider outsourcing salespeople not only for reducing the selling cost but also to grow competitively in the marketplace with quick and high sales turnover. Firms may consider outsourcing salespeople in reference to four factors over time that include the differing roles that internal salespeople and external selling partners should play, the size of the sales force, its degree of specialization, and how salespeople apportion their efforts among different customers, products, and activities. These variables are critical because they determine how quickly sales forces respond to market opportunities, influence sales reps' performance, and affect companies' revenues, costs, and profitability [78].

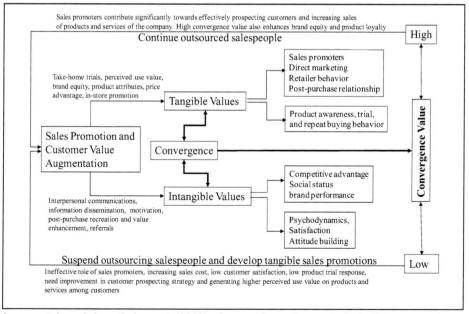

Source: Adapted from Rajagopal (2009), Outsourcing salespeople in building arousal towards retail buying, *Journal of Database Marketing and Customer Strategy Management*, 15 (2), 106-118

Figure 2.4. Direct Selling through Outsourced Salespeople.

At the point of purchase three factors including consumer interest in buying, value for money and product advantages converge, firms must make the most of the sales promotions available to customers to strengthen customer satisfaction, brand value, repeat buying behavior and increase in the sales of products. By making effective pre-purchase interaction with the customers, planning appropriate sales promotions, user friendly packaging, generating adequate shopping arousal, and focusing in-store ambience, marketers can effectively serve various interests of the manufacturer, the retailer, and the customers. A strategic focus considering how customers and competitors will react to any promotional effort, as well as the message delivered and the stature in the marketplace of the brand delivering should be developed by the firms in order to strengthen and streamline the pre-purchase promotions in marketplace. Hence, managers should not only tailor promotion programs successfully to target customers, but also skillfully monitor its implementation through customer culture and competition challenges.

SELECT READINGS

Anderson, R. E., Dubinsky, A. J. and Mehta, R (2005), Personal selling: Building customer relationships and partnerships, Houghton Mifflin, Boston, MA.

Bosworth, M.T. (1994), Solution Selling- Creating Buyers in Difficult Selling Markets, McGraw Hill, New York, NY.

Chang, J. (2003): Born to Sell, Sales *and Marketing Management*, 155 (7), 34-38.

Day, G. (2000), Managing Market Relationships, *Journal of the Academy of Marketing Science*, 28 (1), 24-30.

Direct Selling Women's Association (2005), *Build it big: 101 Insider secrets from top direct selling experts*, Kaplan Business, Birmingham, AL.

Harvard Business School (2009), *Sales and selling*, Harvard Business School Press, Boston, MA.

Malaghan, M (2005), *Making millions in direct sales*, McGraw Hill, Chicago, IL.

Rajagopal (2009), *Consumer behavior: Global Shifts and local effects*, Nova Science Publishers, Hauppauge, NY.

Weitz, B. A, Castleberry, S. B. and Tanner, J. F. (2003) : Selling: Building Partnerships, 5th Edition, McGraw Hill, New York, NY.

REFERENCES

[1]   Shapiro, B. P. (1974), Manage the customer, not just the sales force, *Harvard Business Review*, 63(5), 101–108.

[2]   MacKay, H. B. (1988), Humanizing your selling strategy, *Harvard Business Review*, 66(2), 36–47.

[3]   Macquin, A., Rouziès, D. and Prime, N. (2001), The Influence of Culture on Personal Selling Interactions, *Journal of Euromarketing*, 9(4), 71-88.

[4]   Sharma, A., Iyer, G. R. and Evanschitzky, H. (2008), Personal Selling of High-Technology Products: The Solution-Selling Imperative, *Journal of Relationship Marketing*, 7(3), 287-308.

[5]   Acorn International Inc. (2008), Annual Report. Corporate web site http:// eng.chinadrtv.com

[6]   Ingram, T. N. (1993), The Role of Personal Selling in Direct Sales Organizations, *Journal of Marketing Channels*, 2(2), 57-70.

[7]   Albers, S., Mantrala, M.K., and Sridhar, S. (2008), *A meta-analysis of personal selling elasticities*, Working Paper # 08-100, Marketing Science Institute.

[8]   Sujan, H., Kumar, N., and Weitzby, B. A. (1993), Learning Orientation, Working Smart, and Effective Selling, Working Paper # 93-119, Marketing Science Institute.

[9]   Pelham, A. M.andKravitz, P. (2008), An Exploratory Study of the Influence of Sales Training Content and Salesperson Evaluation on Salesperson Adaptive Selling, Customer Orientation, Listening, and Consulting Behaviors, *Journal of Strategic Marketing*, 16(5), 413-435.

[10]   [Leslie, M. andHolloway, C. A. (2006), Sales learning curve, *Harvard Business Review*, 84 (7), 114-123.

[11]   For details on Amway business model see the corporate web site of the company http://www.amway.com; also see Tan, D. and Tan, J. (2004), *Amway in China (A): A new business Model*, Discussion Case, Ivy School of Business, Canada.

[12]   Tony, C. (2006), The future for 'do it yourself' customer service, *Journal of Database Marketing and Customer Strategy Management*, 13 (4), 324-330.

[13]   Duffy, D.L. (2005), Direct selling as the next channel, *Journal of Consumer Marketing*, 22 (1), 43-45.

[14]   Füller, J., Matzler, K., and Hoppe, M. (2008), Brand Community Members as a Source of Innovation,*Journal of Product Innovation Management*, 25 (6), 608-619.

[15]   Silva, R. V. D., and Alwi, S. F. S. (2006), Cognitive, affective attributes and conative, behavioural responses in retail corporate branding,*Journal of Product and Brand Management*, 15 (5), 293-305.

[16]   Duncan, T. M. (2002), *High trust selling*, Thomas Nelson, Nashville, TN.

[17]   Jholke, M.C. (2006), Sales presentation skills and salesperson job performance, *Journal of Business and Industrial Marketing*, 21 (5), 311-319.

[18]   Elsbach, K.D. (2003), How to pitch a brilliant idea, *Harvard Business Review*, 81 (9), 117-123.

[19]   For details please see the corporate web site of the company http://www. pamperedchef.com; also see Duffy, D.L. (2005), Direct selling as the next channel, *Journal of Consumer Marketing*, 22 (1), 43-45.

[20]   Almquist, E. and Wyner, G. (2009), Boost Your Marketing ROI with Experimental Design, *Harvard Business Review*, 79 (9), 135-141.

[21]   Pederson,C.A.,Wright, M.D., and Weitz, B.A.(1984), Selling: Principles and methods, Richard D Irvin Publishers, Homewood, IL.

[22]   Gupta, S. and Mela, C.F. (2008), What is a free customer worth, *Harvard Business Review*, 86 (11), 102-109.

[23]   Madhavan, P. and Gonzalez, C. (2009), The relationship between stimulus-response mappings and the detection of novel stimuli in a simulated luggage screening task, Theoretical Issues in Ergonomics Science.

[24] For details see Discovery Toys Inc., Corporate web site http://www.discoverytoysinc.com

[25] Havard Business School (2000), Mind Mapping, *Harvard Busines Publishing Newsletter*, November, 01.

[26] Pederson,C.A.,Wright, M.D., and Weitz, B.A.(1984), Selling: Principles and methods, Richard D Irvin Publishers, Homewood, IL.

[27] Simintiras, A. C. andCadogan, J. W. (1996), Behaviourism in the study of salesperson-customer interactions, *Management Decision*, 34 (6), 57-64.

[28] Villanueva, J.,Bharadwaj, P., Chen, Y., and Balasubramanian, S. (2004), *Managing Customer Relationships- Should Managers Really Focus on Long Term*, IESE Business School, Working Paper # D/560, May, pp 1-37.

[29] Anderson, J. C., Narus, J. A., Rossum, W. V.(2006), Customer value propositions in business markets, *Harvard Business Review*, 84 (3), 91-99.

[30] Garver, M. S. and Gagnon, G. B. (2002), Seven Keys to Improving Customer Satisfaction Programs, Business Horizons, 45 (5), 35-42.

[31] Rajagopal (2006), Measuring Customer Value Gaps: An Empirical Study in Mexican Retail Markets, *Economic Issues*, 11 (1),19-40.

[32] Brucks, M., Zeithaml, V. A. and Gilian, N. (2000): price and Brand Name as Indicators of Quality Dimensions of Customer Durables, *Journal of Academy of Marketing Science*, 28 (3), 359-374.

[33] Rajagopal and Sanchez, R. (2004), Conceptual Analysis of Brand Architecture and Relations within Product Categories, *Journal of Brand Management*, 11 (3), pp 233-247.

[34] Williamson, P.J. and Zeng, M. (2009), Value for money strategies for recessionary times, *Harvard Business Review*, 87(3), 66-74.

[35] For details on the corporate profile of the company see web site http://www.blyth.com

[36] IMN Letters (2009), Partylite fans the flames of success with IMN e-newsletters, Waltham, MA http://www.imninc.com

[37] Pfeifer, P. E. (1999), On the use of customer lifetime value as a limit on acquisition spending, *Journal of Database Marketing*, 7 (1), 81-86.

[38] Rust,R. T., Zeithaml, V.A. and Lemon, K. N. (2004), Customer centered brand management, *Harvard Business Review*, 82 (9), 110-118.

[39] Rajagopal (2007), Stimulating Retail Sales and Upholding Consumer Value, *Journal of Retail and Leisure Property*, 6 (2), 117-135.

[40] Lafferty, B. A. and Goldsmith, R. E. (2004), How Influential are Corporate Credibility and Endorser Attractiveness when Innovators React to Advertisement for a New High Technology Product? *Corporate Reputation Review*, 7 (1), 24-26.

[41] Tang, C. S., David, R. B. and Teck-Hua, H. (2001), Store choice and shopping behavior: How price format works, *California Management Review*, 43 (2), 57-74.

[42] Ahmed, S. A. andd'Astous, A. (2006), Product-Country Images in the Context of NAFTA: A Canada-Mexico Study, *Journal of Global Marketing*, 17 (1),23-43.

[43] Grenci, R. T., and Watts, C. A. (2007), Maximizing Customer Value Via Mass Customized e-Consumer Services, Business Horizons, 50(2), 123-132.

[44] Carter, T. (2009), Leadership and Management Performance, *Journal of Hospital Marketing & Public Relations*, 19(2), 142-147.

[45] Ghemawat, P. (2005), Regional strategies for global leadership, *Harvard Business Review*, 83 (12), 98-108.

[46] Mazzei, M. J., Shook, C. L., and Ketchen, D. (2009), Selling strategic issues: Crafting the content of the sales pitch, *Business Horizons*, 52 (6), 539-543.

[47] Larsen, T., Rosenbloom, B., Anderson, R. and Mehta, R. (2000), Global sales manager leadership styles: The impact of national culture, *Journal of Global Marketing*, 13(2), 31-48.

[48] Lane, N. andPiercy, N. (2009), Strategizing the sales organization,*Journal of Strategic Marketing*, 17(3), 307-322.

[49] For details on the company profile see its corporate web site http://www. hellodirect.com

[50] Lane, N. (2009), Searching for strategy in sales, *Journal of Strategic Marketing*, 17(3), 191-197.

[51] Oshagbemi, T. (2008), The impact of personal and organisational variables on the leadership styles of managers, *The International Journal of Human Resource Management*, 19(10), 1896-1910.

[52] Maguad, B. A. and Krone, R. M. (2009), Ethics and moral leadership: Quality linkages, *Total Quality Management & Business Excellence*, 20(2), 209-222.

[53] Jha, K. N. and Iyer, K. C. (2006), Critical Factors Affecting Quality Performance in Construction Projects, *Total Quality Management & Business Excellence*, 17(9), 1155-1170.

[54] For details on the Medfast Inc. direct selling plans see corporate web site http://ir.medifastdiet.com/

[55] Davenport, T. H. (2009), Make better decisions, *Harvard Business Review*,87(11),117-123.

[56] Duffy, D.L. (2005), Direct selling as the next channel, *Journal of Consumer Marketing*, 22 (1), 43-45; Also see details about the Tupperware Brand Corporation at the web site http://www.tupperwarebrands.com/

[57] Walter, A. andGemnden, H. G. (2000), Bridging the gap between suppliers and customers through relationship promoters: theoretical considerations and empirical results,*Journal of Business and Industrial Marketing*, 15 (2), 86-105.

[58] Gómez, M. I.,Maratou, L. M., andJust, D. R. (2007), Factors Affecting the Allocation of Trade Promotions in the U.S. Food Distribution System,*Review of Agricultural Economics*, 29 (1), 119-140.

[59] Davenport, T. H. (2005), The Coming Commoditization of Processes, *Harvard Business Review*, 83 (6), 100-108.

[60] Zoltners, A. A., Sinha, P., and Lorimer, S. E. (2006), Match Your Sales Force Structure to Your Business Life Cycle, *Harvard Business Review*,84 (7-8), 80-89.

[61] Kim, K. I., Syamil, A. andBhatt, B. J. (2007), A resource-based theory of supplier strategy, *International Journal of Logistics Systems and Management*, 3 (1),20-33.

[62] Platz, L. A. and Temponi, C. (2007), Defining the most desirable outsourcing contract between customer and vendor, *Management Decision*, 45 (10), 1656-1666.

[63] Backstrom, K. (2006), Understanding Recreational Shopping, *The International Review of Retail, Distribution and Consumer Research*, 16 (2), 143-158.

[64] Shoham, A. and Brencis, M. M. (2003), Compulsive Buying Behavior, *Journal of Consumer Marketing*, 20 (2), 127-138.

[65] Miranda, M., Konya, L., and Havira, I. (2005), Shopper's Satisfaction Levels are not only the Key to Store Loyalty, *Marketing Intelligence and Planning*, 23 (2), 220-232.

[66] Davies, A. (2004), Moving Base into High-value Integrated Solutions: A Value Stream Approach, *Industrial and Corporate Change*, 13 (5), October, 727-756.

[67] Lafferty, B. A. and Goldsmith, R. E. (2004), How Influential are Corporate Credibility and Endorser Attractiveness when Innovators React to Advertisement for a New High Technology Product? *Corporate Reputation Review*, 7 (1), spring, 24-26.

[68] Backstrom, K. (2006), Understanding Recreational Shopping, *The International Review of Retail, Distribution and Consumer Research*, 16 (2), 143-158.

[69] Rajagopal (2006), Leisure shopping behavior and recreational retailing: a symbiotic analysis of marketplace strategy and consumer response, *Journal of Hospitality and Leisure Marketing*, 15 (2), 5-31.

[70] Anselmsson, J. (2006), Sources of Customer Satisfaction with Shopping Malls: A Comprehensive Study of Different Customer Segments, *The International Review of Retail, Distribution and Consumer Research*, 16 (1), 115-138.

[71] Otieno, R., Harrow,C., and Lea-Greenwood, G. (2005), The unhappy shopper, a retail experience: exploring fashion, fit and affordability, *International Journal of Retail & Distribution Management*, 33 (4), 298-309.

[72] Wakefield, K. L. and Inman, J. J. (2003), "Situational Price Sensitivity: The Role of Consumption Occasion, Social Context and Income", *Journal of Retailing*, 79 (4), 199-212.

[73] Codispoti, M. andde Cesarei, A. (2007), Arousal and attention: Picture size and emotional reactions, *Psychophysiology*, 44 (5),680-686.

[74] Wirtz, J., Mattila, A. S., andTan, R. L. P. (2007), The role of arousal congruency in influencing consumers' satisfaction evaluations and in-store behaviors, *International Journal of Service Industry Management*, 18 (1),6-24.

[75] Menon, S. and Kahn, B. (2002), Cross-category effects of induced arousal and pleasure on the internet shopping experience, *Journal of Retailing*, 78 (1), 31-40.

[76] Ganesh, J., Arnold, M.J., and Reynolds, K. E. (2000), Understanding the customer base of service providers: an examination of the difference between switchers and stayers, *Journal of Marketing*, 64, 65-87.

[77] Rajagopal (2009), Outsourcing salespeople in building arousal towards retail buying, *Journal of Database Marketing and Customer Strategy Management*, 15 (2), 106-118.

[78] Zoltners, A. A., Sinha, P. andLorimer, S. E. (2006), Match Your Sales Force Structure to Your Business Life Cycle, *Harvard Business Review*, 84 (7), 80-89.

Chapter 3

INDUSTRIAL SELLING

Business-to-business (B-to-B) selling strategy is identical with the function of managing sales to industrial clients. This category of selling process involves transactions of goods and services between businesses, such as between a manufacturer and a wholesaler or seller, or between a distributors and a seller. It is estimated that the volume of industrial selling is much higher for the firms as compared to the volume of transactions to the consumers. The primary reason for this is that in a typical supply chain there will be many industrial transactions that can be carried out simultaneously, involving sub-component of the products or raw materials, while business-to-consumer transactions are handled case by case and are limited to the sale of the finished products targeted to the end client. For example, an automobile manufacturer makes several B-to-B transactions such as buying tires, glass for windscreens, and rubber hoses for its vehicles. The final transaction, a finished vehicle sold to the consumer, is a single business-to-consumer transaction. Most firms used price as their key competitive strategy in B-to-B selling process in order to capture market share and cost is controlled by reducing size of the sales force and expenditure on prospecting clients. Hence, most of the industrial sales activities are operated through electronic platform and mobile communications. Most of the industrial selling firms find it more complicated to develop uniform sales strategies than developing different strategies for each client and products. Industrial selling companies need to couple technical know-how with business savvy, as Dell has done. Accordingly, many firms have employed different measures, including raising more cash to stay liquid, adding new target groups and products/services, and going global [1].

Selling industrial goods has become more complicated in recent years due to the mismatch between the pace of technology growth and improvement in the selling skills. It is more likely to be the domain of teams that handle large accounts by coordinating their efforts across product lines. Even without formal teams, greater coordination is often required to land the sale and keep the client. Three concerns have the most serious effect on coordination with an account-sharing strategy that include compensation systems, goal setting, and staffing and training. However, salespeople of industrial and consumer goods can improve the selling process through their complex distribution channels by using various promotional techniques. Sales promotion can be used as a lever to move merchandise through the distribution pipeline by carefully organizing the channels of distribution and determining product objectives. This concept has a focused scope that is highly cost-effective and applicable to a wide range of products and situations to industrial clients [2].

Purchasing decisions in industrial clients frequently requires information on various factors that supports the quality, services and use value of the product under sales. Client decisions on buying are also time bound and purchasing professionals frequently lean on reviewing alternative product information sources, such as the Internet. Traditional direct selling approach is being substituted with a new integrated sales coverage method that incorporates the best practices of direct and database marketing and integrates field sales force into a new communication mix. As industrial clients largely validate their purchase decisions based on the alternative information they obtain from sources other than their salesperson, it is necessary that salespeople equip their prospecting strategy with wide relevant alternative solutions and able to generate stronger value propositions to compete in the marketplace. The sales professionals in selling industrial products and services need also to manage process of selling by controlling a falling outbound call rate of clients, increasing communication clutter, changing sales cycle, chaos among decision makers and influencers, and unclear client knowledge that may cause misleading decisions and feedback [3].

Comtech Telecommunications operates in three divisions: telecommunications transmission, mobile data communications, and RF microwave amplifiers. The company makes equipment used by largely by the US government and related defense contractors. The clients of the company include satellite systems integrators, communications service providers, and oil companies. Its transmission equipment includes modems, frequency converters, high-power amplifiers, very-small-aperture terminal (VSAT) satellite transceivers and antennas, and microwave radios. The mobile data communications segment provides clients with integrated solutions to enable global satellite-based communications when mobile, near real-time, secure transmission is required. A substantial portion of sales in this segment are derived from sales of our integrated mobile data communications solutions to the U.S. military. These solutions include mobile satellite transceivers, vehicle and command center application software, third-party produced ruggedized computers and satellite earth station network gateways and associated installation, training, and maintenance. The company offers products and services on satellite based mobile tracking to fleet operators companies whose vehicles transport oil or industrial chemicals [4].

Many successful firms selling industrial goods have increasingly laid focus on their core competency and perform only certain activities in the value chain, and source part of or the whole product from suppliers before they market the product to their clients. For example Nike focuses on the design and the marketing of athletic shoes and relies on suppliers on the actual manufacturing of the shoes. Instead of simply performing selling activities to their client firms should also develop client ties with the clients to enhance life time value [5]. The tasks of salespeople in business to business selling operations may be categorized as:

- Relationship manager
- Salespeople engaged in concept selling
- Technical sales
- New business selling.

Salespersons engaged in selling activities to the various business-to-business accounts should also focus on delivering quality service, beyond selling products. It is important for salespeople to know that sales firms in general have service orientation either in whole or part, because they create value for clients through performance. Thus, the foundation of industrial sales is based on the client relations and quality of service delivery to achieve sustainable growth. Salespeople should learn that great organizations sell products with great service to their clients and compete in good times as well as bad to attain the lead [6]. All salespeople learn in their basic training program that they should not contradict the client and thrust on them any buying situation as client is always right. A close examination of client behavior, however, reveals that this maxim is not always true. Clients can be not only wrong but also blatantly unjust and such clients take advantage of being always right by demanding unwarranted privileges during the sale negation, which adversely affects companies the performance of salespeople as well as market reputation of the firm [7]. Industrial selling efficiency can be developed by the salespeople through practicing ways to sell new concepts to the clients. In industrial sales the sales cycle is often shorter and clients demand new products frequently. There is a need to reduce sales cost and increase the concept-to-client discussion time, improve quality, and provide innovative technologies that can help clients to enhance user perceptions. By selling right concepts to the right clients, salespeople can increase volume of sales and their market share through efficiently integrating information with the client needs. Leading companies conduct a formal in-depth seller evaluation and risk assessment prior to making purchase decision [8].

Selling new products to the industrial clients is a more challenging task than selling the same to individual clients. New products suffer from high failure rates because of technical shortcomings. Some studies have revealed that timely and reliable knowledge about client preferences and requirements is the single most important area of information necessary for selling the new products. It is often complex for the salespeople to obtain such data, as it increases the cost of selling by ways of investment on time and resources. Some companies integrated the prospecting clients into the sales demonstration process by providing advanced information on the techno-economic features. These firms also ask their clients to show their inclination towards purchasing a new product based on the information made available to them by the salespeople as well as manufacturer. This process is identified as collective client commitment, which can help both manufacturers and selling organizations to avoid product failures, delays in delivery of product and incurring high costs during selling process. Hence, sales of technical products to the industrial clients can be planned with collective client commitment process, which enables clients to make right decisions. Accordingly salespeople can also serve a market segment through direct selling approach [9].

Selling in a new business environment needs an integrated planning. It would be necessary to determine the competitive growth, substitution effects of existing products and services, training to salespeople, investment on sales infrastructure, and role of the government, for planning strategies effectively to sell new products and services in a new business environment. The global selling firms must consider these issues into their selling strategies and preparing salespeople to manage the selling tasks in a new business environment. To successfully negotiate client in a new business the sales managers must rethink their strategic vision and retool their approach to each component of strategy: product and market focus, organizational and supply chain structures, talent management choices, and, of increasing importance, the management of corporate reputation and identity. In particular, selling firms need to work harder to adapt to local differences and pay more attention towards offering competitive advantage to the client and consider every prospect as a launch pad for new products [10].

SELLING TO RESELLERS

Selling products to industrial clients is a bi-dimensional phenomenon, which includes selling to industrial clients by the original equipment manufacturers (OEM) and in turn industrial clients sell the customized products to the users. Such industrial clients who purchase products from OEMs and further resell the products after customization are identified as value added re-sellers (VAR). A value-added reseller (VAR) is a company that adds features to products supplied by OEM and resells it (usually to end-users) as an integrated product or complete turn-key solution to the large accounts. This practice occurs commonly in the electronics industry, where, for example, a VAR might bundle a software application with supplied hardware. Resellers generally have pre-negotiated pricing that enables them to discount more than a client would receive by purchasing directly. Resellers are qualified for higher-tiered discounting due to previous engagements with other clients, and the strategic partnership between the vendor and VAR inherently brings the vendor more business. In addition to selling hardware and software, most resellers also offer strategic planning, system design, implementation, training, asset tracking, technical support, wiring, database development, Web development, consulting and research. In small and medium enterprises, a VAR is generally the vendor of choice for designing, setting up and implementing customized computer systems [11].

A reseller may be defined as a company or individual that purchases goods or services with the intention of adding value to the original product in reference to the consumer preference and engages in reselling them rather than consuming or using them. Such transaction is usually done for enhancing profit as well as establishing brand of the reseller. One example can be found in the industry of telecommunications, where companies buy excess amounts of transmission capacity or call time from other carriers and resell it to smaller carriers. The resellers' product fulfillment-based business model includes a corporate reseller, retail, direct market reseller, and e-retailer that earns less than 10 percent of its revenue from services.

Firms willing to develop new product sales platforms face a formidable challenge as prospective clients do not fully participate until they are confident that the number of end users of the products promoted by the salespeople will be good enough. Traditional strategies used by the sales firms to deal with such situation are found in offering price discounts on products or services to ensure future sales. However, these tactics are expensive and risk averse. Selling to re-sellers strategy can be the less expensive one for building two-sided platforms with the vendors and consumers. However, a firm may start its sales operations as a vendor selling products or services to clients on just one side of the market potential. With the "merchant to two-sided platform" strategy, a firm may start selling as a merchant buying goods from many different suppliers and reselling them, in the process absorbing all the risk of platform failure. In stage two, the firm shifts risk and control back to some or all of its suppliers, giving them more responsibility for managing inventory, pricing, and merchandising their wares [12].

Best Buy Company Inc. is the biggest consumer electronics outlet in the US which works as value added reseller of the products supplied by original equipment manufacturers. The company operates more than 3,900 stores throughout North America, Europe, China, and now Mexico, mostly under the Best Buy and The Phone House banners. Best Buy stores sell a wide variety of electronic gadgets, movies, music, computers, and appliances. In addition to selling products, the stores offer installation and maintenance services, technical support, and subscriptions for cell phone and Internet services. Averaging about 40,000 sq. ft., the big box stores are located in 49 states, Puerto Rico, and all Canadian provinces. While the US accounts for about 75% of the company's sales, it's growing quickly overseas. Generally, U.S. Best Buy stores' and U.S. Best Buy Mobile stand-alone stores' merchandise, except for major appliances and large-screen televisions, is shipped directly from manufacturers to our distribution centers located throughout the U.S. Major appliances and large-screen televisions are shipped to satellite warehouses in each major market. U.S. Best Buy stores are dependent upon the distribution centers for inventory storage and shipment of most merchandise. Contract carriers ship merchandise from the distribution centers to stores. Generally, online merchandise sales are either picked up at U.S. Best Buy stores or fulfilled directly to clients through the distribution centers of the company. The Best Buy stores also offer customized products and services. The company has strong service personnel termed as Geek Squad Agents who offer computer or network solution and also provide a 30-day warranty on all computer services. The service team also offers in-home consultation, HDTV mounting, and calibration [13].

Manufacturing companies in newly industrialized countries contribute substantially to their global innovation networks. These manufacturers perform not only production and physical distribution functions, but also handle design and engineering functions in the processes of bringing new products world-wide. No single selling strategy can fit all original equipment manufacturing (OEM) suppliers and they can sell their own-brand products. A dedicated OEM supplier should emphasize flexibility in manufacturing and selling practices and pursue cross-functional integration as the key for managing industrial accounts successfully [14]. The electronic commerce experience indicates that a good return management process supports client relationship management and enables capturing value by reselling, and redistributing returned products. It also highlights that effectively managing returns in e-business allows recapturing value from products [15].Value added reseller plays the role of a solution provider through careful coordination of four identified supplier's client activity sets in the process of reselling which include warranty services, support and maintenance activities, system extensions, and consulting and optimization services. This coordination target sets specific requirements for key account management organization [16].

A value-based channel management model at Cisco Systems for managing the Value-Added Reseller (VAR) channel can be well illustrated to explain the reselling concept. Cisco initiated a change from a volume-based channel management model that had been driving client value, to a value-based model that tied channel rewards to specific channel value-add activities in early 2001. Critical components of this new model included identifying opportunities for channel value-addition, architecting channel programs to enable value-added channel, tying financial rewards to value-added channel activities including a "holdback system", and exercising significant discipline to manage field pressures for volume-based rewards and diluting certification requirements. This model demonstrated that VARs can serve as a significant sales channel, profitably selling complex solutions to satisfied clients under a value added channel management framework and can also uncover demand opportunities that are incremental to the pull marketing of even a strong brand like Cisco [17].

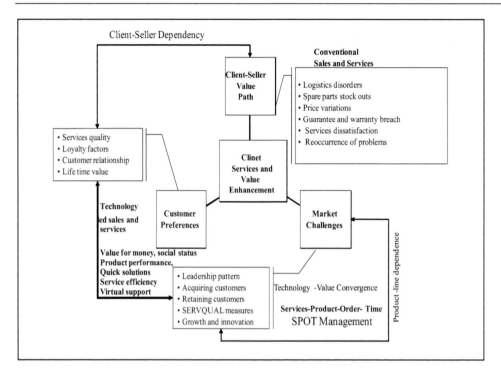

Figure 3.1. Client Services and Value Enhancement Paradigm for Salespeople in Business-to-Business Context.

SALES AND SERVICES INTEGRATION

Measuring and improving service performance becomes an essential strategy for success and survival in today's competitive situation, as service industries are sprouting at an incredible rate. Client satisfaction is perceived to be a key driver of long-term relationships between suppliers and clients, especially when clients are well acquainted with products and markets, and when industries are highly competitive. Services efficiency is one of the principal factors which influence client satisfaction in a business-to-business context and help bridging client-seller-supplier triadic relationship. The key services indicators, which include effective communication, cross-functional teams, and supplier integration, are followed to develop long-term relationships. Client satisfaction has long been considered a milestone in the path towards company profitability. It is widely acknowledged that satisfaction leads to higher market share and stable revenues while relationship between client satisfaction levels and quality of client services influence acquisition of new clients. In fact, there seems to be little guidance for linking company costs to the key elements involved in providing client satisfaction in services, thereby diminishing the ability of a company to manage its activities accordingly [18].

There is a growing competition in the high technology consumer products industry and market orientation has become one of the major factors associated with the success of high technology consumer products manufacturers. Many Asian high technology consumer products manufacturers have developed client orientation by equipping sellers and distributors to deliver quality services to augments client acquisition, satisfaction and retention. Recent research suggests that culture of an organization builds the values, beliefs and assumptions that reflect how services need to be delivered in a competitive market which may be perceived by the clients outside the organization as well [19]. Client retention seems to be the result of a kind of repetitive decision by the clients, but their decision to cross-buy involves a more complicated process. Client satisfaction includes location convenience, one-stop shopping convenience, product services, firm's reputation, firm's expertise in technology, and client relations on both client retention and cross-buying. Trust and satisfaction play different mediating roles in the relationships between service attributes, client retention, and cross-buying [20].

Relationship value is an antecedent to relationship quality and behavioral outcomes and displays a stronger impact on satisfaction than on commitment and trust. Value also directly impacts a client's intention to expand business with a supplier. In turn, its impact on the propensity to leave a relationship is mediated by relationship quality [21]. It is contended that enhanced understanding of this important aspect of business to business relationships leads to the development of more closely aligned strategic plans which may improve return on relational investment as exhibited in Figure 3.1. High technology products sales strategyand customer centric marketing approaches together have significant impact on the performance of retailers and distributors as illustrated in the Figure 3.1. Managers should integrate the high technology products sales and customer services strategies to enhance the customer value in reference to SPOT which indicate the integration of Services, Product-line management, Order-delivery efficiency and Time management as illustrated in the above figure. In developing SPOT coordination, top managers of firms and distribution agencies must pay careful attention to ensure consistent process though there might be variation in the consumer preferences locally. However, the services personnel should also be encouraged to come up with creative process improvements to overcome existing gaps and outperforming competitors [22].

FedEx Corporation hopes its package of subsidiaries will keep delivering significant market share. Its FedEx Express unit was the transportation provider, delivering about 3.4 million packages daily to more than 220 countries and territories. It maintains a fleet of about 655 aircrafts and more than 51,000 motor vehicles and trailers. To complement the express delivery business, FedEx Ground provides small-package ground delivery in North America, and less-than-truckload(LTL) carrier FedEx Freight hauls larger shipments. FedEx Office stores offeres a variety of document-related and other business services and serve as retail hubs for other FedEx units. FedEx offers technology ledground, express, international, freight, and compare services which are negotiated with the large and small accounts. The company followed direct selling strategies in prospecting and developing relationship with their business accounts. FedEx offered top-of-the-line services for unique small business needs through individual business analyses, tailored products and services, and overall supportive customer relations. The customer segment has been determined by the company on the basis of the amount of revenue FedEx obtained from the particular customer during the previous year. At the top of the pyramid were businesses with more than $10 million in annual FedEx billings. At the bottom of the customer-pyramid were small accounts of $6,000 to $40,000 in annual FedEx revenues. The account executives assigned to this segment were responsible for making telephone contact with local offices of the companies in higher levels of the hierarchy (Worldwide Sales, Global Sales, and Local National) which fell within their geographic regions of responsibility. In most of these cases, the account executive would not negotiate agreements with the local office since it is usually best for a customer to negotiate a single contract for all of their FedEx needs. In these cases, the account executive's job is typically to look for ways to increase the use of FedEx at the local office and to report to the World wide account executive any local issues that had arisen. Between them, the FedEx Express sales units geographically covered all 50 states, as well as key global customers abroad. FedEx has developed 'Insight Partners', an international association of professionals and academics that helps companies, governments and individuals negotiate, communicate and mediate more effectively. Their team of conflict management professionals helps Fortune 500 companies, governments and non-profit entities create value and improve working relationships [23].

Value of relationship with the client reveals significant quality and behavioral outcomes in the sales activities. Value displays a stronger impact on satisfaction than on commitment and trust, and also directly impacts a client's intention to expand business with the firm. Perceived strength of the relationship with the clients may be measured by salespeople in reference to technical ability, experience, pricing requirements, speed of response, frequency of client contact, degree of cooperation, trust, length of relationship, friendship and management distance barriers. The success of a supplier firm is interlinked with client value and if they serve clients by offering competitive gains, they will win. Client-seller partnerships are justified only if they stand to yield substantially better results than the firms could achieve on their own. Such co-dependency model elucidates the performance drivers associated with the success of supplier firms. The leading supplier firms are applying new technology, new innovations, and process thinking to augment client value, reduce costs, and enhance client services.

High technology products sales are positively associated with performance of sellers and distributors in terms of client service quality, growth in sales and increase in market share. Manufacturing and services elasticity is widely recognized as a critical component in achieving competitive advantage in the marketplace and improving corporate reputation to augment client value without escalating costs and time overrun [24]. High technology products sale also has a positive influence on measures of cost efficiency, such as productivity and sales per employee. In addition, profitability measures are highly associated with operating effectiveness and cost efficiency. The adoption of a high technology products sale can help supply channels design and offer a service mix that is perceived by core clients as of superior quality, while making a profit and building competitive advantage. Owing to the nature of the dyadic exchange process, the effect of high technology products sale for the supply channels may be more evident because customization may be observed directly by the buyers [25].

Industrial clients are sensitive in buying office durables, hygiene and sanitation products, and office equipments in reference to the country of design (COD) and also value the country of manufacture (COM) of branded products. The transfer of the COD image to brand image affects significantly the buying decision of clients. It is very high in the products of high technology audio visual product category. Brand and COM congruity also plays an important role in buying decisions in determining the product quality and value for money [26]. As prices for office durables and technology products fall over time with firms continually introducing enhanced products, consumers may anticipate these prices and improvements and delay their purchases in the product category. Forward-looking consumers optimize purchase timing in reference to the perceived use value and their expectations on future prices, quality levels, and brand availability. Such forward-looking behavior will result in price dynamics in the marketplace as price changes today influence future purchases [27].

The effect of services performance on relationship quality, in situations characterized by high relative dependence of the buyers on the sellers and distributors, is governed by the efficiency in delivery of services which develop high client satisfaction. Buyer-seller-distributor relationship may have significant effects on the focal firms in reference to the flexibility, responsiveness and client relationship management which would help building capability of supplier firms towards increasing competitive advantage and gaining high client value [28]. Satisfaction plays an important role in relationships, is instrumental in increasing cooperation between channel partners, and leads to fewer terminations of relationships. Client-supplier relationship gets closer and stronger through the information management at both the ends. Distributor information sharing helps to develop quality service oriented client relationships. Interestingly, even if the initial level of trust in the seller is low, the relationship quality substantially improves.

In a more competitive situation, suppliers respond more favorably to seller's information-sharing initiative. In addition, the evidence from previous studies indicates that profitability and market share growth also suffer. Measuring client satisfaction leads to identifying ways to improve product and services quality of the firm, which in turn leads to increasing the company's competitive advantage. The service oriented distributors are applying various listening tools to obtain information about clients' needs, preferences, and perceptions to manage effectively client satisfaction measures [29]. Client centric approaches are practiced efficiently by the call centers to connect the client issues with appropriate and interactive solutions. Call centers not only offer personalized attention to their problems but also help in building client loyalty. Clients rely on call centers which have high value work force towards services scheduling methods such as queuing system models to achieve optimal performance. Most of these models assume a homogeneous population of servers, or at least a static service capacity per service agent. It is observed that specialization minimizes the steady state queue size and reduces time of client services [30].

Many retailing firms are developing sustainable competitive strategies based on client-services relationship. Vertical integration, physical facilities, even a seemingly superior product no longer assures a competitive edge. Sustainable advantage is more likely to come from developing superior capabilities in a few core service skills. Services quality increases the leverage of sales activities in high technology consumer products industry and these changes drive managers towards structuring their organizations and define strategic focus. Most competitors know that a key factor in the success of Japanese network relationships is the practice of dedicating supplier services to the client [31]. Successful service managers pay attention to the factors that drive profitability in this new service paradigm constituting investment in people, technology that supports frontline workers, revamped recruiting and training practices, and compensation linked to performance. The service-profit chain which is developed from analyses of successful service organizations establishes relationships between profitability, client loyalty, employee satisfaction, and productivity [32].

An agent-based model of the online electronics market, consisting of large number of clients and sellers reveals that clients repeatedly interact with sellers, searching for high quality and low price, but only some buyers know seller quality prior to purchase. Clients may learn seller's quality directly or from information obtained from other clients or else estimate quality through market share heuristics [33]. In the growing competitive market client value is perceived in reference to benefits received and price paid. Salespeople can identify any misalignment and use services that add to the benefits of clients to balance the benefit-price equation after measuring clients' perceptions of value [34]. However, in order to meet the increasingly more complex needs of clients and respond to decreasing product margins, sales oriented firms have developed a growing interest in extending their values by combining with sales and service activities. Over the last couple of years, most sales oriented companies have explored the numerous opportunities offered by traditional client service and product-related services. It is observed that cost of client service, risk aversion, tendency to set over-ambitious objectives, and competitive advantage factor play critical role in creating client support services [35].

It has been observed in a study that sales performance is a function of outlet attraction, effective distribution management, services quality, price and promotional strategies as functional factors. Besides, relational variables including personalized client services, leisure support and client convenience also influence the performance of sellers and distributors [36]. The key elements of client-seller-supplier relationships including long-term relationships, communication, cross-functional teams, and supplier integration are followed at different levels of transactional process. Besides, client satisfaction, reliability, and product-related performance are the major drivers that play significant role in the success of such dyadic relationship. Toyota and Honda, leading Asian companies offering sales and services of gasoline based power generators in India, have built taxonomy of client-seller-supplier relationships by consistently following six steps which include understanding functional pattern of distributors, reducing seller-distributor rivalry and providing better opportunity, monitoring closely client relations, enhancing supplier capabilities, sharing information intensively but in a selective way, and help clients continuously improve their association [37].

In the growing competitive markets the large and reputed firms engaged in industrial selling are developing strategies to move into the provision of innovative combinations of products and services as 'high-value integrated solutions' tailored to each client's needs than simply 'moving downstream' into services. Such firms are developing innovative combinations of service capabilities such as operations, business consultancy and finance required to provide complete solutions to each client's needs in order to augment the client value [38]. Manufacturer's high technology products sales increases economic satisfaction of client with the distributor as it reveals competitive advantage in making the buying decision. A strong market oriented strategy of the firm alleviates the possibility of using coercive influence strategies by the competitors and offers advantage to the clients over competitive market forces [39]. High technology products sales is an organization-wide concept that helps explain sustained competitive advantage. Since many manufacturing firms have linked their marketing strategies with services delivery attributes, the concept of high technology products sales is expanding as a system in global corporate settings. The process of high technology products sales to industrial clients contributes to continuous learning and knowledge accumulation by an organization which continuously collects information about clients and competitors and uses it to create superior client value and competitive advantage [40]. In order to cope with increasing competition, high-tech firms need to continuously launch new products. However, adoption of new products may require substantive cognitive efforts from consumers. Therefore, firms engaged in serving to business-to-business accounts should be able to monitor and influence the knowledge base of their clients. A learning roadmap is a relevant instrument for salespeople and indeed a successful concept for implementing customized sales strategies [41].

Firms selling to industrial buyers may routinely pass these costs on to their clients for most of the new products, resulting into high prices. However a less obvious strategy in a competitive situation may be to maintain price, in order to drive the new product in the market with more emphasis on quality, brand name, post-sales services and client relations management as non-price factors. It has been argued that building client value through traditional relationship helps in increasing the demand for products and services of the client and strengthens association with the selling firms [42]. Cognitive and innovativeness attitude among industrial clients enhances the actual adoption of sales offers whereas sensory innovativeness and perceived social and physical risks enhance the propensity to acquire information on alternative sources of buying. Financial risk, on the other hand, has a negative impact on the buying decision of clients. Time, performance, cognitive attributes, and risks of network externalities also affect the decision making of clients during the selling process [43].

INDUSTRIAL SELLING PROCESS

The business-to-business selling process involves a series of steps that may appear to be a straight-line process, but there is a lot of movement back and forth between the different steps that ultimately lead to closing the deal. The selling process will vary by industry and personal preference, but these six essential steps will help you close more deals. The selling process initiates with smile on the face of salesperson to greet the prospect. This initial introduction will set the stage for the rest of the selling process. This is the icebreaker, the time to get the customer to relax and feel comfortable. A salesperson can explore the scope of building rapport with the client or an account to develop appropriate selling strategy that can help in enhancing the perceived value of client. The successful initiation of selling process towards a prospect leads to develop client-seller-supplier triadic relationship to drive the future action on decision making. There are many factor that affect this relationship and selling behavior. The relationship construct is woven around the cognitive variables that include interaction intensity, preferential treatment and tangible treatment, and examines their different effects on three types of client-seller trust, including attribute-based trust, process-based trust, and institution-based trust. It is observed that different relationship selling behaviors indirectly increase the social and functional relationship value in light of different types of trust [44]. The selling process in general includes the following steps:

- Prospecting customers
- Opening the relationship
- Qualifying the prospect
- Presenting the sales message
- Closing the sale
- Servicing the account.

Salespeople should develop skills on qualifying the prospect to proceed ahead with the selling process. At this stage the salespeople explore the customized requirements of client and evaluate his demands, affordability, and potentials of a good referral and repeat buying possibilities. Salespeople may take adequate time to analyze information at this stage. Accordingly sales presentations can be organized with prior exercises on handling the questions during the presentation session. The client may be invited for negotiations on the deal and after managing carefully the win-win negotiations, salespeople may close the deal with the agenda on servicing the account.

Most of the global firms engaged in manufacturing consumer products are focusing on the strategies to enhance consumer value by providing opportunities to experience the products in retail outlets. This strategy has emerged as Do It Yourself (DIY) campaigns for the consumer products which have helped firms to gain consumer loyalty and stay ahead of competition in emerging markets. Consequently, cosmetic and toiletries sales in Belgium had moved at a slightly higher rate in 2007 against the previous year. According to industry sources, this improvement coincides with more favorable economic and consumption context and an exceptionally high level of innovations. Manufacturers focus on more efficient deodorants, sophisticated and/or more natural bath and shower products, more segmented hair care and men's grooming products giving a tanned effect in skin care and color cosmetics. In an increasingly segmented market, they particularly appreciate anti-ageing and firming properties, protective and gentler formulations that often derive from skin care, the main growth driver of the market. All these products have been offered to consumers using DIY promotion at major retail outlets. As the consumers drew satisfaction over the usage of cosmetics and toiletries, Belgian males sought from the companies more convenient blade systems, their own shampoo and shower gel but are still uneasy with openly using personal care products such as skin care [45].

It is necessary for salespeople to be innovative in managing accounts. The just-in-time (JIT) selling strategy is one of the innovative perspectives in this regard. Implementation of JIT-selling strategy requires demonstration of selling tactics that build value during the selling process based on the organization's JIT related capabilities, such as zero-defect production, on-time delivery and quantity precision. The next stage in this process requires development of single-source relationships with customers that result in the full integration of on-site sales representatives into customers' purchasing processes [46]. Among the technology led selling process, digital selling has emerged as most effective solution to clients for accelerating the process of buying. Firms are creating a digitized selling capability by developing Web sites designed to provide information and conduct transactions with customers, replacing many routine sales force activities. This process of sales reveals salesperson effectiveness in managing the selling process to industrial clients. It has been observed that digitization has the positive effect of improving salesperson effectiveness and client satisfaction. Accordingly, managers can improve the technology-enabled multichannel capabilities of the firm by giving priority attention to human capital improvement, sales force control systems, and communication of the digitization strategy [47].

Table 3.1. Stages in Buying and Selling Process

Decision Stages	Buying Process	Selling Process
Stage 1	Anticipation or recognition of the need or problem	Initiating the process to know clients and his needs and setting the stage for sales
Stage 2	Determination and description of the traits and quality of the needed item(s)	Portraying the problems and generating awareness and comprehension on the proposed solutions
Stage 3	Search for and qualification of potential suppliers	Leading the competition, making visual presentations and demonstration of prototypes allowing hands-on experience
Stage 4	Acquisition and analysis of proposals or bids	Developing client-centric proposal, maintaining relationship and carrying periodical follow-up
Stage 5	Evaluation of proposals and selection of suppliers	Engage in discussion on the proposals, set-up negotiation table, adjusting attributes of the deal, placing agenda of client services
Stage 6	Selection of an order routine	Winning the order, setting effective coordination with suppliers and service providers, installation and functional tests
Stage 7	Performance evaluation and feedback	Continuing relationship and client satisfaction measurement

It is necessary for a salesperson to match his selling strategies with the various levels of decision process of the client. A relationship matrix needs to be developed by the salespeople in establishing congruence with the buying process followed by the client. The Table 3.1 exhibits different stages of buying and selling process, and convergence of strategies to be developed by the salespeople.

Sales planners can contribute substantial revenue by developing a percent congruence of actions between client and salesperson during the selling process.Employing relationship selling is crucial for customer centric or service oriented salespeople. It is observed that salesperson's expertise, willingness, and power have a positive effect on selling process to perpetuate the sustainable value added relationship with the clients. In addition, perceived trust in the salesperson and satisfaction with the interactive sessions with the salesperson during the negotiation process also drive the confidence of client in making buying decisions [48]. However, it is important for the salespeople not to allow self-esteem override their personality during the selling process. Such adaptive consequences of pride in personal selling drive the self-regulation during the selling process on one hand and it increases salespersons' performance-related motivations on the other. Specifically, it promotes the use of adaptive selling strategies, greater effort, and self-efficacy.Salespeople, in general, are affected by their emotions, but they are also capable of controlling them to their advantage. Salespeople are capable of self-regulating the expression of these emotions differently toward colleagues and customers via anticipated feelings of fear, shame, and regret [49].A study that investigated the impact of selling strategies on selling effectiveness revealed that when sellers use an agenda selling strategy, target products receive higher evaluations and have higher probabilities of being considered and chosen. In an agenda sales strategy, a salesperson attempts to influence the structure of the buyer's decision by suggesting constraints that eliminate competitive products from consideration and also summarizes the target product's benefits. However the knowledge and expertise of client may moderate this effect. Salespeople in most cases can succeed with the agenda strategy with relatively new clients than expert buyers. The importance of developing appropriate selling strategy to achieve higher selling effectiveness suggests that salespeople can discuss the potential benefit of the target products to influence the structure of the buyer's decision [50].

Salespeople of manufacturing companies who target their products to retail self-service stores prospect them on the rationale of *touch, feel and pick* by demonstrating the products to make buying decisions. The personal selling activities in selling products to large retailers are largely based on the DIY opportunities, which develop the major motivation among the clients and also support them in their decision making process. Motivational factors among prospecting clients play a key influencing role in the driving the buying behavior. Purchasing motives, values and perceived alternatives of clients are often considered as independent inputs into a choice model. Such client behavior shows that buying decisions are influenced by the perception developed over product demonstration [51]. In Business-to-business demonstration process client analyzes alternative brand options, in reference to both service and merchandise quality as it exerts significant influence on store performance measured by sales growth and consumer growth [52]. The business-to-business sales companies should develop congruent strategies to match the broad marketing-mix of clients that helps stores to specialize in certain product categories like *The Home Depot,* which specializes in retailing building construction, interior decoration and gardening equipments. The Home Depot is a US retail self-service chain store operating in North American and Latin American countries. The retail self-service stores operate on a market size effect and a price cutting effect [53].

BUSINESS-TO-BUSINESS SALES RELATIONSHIP MODELS

Industrial selling models are largely built on the relationship networks. The key account management (KAM) is a rapidly growing area of interest in business-to-business marketing. However, unnoticed by marketing, a quiet revolution has taken place in supply chain management (SCM), where the traditional emphasis on least-cost transactions has given way to a focus on long-term relationships with a few key suppliers. It is thus apparent that the two disciplines are converging [54]. Fiocca (1982) suggests a number of mechanisms for assessing the proposed axes: 'Difficulty in managing the customer' is a function of the level of competition for the customer, customer buying behavior and the characteristics of the product bought by the customer [55]. 'Strategic importance' is determined by the value/volume of purchases, the potential and prestige of the customer, customer market leadership, and process improvements and adaptation to customer specifications. This combination of subjective and actual values makes such calculations difficult especially when the main point of using such analysis is surely to produce data which can be used for comparison. 'Business attractiveness' is determined by considering a number of factors that are related to the customer's market (growth rate, competition, maturity, changes in the environment, etc.) and the status/position of the customer's business within the market. The strength of buyer-seller relationships is measured by applying sales objectives to judgmental or subjective factors, which include:

- Length of relationship,
- Importance of the customer,
- Friendship,
- Co-operation in product development,
- Social distance

Perceived strength of the buyer-seller relationship can be measured using the variables technical ability, experience, pricing requirements, speed of response, frequency of contact, degree of cooperation, trust, length of relationship, friendship and management distance (frequency of contact). A customer portfolio for retail operations includes new-primary, loyal-primary, and defector (formerly primary) customers, as well as occasional customers who shop primarily at competing stores. The attitude-accessibilities dyad of business-to-business clients may offer useful information about the reasons of their preferences for sales or supplier agencies. A study revealed that the attitude and accessibility of the major retail clients differ substantially in reference to loyalty. More specifically, for a given store, greater shares of loyal and new clients will exhibit positive, store attitude-accessibility profiles compared to non-customers and defectors. A greater share of defector clients exhibited a negative attitude-accessibility profile compared to each of the loyal and new buying segments [56].

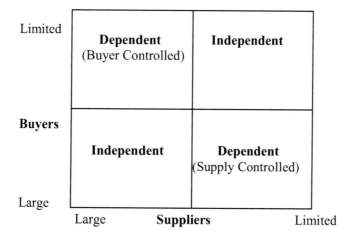

Figure 3.2.Power Judgement in Portfolio Decision.

Reviewing back, Campbell and Cunningham (1983) proposed a three step portfolio analysis strategy for marketing management. The Figure 3.2 exhibits the power balance factors and their impact on making the portfolio decisions. Using a case study of a major packaging supplier, they suggest a three step analysis using two variables at each stage. The *first step* focuses on the nature and attractiveness of the customer relationship using customer life cycle stage on one axis and various customer data on the other. The customer life cycle stage is divided into tomorrow's customers, today's special customers, today's regular customers and yesterday's customers.

Another dimension of this analysis is multivariate, involving sales volume, use of strategic resources, age of relationship, supplier's share of customer's purchasers and profitability of customer to supplier. This type of categorization will facilitate the understanding of how strategic resources, which will ensure the future health of the business, are allocated among customers [57]. First step is the conceptual validity and practicality of using a life cycle approach to customer analysis can be challenged. Secondly, the choice of appropriate variables for analysis can be difficult; obtaining the required data on the variables can also present major problems. The *second step* of analysis focuses on the customer's own performance as an important aspect of customer portfolio planning. The third and *final step* involves the selection of the key customers for analysis. Another two-dimensional grid is proposed for this stage with growth rate of Customer's market (high, medium, low and decline) on the vertical axis and competitive position (relative share of customer's purchases) on the horizontal axis. Companies are placed on the matrix and are represented by a circle that represents their sales volume. However, such a framework provides a useful conceptual starting point for undertaking strategic analysis of an organization's customer portfolio. The benefits of consumer-firm relationships have been recently addressed and include increasing efficiency and effectiveness in maintaining current customers rather than prospecting new customers, and improved competitive advantage. The consumer benefits through consumer learning, in such situations can be stored, processed and retrieved to use in subsequent situations. This leads to an ability to manage future decisions based on simplifying problem-solving situations and reducing risk [58].

It is necessary that salespeople understand the difference between sales efficiency and sales effectiveness. Sales efficiency can be measured as ratio of output i.e. the volume or value of sales realized, and input i.e. selling costs; whereas the sales effectiveness is measured as the degree of variation between actual and desired output. Hence the client portfolios need to be carefully developed integrating the attributes that help maintaining the long-term relationship. Conventional relationship marketing has focused on long-term, committed and partnerships relationships in both business-to-business settings and also firm - to - consumer settings. It has also been assumed to some extent that relationships are always desirable.

Buyer-seller Relationship Theories

It is important for the salespeople to consider a client as a profit centre in business-to-business environment while developing a client perception matrix. Among many, three major variables including costs to serve suppliers, customer behaviour and management of customers can be used to investigate the profit dispersion by the salespeople within the client portfolio. Four types of costs comprising pre-sale, production, distribution and post-sale service costs determine the cost to serve axis. It is argued that while many suppliers believe

that if they analyze the breakdown of their accounts, most accounts will fall into the'carriage trade' and 'bargain basement' quadrants. Yet, when analysis is actually performed, it will usually show that over half a suppliers' accounts fall into the 'passive' and 'aggressive' quadrants as exhibited in Figure 3.3. They contend that four aspects of the customer's nature
      Krapfel, Salmond and Spekman [60] also used a portfolio approach to analyse buyer-seller relationships and proposed a relationship classification matrix based upon the concepts

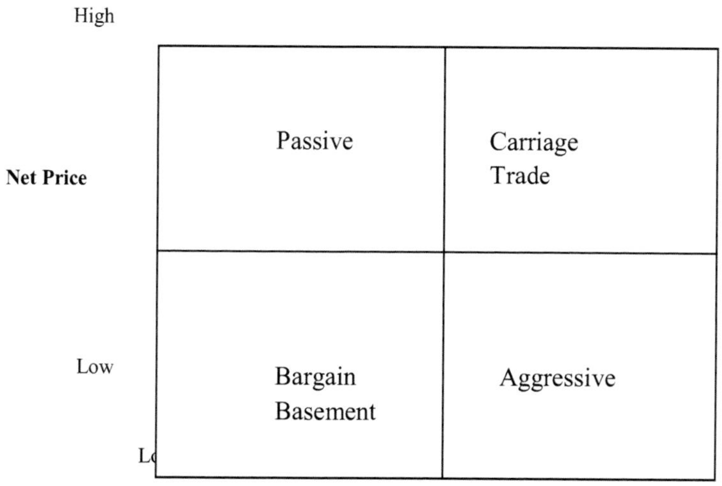

High

Net Price	Passive	Carriage Trade
Low	Bargain Basement	Aggressive

Figure 3.3.Customer Perception Matrix.

of 'relationship value' and 'interest commonality'. They defined relationship value as a function of four factors consisting of criticality, quantity, substitution and slack. This relationship can be expressed as:

$$RV_i = f\left[C_j Q_j R_j S_j\right]$$

Whereas,

RVi is the value of the relationship to the seller
Cj is the criticality of the goods purchased by the buyer
Qj is the quantity of the seller's output consumed by this buyer
Rj is the subsitution of this buyer (i.e. the switching cost of accessing other buyers)
Sj is the cost savings resulting from the buyer's practices and procedures

They argue that three-dimensional analysis based upon cost to serve, net price and relationship value, is appropriate when segmenting the customers of any firm, especially because such an analysis provides a more comprehensive overview than can be gained from simply using two variables. These portfolio theories have been tested over the period with variety of data sets and improved subsequently. The major criticisms on the current models include as following:

- Is it viable to transpose product life cycle concepts into a 'customer life cycle' and then use this as a basis for planning? While a number of authors have discussed this concept at length, its application to this sort of analysis can be problematic.
- There are a wide range of variables and potential ways to calculate the dimensions of analysis, which mitigates against easy comparison of analyses.
- The actual analysis may be easily distorted by a number of factors, including:
  - lack of accurate data
  - suppliers being reliant on one or two major customers
  - data being collected over too short a period
  - the subjective basis of many of the variables.
- Many of the models do not explicitly include customer profitability; experience shows that customer profitability data is difficult to collect: although direct costs should be apportioned directly on a customer-by-customer basis, many companies do not have adequate mechanisms for allocating indirect costs.
- When matrix positioning involves a mixture of actual and subjective data, the results may prove unsuitable for use in future comparisons. Although weighing of variables may go some way to alleviating this.
- Generally, the scales proposed for axes are imprecise; for instance, what are low and high values? Again, such values implicitly involve subjective judgements and therefore become more difficult to assess. However, they can be very useful if it is accepted that they simply provide a rough conceptual guide to sorting out the major customers from the mass of customers, especially when it is not very clear what to do because the majority of customers occur in a large cluster.

In a dynamic framework where franchise value is determined endogenously by the clients in business-to-business environment, different techno-economic determinants affect the risk taking behaviour of the client during the decision making process, the probability of default and the value of deposit insurance liability. On the contrary a dynamic active portfolio management problem can be considered in selling products and services to the industrial clients, where the objective is related to the trade-off between the perceived use value and the risk of economic disadvantages. This allows sales managers to make quantitative analysis of the risk/return trade-off, with risk defined directly in terms of probability of shortfall relative to the benchmark and return defined in terms of the expected time to reach investment goals relative to the benchmark [61]. The market environment related factors affect the customer portfolio decisions to a large extent in a given market conditions. The client-organization (C-O) fit has been exhibited in the model and the major attributes of the factors involved in buying decisions [62]. The model may be considered as an instrument to analyze the client attraction in reference to competition, brand life cycle, diffusion and adaptation of innovation and technology that determine the strength of industry attractiveness. The construct of the model has been built around the market environment factors between selling and buying organizations on five indicators that include behavioral dimensions of the clients, sales and service attractiveness, competition, economic variables and brand performance. The risk factor is predominant in marketing and is associated with the product attractiveness in making buying decisions. The strategic sales perspectives in reference to portfolio of industrial clients are exhibited in Figure 3.4.

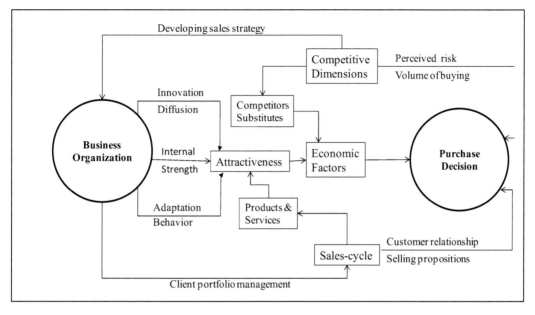

Figure 3.4.Client Portfolio and Sales Paradigm in Business-to-Business Evironment.

The strength of the business organizations in effective diffusion of innovations and technology and inducing the responsive behaviour towards its adaptation would help in building the industry and product attractiveness. The Figure 3.4 illustrates that the industry attractiveness is measured in terms of its competitive gains that reflect in terms of the relative market share, growth and sales. The sales cycle in association with product and services life cycles determines the product attractiveness factors. The determinants that affect the buying decision of clients include 4As comprising accessibility, approachability, affordability, and adaptability; and 4Cs encompassing convenience, comprehension, cost to customers and care along with price and non-price factors leading to quality of sales and services. The risk factor in buying decisions may drive the clients towards higher prices and lower risk premiums for an isolated client portfolio; while for the repeat buying lower prices with low premiums may turn out to be favourable [63].

The value concept in the above relationship governs the client portfolio in terms of formulation of recursive utility over time in business-to-business sales operations. It shows that the optimal portfolio demand for products under competition varies strongly with the values associated with the brand, industry attractiveness, knowledge management and ethical issues of the organization. The extent of business values for both selling and buying firms determines the relative risk aversion in terms of functional and logistical efficiency between the organization and supplier, while the switching attitude may influence the clients if the organizational values are not strong and sustainable in the given competitive environment. The model assumes that a high functional value integrated with the triadic entities would raise the market power of organization, sustain decisions of customer portfolios and develop long-run relationships thereof. The clientvalue concept is utilized to assess product performance and eventually to determine the competitive market structure and the product-market boundaries. The value driven client portfolio management process is exhibited in Figure 3.5.

The model exhibited in Figure 3.5 illustrates that the value based customer portfolios would enhance the customervalue in reference to the ratio of outputs (e.g. reuse value, reliability, safety, comfort), which customers obtain from product related inputs (price, running costs). The derived efficiency value of product or service targeted to clients for sales can be understood as the return on the client's investment. Products offering a maximum customervalue relative to all other alternatives in the market are characterized as efficient. The value based portfolio model illustrates the client portfolio management (CPM) within the triadic relationship of the organization-supplier and customer. The client values are reflected in their competitive gains, perceived use values, volume of buying and level of quintessence with the customer relationship management services of the organization. If these variables do not measure significantly, there emerges the development of switching attitude among the business-to-business buyers. If the organizational values are low the relationship of clients with sellers may be weak and myopic.

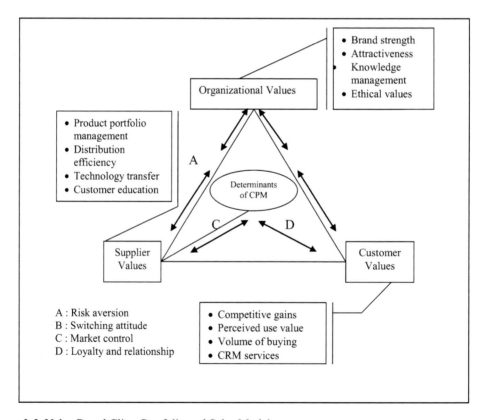

Figure 3.5. Value Based Client Portfolio and Sales Model.

An appropriate client portfolio model can be determined by the salespeople for optimizing returns on sales and managing value-at-risk measures by following the low selling high customer value strategies. The application of these models needs adequate training to salespeople, strong support of data base and communication skills. These models can also be used by the corporate sales managers to forecast the client preferences, demand, brand switch probabilities, response to pricing strategies, and impact on the innovations to equip salespeople to develop appropriate strategies in selling. This analysis of the client data can be done through the following process:

- Identifying the corporate needs
- Defining the problem for analysis
- Analyzing the nature of data-discrete, clustered, time series etc
- Choosing the appropriate model with assumptions
- Testing the model
- Analyzing results
- Undertaking repeated analysis of the data
- Preparing appropriate action plans
- Implementing the plans
- Monitoring and evaluating the outcomes of plans, and
- Input the results back into the planning process.

On a tactical level managers need to consider the optimum spread of clients the sales territory. This needs careful attention and the application of managerial judgement and experience. Salespeople should also be prepared to face the varied management style in response supplier-customer relationships. Client portfolio management is a multi-level process that encompasses understanding their business portfolio, developing a plan, managing implementation of the plan and evaluation thereof. The sales managers of the company may identify high-potential clients and in order to increase their satisfaction through target offers should drive personalized services to ensure loyalty and long term relationship.

Miles and Snow Competitive Sales Model

Miles and Snow [64] argued that companies have different strategies to address three fundamental problems emerging from entrepreneurial, operational, and administrative points of view. Managing effective market share is identified as an entrepreneurial problem, while implementing an appropriate solution to a given market situation is considered as an operational issue. It has been postulated by Miles and Snow that there are four general strategic types of organizations including prospector, defender, analyzer, and reactor organizations. The typology of Miles and Snow guides the sales organizations about the attributes of client organizations and to enable them to develop appropriate selling strategies by understanding the client orientation on buying products and services. The implications of Miles and Snow typology for industrial sales organizations are exhibited in Table 3.2.

Table 3.2. Organizational Typology and Management of Industrial Sales

Miles and Snow's typology	Sales force implications
Prospector	
Attempts to lead in developing new products Innovates and leans on introducing improvements frequently Ignores short-term profits to gain strategic advantages in the long-term Looks for strong-hold in their markets and sustainable growth.	Primarily focuses volume of sales and growth Manages territory with emphasis on prospecting Develops value chain based industrial customer networks
Defender	
Offers a narrow product-mix with sustainable product line Avoids penetration into new markets Operates in a predictable market Moves with slow growth or enters in early maturity phase of the product life cycle Low-cost manufacturing with high product turnover	Maintains the current customer base Puts less effort on prospecting for new customers Links customer service to account attributes
Analyzer	
Chooses high-growth markets Shows attributes of an intermediate type of firm Manufactures fewer product Slow reaction to product market changes Exhibits, weak stability and efficiency than defenders	Leans on offering multiple roles Prioritizes servicing existing customers Prospects for new customers Manages distribution of mature products Organizes campaigns for new products.

Prospector organizations thrive in improving business environment in search of new opportunities for salespeople. Moreover, prospector organizations look for broad product or service lines and often lean on creativity for higher efficiency and effectiveness. Consequently, prospector companies prioritize suppliers, and sales and services companies to new and changing needs for creating new demands. The administrative problem of these companies is how to coordinate diverse business activities and promote innovation. Prospector organizations solve this problem by being decentralized, employing generalists (not specialists), having few levels of management, and encouraging collaboration among different departments and units. It is argued that generic sales strategies are of considerable value in the regulated industries [65]. Defender organizations stress on cost leadership and so these organizations achieve success by specializing in particular areas and using established and standardized technical processes to maintain low costs. In addition, defender organizations tend to be vertically integrated in order to achieve cost efficiency. Defender organizations face the administrative problem of having to ensure efficiency, and thus they require centralization, formal procedures, and discrete functions.

Analyzer organizations share characteristics with prospector and defender organizations; thus, they face the entrepreneurial problem of how to maintain their shares in existing markets and how to find and exploit new markets and product opportunities. These organizations have the operational problem of maintaining the efficiency of established products or services, while remaining flexible enough to pursue new business activities. Consequently, they seek technical efficiency to maintain low costs, but they also emphasize new product and service development to remain competitive when the market changes. Reactor organizations, as the name suggests, do not have a systematic strategy, design, or structure. Their new product or service buying process fluctuates in response to the way their managers perceive their environment. Reactor organizations do not make long-term buying plans or making association with selling organizations, because they see the environment as changing too quickly for them to be of any use, and they possess unclear chains of command.

This model argues that companies develop their adaptive strategies based on their perception of their environments. Hence, as seen above, the different organization types view their environments in different ways, causing them to adopt different strategies. These adaptive strategies allow some organizations to be more adaptive or more sensitive to their environments than others, and the different organization types represent a range of adaptive companies. Since sales strategies vary from organization to organization, having a less common strategy may be beneficial to most of the industrial clients. On the other hand, prospector organizations clearly have an advantage over the other types of organizations in business environments with a fair amount of flux. Sales companies operating in mature markets in particular benefit from introducing new products or services and innovations to continue expanding.

Table 3.3. Supplier Strategies in Business-to-Business Environment

Porter's typology	Sales force implications
Supplier attributes	
Aggressive selling	Sales and service to large accounts
Linking performance with efficiency-scale	Low operational costs
Focus on reducing selling cost	Price is considered as the decisive factor
Non-flexible economic budgets and overhead control,	Driving high order sales
Aiming at high relative market share	
Sales differentiation	
Setting up unique selling proposition	Selling on non-price benefits premise,
Protecting accounts against competitive	Generating quick orders
Developing brand loyalty	Providing high quality of customer service
Lower sensitivity to price	Staying responsive to buyers,
	Targeting low prices sensitive accounts
	Selling through high-quality sales force.
Niche selling	
Market orientation	Becoming experts in the operations
Developing market led functional policy	Exploring opportunities within target market
Limited market share	Focusing customer attention on non-price
Dominates business in a small market segment	benefits Allocating selling time

Porter's Typology

Some manufacturers in the business-to-business environmenthold an incremental profitability due to innovative marketing strategies in multi-domestic strategy, which allows them to state their manufacturing requirements independently and choose suppliers of raw material and ancillary services in different domestic markets. Before developing a global procurement strategy a company recognizes its business as potentially global look forward to establish long term buyer-seller relationship [66]. The porter's typology on industrial selling is exhibited in Table 3.3.

Selling industrial goods has become more complicated in view of the growing competition. The firms engaged in selling products and services to industrial clients coordinate large accounts through their efforts across product lines. However, sometimes with small and specialized sales teams industrial products selling firms drive greater coordination and provide high customer value. In the porter's typology of industrial sales three concerns have the most serious effect on coordination with an account management strategy, which include value systems, goal setting, and sales administration [67].

EXTERNAL AND INTERNAL FIT

Salespeople should know that the demand for products in the business-to-business marketplace is usually derived from the demand of products at the end user level. The manufacturing companies first assess the consumer preferences for the finished products and respond accordingly to the stimulus of products being prospected by the sales people. Such derived demand in industrial marketplace can be illustrated with an example of selling raw material to an automobile company. Different parts assembled in an automobile are the end result of a supply chain consisting of many other components and raw materials. The dashboard is usually molded from a chemical compound Acrylonitrile butadiene styrene [68], commonly known as ABS plastic. The ABS plastic is made through a series of chemical processes derived at three different levels generating opportunities for selling root chemicals. ABS plastic resin is manufactured from styrene, while styrene is made from ethylene, and the ethylene is extracted from petroleum. Thus the demand for each of these intermediate goods is influenced by the consumer demand for automobiles and during the recession in automobile market, manufacturers of the ABS pellets are directly affected. Many manufacturers have sales forces for "missionary selling" to locate potential customers for their intermediaries, consumer preferences, economic cycles, and social trends affect industrial markets by influencing final consumer demand. Hence, it is argued that industrial selling has a long term perspective unlike selling consumer goods, and salespeople lean on strategic selling approaches than applying tactics to reach quick results [69].

Henkel manufactures adhesives, sealants, and surface treatments for industrial use. It supplies these products to many different manufacturing industries with advanced products and systems solutions. The company operates with high customer focus, customized quality, drives for innovation and sustainable business practices have contributed to its success in competitive marketplace. Managing close relationship with customers, Henkel also plays a major role in the continuous improvement of industrial products and manufacturing processes. The company offers high standards of research, product development, application testing and sales support through the seamless global network of the company. The industrial adhesives of

Henkel are used by the automotive industry, and manufacturers of trains and trams, and the shipbuilding and aerospace sectors all profit from Henkel's extremely high performance adhesives and sealants. These materials create the opportunities to design and build advanced lightweight constructions and to enable seemingly conflicting aims such as higher speeds and lower fuel consumption to be reconciled. The company also makes an important contribution to safety, comfort and value retention. Components made of highly rigid structural foams absorb impact energy in the event of a crash (improved safety); innovative sound-deadening materials improve vehicle acoustics (more comfort), and special pretreatment systems for car bodies optimize corrosion protection and paint adhesion. At the same time, Henkel's know-how helps customers to conserve environmental resources and increase the efficiency of their production processes[70].

Among the industrial clients, the decision making process is complex that involves a chain process involving many departments within the organization. Purchasing a small volume of industrial raw material might involve a host of departments, such as purchasing, engineering, finance, and manufacturing in seeking approvals at various stages buying process. The industrial buying process is generally driven by the following factors:

- Influence of organizational culture
- Long-term importance of the item being purchased,
- Cost advantage in purchase
- Experience of selling organization, and
- Potential of effective servicing.

The purchase decisions among industrial clients become more complex due to hierarchical structure of the organization. Accordingly, buying decision increases the time and cost. The experience of the companies selling products and services to industrial customers also contributes as a significant determinant in the making purchase decisions by the client organizations. The delay in decisions of industrial clients is also attributed by the practice of simultaneous buying operations from multiple locations. Each purchase unit may use the same basic product, but unique requirements of each production unit may necessitate approval of corporate headquarters as well as service requirements. Business-to-business clients have economic requirements that are different from the needs of common customers. In most consumer purchases, the buyer-seller relationship usually ends with the sale while industrial selling often causes the buyer to become dependent on suppliers for the consistency in quality supplies and services. Consequently, many industrial customers evaluate selling firms and suppliers as potential contenders for long-term sustainable relationships. Unlike consumer goods selling organizations, industrial selling firms generally have a small number of potential clients and in such markets even a small number clients represent a higher buying potential. Industrial sellers practice direct marketing and personal selling as an economically viable proposition.

Selling products and services in business-to-business marketplace high level of customization stimulates the response for industrial products. A large variety of industrial products are made to order. Even if produced to standard specifications, the actual manufacturing may commence only after the customer order is received. Industrial firms in many cases may actually sell standard products, but the complexity of the machinery or system may cause the accompanying information and service support to be customized. Consequently, when a client initiates an order for industrial products, the order processing chain is linked with the manufacturing operations. Industrial selling companies should select their accounts in reference to the product requirement, buying power and nature of decision making process. Industrial accounts should typically require a large, long term commitment of often specialized resources from their suppliers. Such effort drives a sizeable opportunity cost for the selling company that indicates attending large volume of sales calls and effective customer relationship. The industrial products selling companies can categorize the accounts as major, small and casual type depending on the size of their order and customization needs. Account selection is a crucial decision for the selling company, and clarifying account-selection criteria a necessary component of effective account-management tactics.

SELECT READINGS

Coe, J. (2003), *The Fundamentals to Business-to-Business of Sales and Marketing*, McGraw Hill, Chicago, IL.

Hutt, M.D. and Speh, T.W. (2009), *Business Marketing Management: B2B*, South-Western College Publishers, Florence, KY.

Rigby, D. (2009), *Winning in Turbulence*, Harvard Business School Press, Boston, MA.

REFERENCES

[1]   Sharma, A. (2001), E-Coms and their marketing strategies, *Business Horizons*, 44 (5), 14-20.

[2]   Shapiro, B. P. (1977), Improve distribution with your promotional-mix, *Harvard Business Review*, 55 (4), 104-114.

[3]   Coe, J. M. (2004), *The Fundamentals of Business-to-Business Sales and Marketing*, McGraw-Hill, New York, NY.

[4]   Comtec Telecommunications Corporation (2009), Annual Report. For details see http://www.comtechtel.com/annual.cfm?Year=2009

[5]   Kotabe, M. and Murray, Y. (2004), Globalization; Procurement; Services; Management strategy, *International Marketing Review*, 21 (6), 615-633.

[6]   Berry, L. L. (2009), Competing with quality service in good times and bad, *Business Horizons*, 52 (3), 205-207.

[7]   Berry, L. L. and Seiders, K. (2009), Serving unfair customers, *Business Horizons*, 51 (1), 29-37.

[8]   Handfield, R. B., Ragatz, G. L., Peterson, K. and Monczka, R. M. (1999), Involving suppliers in new product development, *California Management Review*, 42 (1), 59-82.

[9]   Ogawa, S. and Piller, F. T. (2006), Reducing the risk of new product development, *Sloan Management Review*, 47 (2), 65-71.

[10]   Ghemawat, P. (2010), Finding your strategies in the new landscape, Harvard Business Review, March (in press).

[11] Surhone, L. M., Timpledon, M.T. and Marseken, S. F. (2010), *Reseller: Company, Good, Service, World Wide Web, ICANN, Domain Name Registrar, Retailing, Web Hosting Service*, Betascript Publishing, Beau Bassin, Mauritius.

[12] Eisenmann, T. and Hagiu, A. (2007), *Staging two-.sided platforms*, Background Note, Harvard Business School Press, Boston, MA.

[13] Best Buy Company Inc. (2009), *Annual Report*. For details about the company see corporate web site of the company http://www.bestbuy.com

[14] Lin, B. W. (2004), Original equipment manufacturers (OEM) manufacturing strategy for network innovation agility: the case of Taiwanese manufacturing networks, *International Journal of Production Research*, 42(5), 943-957.

[15] Dissanayake, D. and Singh, M. (2008), Managing Returns in E-Business, *Journal of Internet Commerce*, 6(2), 35-49.

[16] Helander, A. and Möller, K. (2008), How to Become Solution Provider: System Supplier's Strategic Tools, *Journal of Business To Business Marketing*,15(3), 247-289.

[17] Kalyanam, K. and Brar, S. (2009), From Volume to Value: Managing the Value-Add Reseller Channel at Cisco Systems, *California Management Review*, 52(1), 94-119.

[18] Cugini, A., Carù, A., and Zerbini, F. (2007), The Cost of Customer Satisfaction: A Framework for Strategic Cost Management in Service Industries,*European Accounting Review*, 16 (3), 499-530.

[19] Macintosh, E. and Doherty, A. (2007), Reframing the service environment in the fitness industry,*Managing Leisure*, 12 (4), 273-289.

[20] Liu, T. C. and Wu, L.W. (2007), Customer retention and cross-buying in the banking industry: An integration of service attributes, satisfaction and trust,*Journal of Financial Services Marketing*, 12 (2), 132-145.

[21] Ulaga, W. and Eggert, A. (2006), Relationship value and relationship quality: Broadening the nomological network of business-to-business relationships, *European Journal of Marketing*, 40 (3-4), 311-327.

[22] Rajagopal (2010), Bridging Sales and Services Quality Function in Retailing of High Technology Consumer Products, *International Journal of Services and Operations Management*, 6 (5), (in press).

[23] For details see corporate web site of FedEx http://www.fedex.com; also see Godes, D.B. (2008), *Sales force integration at FedEx (A)*, Discussion Case, Harvard Business School Press, Boston, MA.

[24] Oberoi, J. S., Khamba, J. S., Sushil, and Kiran, R. (2008), An empirical examination of advanced manufacturing technology and sourcing practices in developing manufacturing flexibilities,*International Journal of Services and Operations Management*, 4 (6), 652- 671.

[25] Chang, T. Z., Mehta, R., Chen, S. J., Polsa, P., and Mazur, J. (1999), The effects of market orientation on effectiveness and efficiency: The case of automotive distribution channels in Finland and Poland, *Journal of Services Marketing*, 13 (4-5), 407-418.

[26] Essoussi, L. H. and Merunka, D. (2007), Consumers' product evaluations in emerging markets: Does country of design, country of manufacture, or brand image matter?*International Marketing Review*, 24 (4), 409-426.

[27] Song, I. and Chintagunta, P. K. (2003), A Micromodel of New Product Adoption with Heterogeneous and Forward-Looking Consumers: Application to the Digital Camera Category,*Quantitative Marketing and Economics*, 1 (4), 371-407.

[28] Squire, B., Cousins, P. D., and Brown, S. (2005), Collaborating for customization: an extended resource-based view of the firm, *International Journal of Productivity and Quality Management*, 1 (1-2), 8-25.

[29] Maguire, S., Koh, S. C. L., and Huang, C. (2006), Managing customer satisfaction through efficient listening tools: an evaluation of best practice in four world-class companies,*International Journal of Services and Operations Management*, 2 (1), 22-41.

[30] Ryder, S. G., Ross, K. G. and Musacchio, J. T. (2008), Optimal services policies under learning effects, *International Journal of Services and Operations Management*, 4 (6), 631-651.

[31] Dyer, J. H. (1994), Dedicated Assets: Japan's Manufacturing Edge, *Harvard Business Review*, 72 (6), 174-178.

[32] Heskett, J. L., Jones, T. O., Loveman, G. W., Sasser, W. E. Jr., and Schlesinger, L. A. (1994), Putting the Service-Profit Chain to Work, *Harvard Business Review*, 72 (3), 164-174.

[33] Lundquist, D. and Bhattacharyya, S. (2008), Finding seller quality and quality sellers in online markets: an agent-based analysis,*International Journal of Electronic Business*, 6 (1), 47-66.

[34] Bala, V. and Green, J. (2007), Charge What Your Products Are Worth, Harvard Business Review, 85 (9), 22-25.

[35] Heiko, G., Regine, K. and Elgar, F. (2008), Exploring the Effect of Cognitive Biases on Customer Support Services,*Creativity and Innovation Management*, 17 (1), 58-70.

[36] Rajagopal (2007), Optimising franchisee sales and business performance,*Journal of Retail and Leisure Property*, 6 (4), 341-360.

[37] Rajagopal (2008), Consumer Response and Cyclicality in New Product Management, *Journal of Customer Behaviour*, 7 (2), 165-180.

[38] Rajagopal (2007), Sales Management in Developing Countries: A Comparison of Managerial Control Perspectives, *Journal of Asia Pacific Business*, 8 (3), 37-61.

[39] Chung, J., Jin, B. and Sternquist, B. (2007), The role of market orientation in channel relationships when channel power is imbalanced,*The International Review of Retail, Distribution and Consumer Research*, 17 (2), 159-176.

[40] Narver, J. C. and Slater, S. F. (1990), The effect of market orientation on business profitability, Journal of Marketing, 54 (4), 20-35.

[41] Hanninen, S. and Sandberg, B. (2006), Consumer learning roadmap: a necessary tool for new products,*International Journal of Knowledge and Learning*, 2(3), 298-307.

[42] Reichheld, F. F. and Sasser, W. E. (1990), Zero Defections: Quality Comes to Services, *Harvard Business Review*, 68 (5), pp 105-111.

[43] Hirunyawipada, T. and Paswan, A. K. (2006), Consumer innovativeness and perceived risk: implications for high technology product adoption,*Journal of Consumer Marketing*, 23 (4), 182-198.

[44] Chen, T., Chen, C. and Tsung, C. (2007), Promoting Relationship Selling Behaviors to Establish Relationship Value: The Case of International Airlines, *Journal of Relationship Marketing*, 5(4), 43-62.

[45] Euromonitor International (2008), *Cosmetics and toiletries in Belgium*, Euromonitor International, Market Report, August.

[46] Green, K. W. and Inman, R. A. (2005), Using a just-in-time selling strategy to strengthen supply chain linkages, *International Journal of Production Research*, 43(16), 3437-3453.

[47] Johnson, D.S. and Bhara, S. (2005), Digitization of Selling Activity and Sales Force Performance: An Empirical Investigation, *Journal of the Academy of Marketing Science*, 33 (1), 3 - 18.

[48] Wagner, J. A., Klein, N. M., and Keith, J. E. (2001), Selling Strategies: The Effects of Suggesting a Decision Structure to Novice and Expert Buyers, *Journal of the Academy of Marketing Science, 29 (3),* 290 – 307

[49] Verbeke, W., Belschak, F., and Bagozzi, R. P. (2004), The Adaptive Consequences of Pride in Personal Selling, *Journal of the Academy of Marketing Science, 32 (4),* 386 - 402.

[50] Wagner, J. A., Klein, N. M., and Keith, J. E. (2001), Selling Strategies: The Effects of Suggesting a Decision Structure to Novice and Expert Buyers, *Journal of the Academy of Marketing Science, 29 (3),* 290-307.

[51] Morschett, D., Swoboda, B. and Foscht, T. (2005), Perception of store attributes and overall attitude towards grocery retailers: The role of shopping motives, *The International Review of Retail, Distribution and Consumer Research*, 15 (4), 423-447.

[52] Babakus, E., Bienstock, C. C. and van Scotter, J. R. (2004), Linking Perceived Quality and Consumer Satisfaction to Store Traffic and Revenue Growth, *Decision Sciences*, 35 (4), 713-737.

[53] Konishi, H. (1999), *Concentration of Competing Retail Stores*, Boston College, Working Papers in Economics, No 447.

[54] Ryals, L. J. and Humphries, A. S. (2007), Managing Key Business-to-Business Relationships: What Marketing Can Learn From Supply Chain Management, *Journal of Service Research*, 9 (4), 312 – 326.

[55] Fiocca, R. (1982) "Account Portfolio Analysis for Strategy Development", Industrial Marketing Management, 11, pp. 53-62.

[56] Woodside, A. G. and Trappey, R. J. (1996) Customer Portfolio Analysis among Competing Retail Stores, *Journal of Business*, 35 (3), pp 189-200.

[57] Campbell, Nigel C.G., and Malcolm T. Cunningham (1983) Customer Analysis for Strategy Development in Industrial Markets, Strategic Management Journal, 4 (4), 369-380.

[58] Sharma, A.,and Jagdish, N. S. (1997), Relationship Marketing: An Agenda for Inquiry, *Industrial Marketing Management*. 26 (2), 87-89.

[59] Shapiro, B. P., Kasturi Rangan, V., Moriarty, R. T. and Ross, E. B. (1987), Manage Customers for Profits (not Just Sales*), Harvard Business Review*, 65 (5), 101-108.

[60] Krapfel, R, E., Salmond, D. and Spekman, R. (1991), A Strategic Approach to Managing Buyer-Seller Relationships, *European Journal of Marketing*, 25 (9), 22-37.

[61] Brone Sid (2000), *Risk Constrained Dynamic Active Portfolio Management*, Columbia University, Working Paper, June.

[62] Rajagopal and Sanchez, R. (2005), Analysis of Customer Portfolio and Relationship Management Models : Bridging Managerial Gaps, *Journal of Business and Industrial Marketing*, 20 (6), 307-316.

[63] *ibid.*

[64] Miles, R, and Snow, C. (2003), *Organizational Strategy, structure and process*, Stanford Business Books, Stanford, CA.

[65] Ghobadian, A., James, P., Liu, J. and Viney, H.(1998), Evaluating the Applicability of the Miles and Snow Typology in a Regulated Public Utility Environment, *British Journal of Management*, 9 (1), 71-83.

[66] Hout, T. M., Porter, M. E. and Rudden, E. (1982), How global companies win out, *Harvard Business Review*, 60 (5), 98-108.

[67] Cespedes, F. V., Doyle, S. X. and Freedman, R. J. (1989), Teamwork for today's selling, *Harvard Business Review*, 67 (2), 44-58.

[68] Acrylonitrile is a synthetic monomer produced from propylene and ammonia; butadiene is a petroleum hydrocarbon and styrene monomer is made by dehydrogenation of ethyl benzene which is a hydrocarbon obtained in the reaction of ethylene and benzene.

[69] Rangan, V. K. and Issacson, B. (1991), *What is industrial marketing*, Background note for classroom discussion, Harvard Business School Press, Boston, MA.

[70] For details about the company see corporate web site of Henkel International at http://www.henkel.com

Chapter 4

SALES FORCE ORGANIZATION

Salespeople play a key role in the formation of long-term relationships with business partners including buyers and suppliers. As the primary link between the buying and the selling firm, they have considerable influence on buyer's perceptions of seller's reliability, the value of seller's services, and consequently the buyer's interest in continuing the relationship. Buyers often have a greater loyalty to salespeople than they have to the firms employing the salespeople, as salespeople develop buyers' perceptions [1]. Sales management is a continuous process in a firm. The more a company learns about the sales process, the more efficient it becomes at selling, and the higher the sales yield. As the sales yield increases, sales learning process unfolds in three distinct phases including initiation, transition, and execution. Each phase requires a different size and type of sales force and represents a different stage in a company's production, marketing, and sales strategies [2].

Effective management of sales forcetypically requires both planning and action to deliver powerful aspiration, establishing performance measuring and rewarding systems to reinforce specific goals of sales force, and developing mind-sets of salespeople by creating a sense of shared responsibility in performing tasks. Although companies devote considerable time and money to administer their sales forces, it is necessary to focus management strategies on how the structure of the sales force needs to be closely associated with the performance indicators of a firm's business. Firms must consider the relationship between the differing roles of internal salespeople and external selling partners, the size of the sales force, its degree of specialization, and how salespeople share their efforts among different customers, products, and activities. These variables are critical because they determine how quickly sales force of a firm responds to market opportunities, influences performance of sales people, and affects revenues, costs, and profitability of the firm [3]. The task of a salesperson changes over the course of the selling process. Different abilities are required in each stage of the sale including identifying prospects, gaining approval from potential customers, creating solutions, and closing the deal. Success in the first stage, for instance, depends on the salesperson acquiring precise and timely information about opportunities from contacts in the marketplace. Managers often view sales as a set of stereotype activities which begins with prospecting and terminates with closing the deal. But in practice it is a diligent task which involves variety of conflicts and high risk. Hence, sales performance requires effective administration in a phased manner to resolves conflicts and ambiguities during the negotiation and transactional process. The organizational design of sales force should be considered by the companies on the following lines:

- Organizational structure should reflect a marketing orientation
- Organization should be built around activities, not people
- Responsibility and authority should be related properly
- Span of executive control should be reasonable
- Organization should be stable but not flexible
- Activities should be balanced and coordinated.

The contemporary practices of organizing sales forces are based on a multilayer model of market oriented organizational culture. This new perspective distinguishes between market-oriented values and norms that is built upon the various layers of a market-oriented culture and the behavior of sales-force personnel. It is observed that market-oriented values affect role conflict, role ambiguity, organizational commitment and job satisfaction, and that market-oriented norms drive the sales people competitive to take a sales lead among competitors. However, firms that have developed market-oriented sales force structure could not set strong impact on the customer relationship and enhancing their value [4].

Many companies competing in the global marketplace thrive in organizing their sale force in order to effectively respond to the three major operational factors including time, target and territory. These firms also invest considerable time and money to managing their sales forces. The firms should concentrate their managerial focus on how the structure of the sales forces needs to be linked to the life cycle of a product or a business. However, the organization and sales objectives have to be initiated with the businesses start up, and pushed to grow, mature, and decline if a company wants to keep winning the race for customers. During organizing sales force firms should consider the differing roles to be played by the internal salespeople and external selling partners, the size of the sales force and its degree of specialization. These variables are critical in organizing sales force because they determine how quickly sales forces respond to market opportunities, influence sales reps' performance, and affect companies' revenues, costs, and profitability. It is observed that during the initial stages of the market penetration, firms focus the size of their sales staff to reach customers effectively. Firms also consider the extent of dependence of salespeople with their outsourced selling partners while setting the structure of the sales force. In the growth phase, firms should concentrate on getting the sales force's degree of specialization and right size. When firms reach the stage of maturity, they should better allocate existing resources and hire more general-purpose salespeople. Finally, as organizations go into decline, wise sales managers shrink the size of sales force and use partners to keep the business afloat for as long as possible [5].

The sales planning and pricing team manages the demand and supply of Toyota and Lexus models across more than 25 countries in Europe. They implement market-oriented prices to maximize volume and profit. The Toyota Sales team works closely together with the national distributors to increase sales profit. Toyota offers the team members experience in selling activity and development opportunities in an environment close to the retailers and end customers for the products and services of the company. Toyota sales strategywas to sell Lexus with an identity of a separate company with almost no references to Toyota.

Consequently, a heavy emphasis towards quality customer service and it had a separate dealership network from Toyota. This marketing strategy has allowed Lexus to become one of the best selling luxury cars in the US by 2000 and it encouraged Nissan to sell luxury cars with the Infiniti brand. The lean operating principles are based on the systems of Toyota Japan. Companies around the world, in all sectors of the economy, now embrace these approaches to improve quality, cost and productivity. Different purchasing, selling, and outsourcing activities are practiced in lean organizations that are oriented towards winning sales by introducing critical changes to sales strategy and practice. The role of this future-sales-force is considered in relation to its shift towards a marketing-orientated rather than sales-push strategy [6].

Many leading companies are taking a scientific approach in organizing sales force in view of their productivity, contribution to profit of the company and involvement in creating customer lifetime value. Sales force of a company is accounted for acquiring new customer and retaining those currently associated with the company. The skilled members of the sales force are expected to reach out to new customers in innovative ways and show increasing productivity by helping the field salespeople with advanced skills and resources. In order to enhance the effectiveness of the sales force, firms need to systematically target their offerings to the customers, optimize automation process in the sales, and pay close attention to sales force deployment in reference to marketplace environment. If the sales force is properly organized and equipped with competitive skills, the overall effect of increasing the average sales per employee can be exponential and company can stay competitive [7]. A well trained sales force may drive its operations on the premise of the following objectives:

- Calling on the right customer at the right time in the right way
- Setting contacts of an individual salesperson to buyer
- Negotiation plan of salesperson to buyer team
- Carrying selling process of a sales team to buyer team
- Organizing sales conference and seminar to stimulate prospects
- Prospecting through outsourced salespeople.

Asian automobile companies like Toyota and Honda have built sustainable buyer relationships by consistently following six steps in managing their sales force- understanding functional pattern of distributors, reducing distributor rivalry and providing better opportunity for clean sales, monitoring closely buyer relations, enhancing distributor capabilities, sharing information intensively but in a selective way, and helping buyers continuously improve their association with the distributors. Distributors of automobiles who also have to meet the sales targets, attempt to minimize the inherent risk associated with operating in a given business environment and increasing the customer value at the same time. Supplier collaboration might have a positive effect on buyer performance both in terms of innovative capability and financial results driven by trust and dependence which play an important role in building mutual relationships. However, there exists significant gaps between buyer and seller expectations concerning how relationships should be evolved and the issues of power and trust need to be explored in greater depth if relationships are to be optimized for achieving higher functional efficiency [8].

A firm may designate the sales force as a profit center with an objective of reducing cost, increasing volume of sale, enhancing customer life time value and increasing profit contribution to the company. A profit center is held accountable for both revenues, and costs (expenses), and therefore for profits. In terms of managerial responsibilities a profit center approach of a firm aims at driving the sales revenue by generating multiple selling activities for new customer which leads to cash inflows and at the same time control the cost (cash outflows) causing activities. This makes the profit center management more challenging than cost centre management. A profit center business unit aims at lowering the operational costs and increasing the volume of sales. On the contrary the cost center orients sales force with an objective of reducing cost by observing austerity measures in selling and reducing human resources in the sales force, though increase in sales revenue is not guaranteed. The sales force takes up the core task of reducing the operational cost and maintaining status quo of revenue generation, in case augmenting the revenue is not possible.

SALES FORCE DESIGN

Sales force is considered as one of the most valuable resources for attaining competitive growth in a business organization. The sales force should be designed in reference to long term goals of the firm and carefully evolving it according to market conditions requires a thorough understanding of customer needs and potential. The design of the sales force should fit into the cost-effective mix of sales channels and sales teams need to be given precise directions to reach larger segment of customers and optimize market coverage. The sales force of a company is also accountable for individual and industrial customer accounts, products and, services to be offered. A cost effective sale force design in a company should include optimal number of salespeople in each sales team and equipped with technical support services. A sales design may function efficiently if it possesses an integrated information flow pattern comprising dissemination of policy directives as top-down pattern, and streaming of market information through salespeople as bottom-up model. Most companies have weak bottom-up information flow, which causes mismatch of corporate marketing and selling strategies with those desired for competitive lead in the market. In the global marketplace companies have experience across a broad range of sales models including global, strategic, and key account teams, generalist and specialized field sales teams; telesales teams, and business partners. It is necessary for a company to design the sales force by considering the following attributes:

- Sales force design should be subjected to pilot tests in reference to the strength of sales models implemented by the sales teams to increase market share,
- Sales force design should comply to business growthmodel of the company and ensure sales effectiveness by minimizing customer switching,
- Sales force design should be linked to marketing plans to drive long term and sustainable sales related decisions,
- Sales force design should be able to improve the skills of salespeople to adapt to
- sales force automation,

- The design of sale force should be able to develop competency of sales teams to work with advanced information technology tools, and market insights to strengthen decisions on the sales and acquiring new accounts, and
- Sales force design should develop skills towards analytical client insights pertaining to demand for products and services, and customer value.

Sales force design should be linked with the anticipated service area of the market and skills to the salespeople need to be inducted in the design accordingly to enable periodic training to the members of the sales teams. Most advantageous strategies of a firm can be built on sales and services differentiation offering customers something they value against brands of competing companies. But most companies concentrate only on their products or services without leveraging sales force with updated knowledge and skills. In fact, the salespeople can help a company differentiate against competitors at every point where it comes in contact with its customers from the moment customers realize they need a product to service, to the time when they dispose it off. It is observed that if companies open up their thinking to their customers' entire experience with a product or service, the sales value chain can uncover opportunities to position their offerings in ways that neither they nor their competitors thought possible [9]. However, most companies lean on designing own sales force to reduce cost and exercise better control over their performance. In a standard analysis companies consider sales cost as the major guiding factor in organizing or outsourcing salespeople. It is assumed that the sales force of the company has a fixed cost and that the cost of outsourced sales force varies with sales. The sales volume, at which the cost of company's sales force equals the cost of outsourced sales force, suggests that for sales volume above that quantity, firms should use a direct sales force. However, a company should also consider the issues of sales effectiveness for the types of sales force designs [10].

A healthcare sales team of a business consulting company began working with the corporate client to develop an inbound and outbound sales program for new pharmaceutical products. This program was supported by a trained sales force responsible for contacting over 1,700 physicians each month. The members of the sales force were trained on the company's full line of products so that they could change their call strategy based on the physicians' preferences. In addition, the client switched their featured products every three months and due to the flexibility of the program, the sales force design was kept flexible. So, new sales messages were incorporated with virtually no transition time which ensured that the message being delivered to physicians was aligned with the company's focus. The member of the sales force attended the client's field training prior to launch and new training procedures were implemented as the program evolved. The sales force provided by the consulting company in the diagnostics division was responsible for gaining and retaining business, fulfilling patient literature and handling all aspects of orders. Upon implementing the program the client's field representative visited the physician's office to discuss the new diagnostic product and set up a nurse training session. The sales force was then responsible for coordinating the nurses with the physician's office, placing reminder calls, shipping the training kits and ensuring that everything was in place for the appointment. In view of the high operational costs of the nurse training program, the client developed self-training audio-visual aids which were less costly, more efficient and allowed the physician to complete the training at their own convenience. Consulting firm implemented new sales procedures in short notice and the sales force took over the responsibility of distributing the CDs, following-up and handling inquiry calls from physicians [11].

Successful companies must devote significant effort to designing the sales force they need and have robust processes in place to make sure that it is implemented successfully. Designing sales force in a company should also be subject to sales systems analysis that may increase the productivity of a sales department by revealing the right combination of products, the right promotional methods, and the right locations. Sales managers implement the system by setting improvement objectives and members of sales teams then apply output and input variables to determine the control variables that need refinement. Efficiency of sales force can be measured in reference to contribution to profit, returns on assets, sales to cost ratio, market share, and achievement of marketing and sales objectives of the company [12].

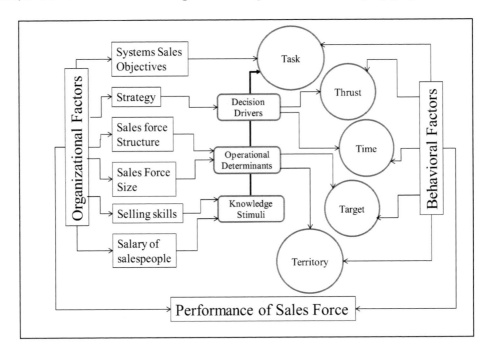

Figure 4.1.Sales Force Design and Management.

Sales force design should be integrated with the organizational factors consisting of system objectives, strategy, structure, size, skills, and salary (6-S), and behavioral attributes of salespeople affecting task, thrust, time, target and territory (5-Ts). The relationship among these variables in strategic sales and operations activities is exhibited in Figure 4.1. Sales force design in reference to a particular territory is also considered as an important factor, which has received little analytical attention in the traditional literature, but which appears to be an important influence on the effectiveness of the sales operation. It has been argued that organizational effectiveness is determined by outcome performance and behavioral performance of sales team, as well as through systems control approach. Although conventional theory has suggested that behavior performance and outcome performance result from different stimuli, behavior-based control is positively associated with both behavior performance and outcome performance [13]. Members of the sales team have the promising behavior intending to offer a pleasant, positive and rewarding scenario of the situation under discussion and the unpleasant consequences are kept undisclosed. The members of the sales team avoid any negativity in their conversation as far as possible. Sales force culture in general is embedded with 5-T power-grid which comprises a synergy of task (commitment), thrust (driving team) and time (punctuality), target (right customers), and territory (marketplace) that affect their overall performance [14]. While designing the sales force it is important for the managers to see that performance of the team is measured on the predetermined key indicators. A sound pricing policy lets customers know that the price they are paying is related to the value they receive, and keeps salespeople from dealing with long, complicated negotiation processes. While some managers worry that their sales force do not accept price as an indicator of their performance, most salespeople will adopt it if the new policy is implemented well. The key is to reward salespeople for driving profitability instead of just revenue and sales volume [15].

Territorial Sales Force Design

Sales force can be organized in many ways subject to the product type, coverage and buyers prospects. The different ways of organizing the sales force are stated below:

- Territorial structure
- Product structure
- Customer structure
- Matrix Structure
- Complex structure

The region based sales force (territorial structure) may be organized for vertical and horizontal coverage of different areas spread across the operational area. The vertical hierarchy of sales force may be at the central level, zonal level, area level, and down to the markets located in various towns. The horizontal area of operation may be carried in a given area across the towns or villages. The size of the sales force depends on the geographical spread of the sales operation, volume of sales and the density of buyers. Sales territory design is also considered as a particularly important managerial variable and analytical model on design process suggests that territory design has a large effect on sales organization effectiveness both directly, and indirectly through its relationship with sales force behavioral performance [16]. The territory design model illustrates that organization effectiveness of sales is determined by sales force outcome performance and behavioral performance, as well as by the use of a behavior-based control approach. Although conventional theory has suggested that behavior performance and outcome performance result from different stimuli, behavior-based control is positively associated with both behavior performance and outcome performance [17].

The importance of designing effective sales territory is widely supported in the process of organizational restructuring. Despite this importance, the impact of sales territory designs on salesperson and organizational consequences has not gained significant research attention [18]. However, the role of designing effective sales territories in the multinational sales organization is growing substantially which includes the challenges towards integrating the local market differences in sales infrastructure and other competitive market attributes. Organizational commitment and sales territory design are significantly related to sales force performance and the study highlights the growing emphasis on building long-term, collaborative buyer-seller relationships that favor the use of behavior-based control systems in many sales management situations, and suggests a new agenda for management attention in improving sales force effectiveness.

The territory design also helps in exploring the new opportunities for salespeople towards acquiring new accounts and developing long-term buyer relationship. A convenient territory augments the task performance and encourages them to earn incentive pay where incentives are linked directly to territory-level individual performance [19]. It has been observed that organizations may not define salespeople's territories based on individual conveniences. However, the effective sales organizations place more emphasis on tsheir sales territory design, and, additionally, their sales forces show significant differences in both personal characteristics and performance dimensions. Salespeople in the more effective sales units display higher levels of intrinsic and extrinsic motivation, sales support orientation, and customer orientation. Experienced sales managers understand that poor territory designs will have a negative impact on salesperson morale, result in inadequate market coverage, and complicate management evaluation and control [20]. Several factors beyond control of the salesperson often guide the territory design decision including the size of the sales force, buying power of the accounts, geographical dispersion of accounts, time required to service each account, and competitive intensity. The attributes of different sales force designs are illustrated in Figure 4.2.

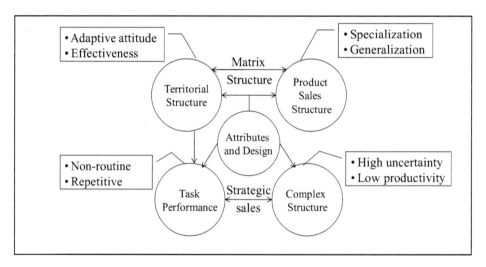

Figure 4.2.Sales Force Design and Performance Attributes.

Major attributes of the territorial sales force design are as listed below:

- Salespeople deal with less diverse product lines
- Territorial sales force target more on consumers than accounts
- Persuasion is considered as a critical variable
- Sales force depends on limited channels for selling different products
- Products are based on the local consumer needs
- Responsibility and delegation of line of authority.
- Consumer demand and sales coordination

- Hierarchical structure and generally possess large number of executives
- Conflicts of interests and responsibilities
- Lack of specialists in the product sales line.

An important disadvantage in territorial sales force design it that large number of employees involve in operational tasks which lead to the conflict in power play and command execution in the organization. Besides the agglomeration of top management personnel, the individual products may suffer, as responsibilities cannot be fixed easily on the operational executives.

Product Bases Sales Force Structure

Sales force based on the product structure generally belongs to a company and functions as a team to promote the sales of the product in the given area. The sales force needs to be specifically organized to be in tune with the functional requirements of the market. In sales of computations products, a group of professional sales and service engineers are required to boost-up the market for the principle product as well as for the service packages of the ancillary unit. The product based sales force design incorporates two sales lines that include generalists and specialists. Generalist sales force operates on simple and commonly used products while specialist salespeople are trained for the product line and deal with complex products. The coordination of activities in a geographic area is handled through specialists at the corporate staff level but in the product based sales force focus is laid on the performance of product-mix in a given area. Multinational companies which operate within this structure have variety of end users, handle diversified product lines with high technological capability and logistics cost are diverted to the local manufacturers. This type of sales force design has several benefits including:

- Decentralization of authority
- High motivation to salespeople
- Sales strategies developed in congruence to product life cycle.

In this type of sales structure, a firm is segregated along product lines considering each division as a separate profit center with the division head directly accountable for profitability. Decentralization of sales operations is critical in this structure and more decisions are likely to be left to the local sales manager. Decentralization of authority is a prime advantage of this structure where product division heads, responsible for sales are highly motivated. This structure allows the product managers to add new products and product lines and withdraw old ones with only marginal effect on overall operations. Another advantage of this structure is that the control of a product through the product life cycle can be managed promptly in a secured manner. However, firms following product sale force design often face the problem of coordination among product and territory sales managers. In addition, it is felt that executives quickly get biased towards the regional and corporate staff in managing any product selling process. Product sales force structure largely operates on the coordination of customer needs and products offered in a marketplace. Salespeople derive various sales strategies according to product-market relationships. Salespeople apply psychological intelligence to acquire and retain consumers for selling the broad range of products offered by the firm. However, there is no formula for placing sound psychological analyses magically in the hands of the salespeople. Product led sales forces operate on the three broad guiding determinants to make sure that sales calls are highly productive and informative. These guidelines include, driving out-bound calls with customer determination, listening to the sales force, and rewarding for secret information collection, analysis and execution to help increase sales effectiveness [21]. The Figure 4.3 exhibits the ways of strategy development in product sales force design.

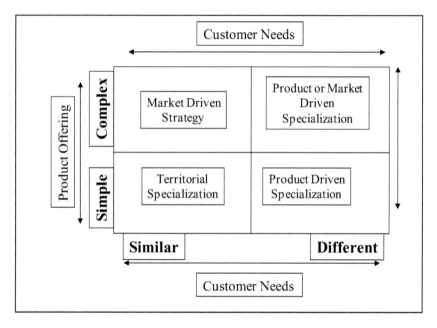

Figure 4.3.Customer and Product determinants for Sales Force Organization.

Many companies face large, complex selling situations with the product led sales force design. Such companies might be engaged in selling expensive equipment that affects many parts of a client's manufacturing or packaging process for onwards sales. The sales of such high value and complex products need special handling skills by the salespeople. Sales people engaged in dealing complex product categories should realize that they are not smaller transactions as their potential profit is larger, and they have a more lasting effect on both buyer and seller. In case of similar customer need across markets for complex products, salespeople may follow market driven sales strategylearning from their competitors, while geographic specialization can be developed across market segments for selling the products of similar needs. However, when there are different customer needs for the simple products it would be right for the salespeople to follow product specialization to drive higher confidence among customers. This strategy would also enhance the customer value and build loyalty in the long run. Salespeople need to develop a profile of a company's needs and key personnel, justify the purchase to the buyer, make the sales pitch, coordinate company resources, close the sale, and maintain the account. Before they can engage in strategic selling, companies require revising the approach of their sales forces according to the kind of sales they want to make, which may include different types of non-recurring sales [22].

Matrix Sales Force Structure

In recent years, synergies of all the above sales force organization designs have emerged among the multinational companies which is defined as *matrix structure*. The matrix structure offers greater flexibility than the single lineof-command sales force designs already discussed, and reconciles this flexibility with coordination and economies of scale to keep the strength of large organizations. The attributes and advantages of matrix sales force design include:

- Multiple command line
- Product and geographic sale force coordination
- Organization design reacts quickly to the local environment demand

For the multinational firm, the matrix organization is a solution to the problem responding to both economic and political environments. General Electric Company in Asia operates with matrix structure and has been successful. A matrix organization can encompass geographic and product based sales management components. However, some of the disadvantages in following this organizational design include international competition on sales, power struggles among the mangers and parallel decision making.

BOTTOM OF THE PYRAMID SALES FORCE DESIGN

Global companies are deploying their sales forces to target consumers in the bottom of the pyramid (BoP) market segments comprising large consumer communities with small per capita purchases. These companies are developing BoP sales strategies based on educating consumers on global brands through inter-personal relationships with buyers, local retailers and distributors in the region. BoP sales segments have been identified as potential outlets for global brands when the semi-urban and rural markets are modernized. Globalization has segregated the consumer behaviorin rural and semi-urban marketplaces and influenced by the urban marketplaces [23].

Intensive competition of global firms has not only decreased the market share in the premium and regular consumers' market segments but also created price wars reducing profit margins and limiting market growth of firms. This situation motivated companies to consider positioning brands in the sub-urban and rural segments, which are unexplored. By targeting these segments with products in small packs at lower price points, companies have experienced great success [24]. The BoP sales segments which constitutes large number of small consumers has become the principal target of most of the consumer brands emerging from multinational firms. The products and services penetrating to the BoP should provide constancy and agility at the same time. Constancy is required if the brand is to build awareness and credibility while agility in the brand builds perceived values among consumers. Agility is required if the brand is to remain relevant in a free marketplace [25].

Sale force assumes principal task of building consumer-based brand equity by developing awareness about the products and services among consumers in BoP market segment. Hence, sales strategies may not be implemented uniformly in BoP consumer segment without understanding the cognitive dimensions of consumer. Salespeople should carry out the consumer mindset mapping may be as a basis for spatial segmentation of markets in urban areas. Companies holding strong brands define market access strategies, consumer value propositions based on product features, and brand relations among consumers in terms of perceived brand values across market segments within urban habitats. In the process companies determine markets first and then build sales strategies around identified market segments [26]. The Figure 4.4 exhibits the sales strategies to be developed in the BoP consumer segment.

Source: Adapted from Rajagopal and Rajagopal, A. (2010), Measuring performance of sales force: administrative and behavioral parameters, International Journal of Economics and Business Research, 2 (4), in press.

Figure 4.4.Bottom of the Pyramid Sales Force Management.

Sale force deployment should be planned according to the target markets and customers attributes. The sales targets are always higher in the bottom of the market pyramid as illustrated in Figure 4.4. Managers should understand the lying market conditions while demarcating the sales territories in up-market, regular market and bottom-line market. Although the per capita response to sales may be lower in the bottom-line markets, the aggregate buying power of customers is actually quite large, representing a substantial variety of goods and services. These markets indicate sales potential that there exists long-run sales advantage for large number of companies. Thus, managers must shift their thinking towards the bottom-line market which holds value of high-volume but high-margin businesses in the long run.

Consumers' preference for utilitarian brands are significantly higher in the BoP markets while opulent brands build high equity in the premium markets attracting consumers to make personal visits to evaluate the brand. Merchandise-related attributes of store formats have a higher impact on the utility formation of consumer brands, and service and convenience related attributes of the brands add value to the perception of consumers in bottom line markets. Salespeople in the BoP market segment should prospect consumers and retailers simultaneously [27]. A study revealed that rural residents in India found that packaging is more helpful in buying, that better packaging contains a better product and that they are more influenced by the ease of storing a package than their urban counterparts. Easy to carry size of package, gross weight, simplicity, transparency and similarity of packaging have also emerged as critical brand identity factors among the consumers of BoP market segments in urban areas [28].

Hindustan Unilever Limited (HUL), one of the largest consumer goods manufacturing company in India has pioneered the bottom of the pyramid marketing strategies moving its action sales force to rural consumer segments. *Shakti* (means strength in Hindi, the national language of India) was initiated as a bottom line action program by the company to reach the massive un-served and under-served markets that cannot be economically and effectively serviced through regular methods. HUL identifies underprivileged women in villages and traines them to become *Shakti* Entrepreneurs (SEs) to play the role as distributors of HUL products in order to earn a sustainable income. The company invested resources in training SEs by helping them become confident and independent and stand as a source of inspiration for the other women in the community. Hence, besides being sales, distribution and communication initiative, *Shakti* has become a micro-enterprise initiative that created livelihoods and improved the standard of life of rural consumers at the bottom of the pyramid. This bottom line initiative also enabled rural consumers to access world class products, thereby reducing the menace of spurious products. This program has benefited the consumer goods business by significantly enhancing direct l reach of the company to the bottom of the pyramid market segment and enabled communication of its brands effectively in the regions which were not enlightened by the television and print media. The products distributed through project*Shakti* are some of the popular brands of consumer goods which have specifically relevant to rural consumers. It has been observed that the popularity of this bottom line marketing program was reached from 17 SEs in 2001 to more than 45,000 SEs in 2008. It is realized that *Shakti* has come a long way in impacting lives of rural consumers in India [29].

The concept of sales leadership has been driven by some companies; however a few companies could maintain the leadership with sustainable growth in sales. Multinational companies optimize their sales performance in the BoP market by building strong buyer-seller relationship and several other activities as discussed below:

- *Consumer Knowledge*: Consumer goods retailing firms study their consumers and channel partners through periodical surveys and intelligence gathering. The suggestions of the consumers and the distributors are screened and due consideration is given to those worth reengineering market strategies.
- *Long Term Vision*: Long term implications in developing brand extensions are drawn by the firms intending to penetrate in the BoP market segments. Such strategies help salespeople in prospecting consumers from the up-stream markets and drive high

perceived use value among consumers. However, sales of products and services of multinational firms may struggle with the local brands existing in the market in terms of volume and price sensitivity factors affecting buying behavior of consumers.

- *Quality Confidence*: Salespeople should also bank on the quality platform to maintain its brand share and expand in the new market segments. This would increase confidence on quality among the consumers and the channel partners, and enhance brand equity.
- *Product Line:* Firms should extend product line covering the class and mass market segments to provide more shelf space to brands and narrow down competition to emerge as brand leader in the BoP market segments. In these markets the brand equity is determined by the consumers in reference to perceived use value and value for money measures. Therefore in a large product line, brand switching and inter-temporal substitution within and outside the product line is very common [30].
- *Multi brand selling strategy:* In order to sell multi-brands in the BoP market segments, firms should operate on a two-dimensional matrix structure with products and geographic platforms for effective diffusion of brands. Such selling strategies would enable firms to promote products and services to different consumers in the BoP segments and compete against the local sellers.
- *Aggressive selling:* BoP market strategy demands intensive advertisements, sales schemes and attacking sales force. Salespeople should employ tactical approaches like develop effective point of purchase displays, sales promotion schemes and consumer loyalty programs to achieve quick results. Price premium on the consumer products in the BoP market segments delineates the role of uniqueness, together with the awareness, qualities, associations and loyalty as principal dimensions of products and services.

The concept of developing sales in the bottom line consumer segments of the society argues that moving beyond decades of mutual distrust and animosity, competitors are learning to cooperate with each other. Realizing that their interests are converging, firms are working together to create innovative business models that are helping to grow new markets, also accelerate the eradication of poverty. People in China, India, Indonesia, and Brazil are eager to enter the global marketplace. Yet multinational companies typically pitch their products to the tiny segments of affluent buyers in the emerging markets, and thus miss out on much larger markets further down the socioeconomic pyramid [31]. The BoP philosophy of business further argues that by stimulating commerce and development at the bottom of the economic pyramid, multinationals could radically improve the lives of billions of people and help create a more stable, less dangerous world. Achieving this goal does not require MNCs to spearhead global social-development initiatives for charitable purposes but rationally establish business by positioning global brands also in the BoP markets [32].

TYPOLOGY OF SALES ORGANIZATIONS

The characteristics of sales organizations are often stubborn as their work culture is laid on the maxim of coping with challenges. However, sales organizations fall into one of five natural configurations, each a combination of certain elements of structure and situation. These configurations include the simple structure, Process automation hierarchy, professional bureaucracy, compartmentalization of activities, and outsourcing of services. These distinct configurations serve as an effective tool in diagnosing the problems of organizational design [33]. Organization design has become a cornerstone of competitive advantage in increasingly complex companies which derived powerful synergies among their capabilities, among sales opportunities, and between the two. But the way they recognized and realized these synergies was through their multidimensional organizational designs. Sales organizations should be built with good sales leaders who spend a lot of time evaluating and occasionally redesigning the sales force structure to ensure that it supports corporate strategy. Often, this involves finding the right balance between specialized and generalized sales roles. Organization design involves the creation of roles, processes, and formal reporting relationships in an organization. The design process of organization has two phases encompassing strategic grouping, which establishes the overall structure of the organization with levels of internal relationship with different functional areas, and operational design, which defines the more detailed roles and processes. Organizational design for sales activities should be developed considering the following attributes:

- Specialization of labor, activity and control
- Stability and continuity
- Coordination between the lines of activities and within a line of activity
- Levels of reporting
- Length of the sales organization
- Decision making personnel and accessibility
- Flow of directions

A multiunit sales workplace is a geographically dispersed organization built from standard functional units like distributors, stores and vendors that are aggregated into larger geographic groupings comprising sales units, districts and regions. Sales managers are responsible for meeting targets at different functional units, set by corporate headquarters and implementing customer driven sales strategy. To do so, salespeople and managers should adhere to following principles of organizational design [34]:

- The responsibility of sales personnel at the field level need to be fixed to avoid any overlapping of tasks and responsibilities
- A multilayered net to catch any problems in sales process need to be created earmarking time, target and territory related responsiveness
- Managers and leaders of sales teams at all levels should be trained to serve as integrators, coordinating diverse activities and optimizing the efforts of the whole organization rather than its parts.
- Higher-level sales manager who serve as decision makers in the organization should filter market data and reported sales statistics from salespeople to support situation based sales strategies.
- In a multi layered sales organization heads of sales units in particular should act as motivators of salespeople in rolling out new initiatives, and

- Each salesperson and managers should be tuned to share responsibility for the development of competitive skills to enhance the sales performance in the marketplace.

Organizational design is thus considered as a hierarchical process which identifies various aspects of work flow, procedures, structures, systems, and realigns them to fit into current sales goals for developing strategic plans implement the continuous changes. For most companies, the design process leads to a more effective organization design, significantly improved results reflecting in profitability, customer service, and internal operations [35]. The principal attribute of the design process is a comprehensive and holistic approach to organizational improvement that touches all aspects of organizational life and enables to achieve:

- Effective customer services with customer value-chain
- Increase in profitability and sustainable growth
- Low operating costs
- Improved sales efficiency and short sales cycle time
- An adaptable sale culture in the organization, and
- Innovative sales approaches for competitive advantage.

The growing demand for new approaches to support the changing nature of work and organizational structure has prompted innovative ideas. New relationship paradigms, technologies, and routes to markets are frequently implemented by competing firms to enhance customer value and refine goals of sales organizations. The organizational designs are also influenced by the physical attributes of the workspace and the physical space attributes that are driven by the work processes, perceptions, and attitudes of people at work [36]. A multi-level sales organizational design need to be considered in reference to the following functional attributes:

- Multi level persuasion
- Flexibility and task coordination
- Logistic and administrative support
- Account serving and prospect management
- Networking organization sales with key persons in buying organizations.

There are significant differences in multilevel sales force design on the behavioral and attitudinal variables of salespeople in carrying the sales process. It is observed that the relationship between some of these variables and quitting intentions differed substantially between multilevel salespeople. These variables include job satisfaction, organizational commitment, perceived image of direct selling in the marketplace, and the importance of the job characteristics of work rewards and career growth [37].

Mary Kay Inc. is a multi-level selling company that engaged in selling beauty products in the US. It offers more than 200 products in six categories including body care, color cosmetics, facial skin care, fragrance, nail care, and sun protection. The company has over 1.8 million independent sales consultants demonstrate Mary Kay products in the US and some 35 other countries. Consultants compete for awards each year, ranging from jewelry to the company's trademark pink Cadillac that was first awarded in 1969. The company also founded the Mary Kay Ash Charitable Foundation in 1996 to fund cancer research and domestic violence programs. The family of founder Mary Kay Ash owns most of the company. Direct-sales businesses that rely on home-based representatives to peddle their goods in a network that grows by inducting continuously new salespeople. In this process many buyers who realize the benefits of direct selling also joins the network. The multilevel selling companies have observed that sales force in this segment are rapidly expand as the unemployment rate soars in the society and those who lost jobs look for new ways to make money. Avon Cosmetics Company launched its TV commercials in mid 2009 with a 30-second spot during the Super Bowl pregame show to attract new entrepreneurs to drive direct selling. This cosmetics company also participates in the job fairs this year, scouting for new salespeople at more than 140 such events. In multilevel selling organizations compensation systems vary, but representatives primarily earn money from commissions on product sales or by purchasing the products wholesale and selling them at retail prices [38].

Multilevel sales organizations also often call themselves as direct selling organizations (DSO) or network marketers. The major difference between DSOs and a true multilevel sales organization (MSO) is network selling. Social network ties are particularly necessary in increasing sales force behavioral performance while poor family background coupled with lack of career planning and limited professional sales presentation, impact negatively on the performance of the sales career. A matter of chance, development networks, social dyadic interactions with customers and key stakeholders are found necessary in generating, building and retaining customers [39].The product is predominant in the sales process of DSO whereas in the case of MSOs, the product often is irrelevant or secondary and the emphasis is not placed on selling product but on recruiting other salespeople to in turn recruit still others. A direct selling organization turns into a multilevel selling organization when the emphasis switches from selling product to recruiting other salespeople [40]. MSOs expand horizontally by recruiting salespeople continuously and expand horizontally. Many MSOs also encouraged building sales teams to manage their respective accounts. These companies need to consider the following attributes in recruiting salespeople for taking up direct selling activities:

- Economic criteria
- Control and strategic criteria
- Transaction costs
- Operational flexibility
- Innovative and organizational commitment
- Aptitude and conduct
- Customer relationship management.

It has been recognized by most direct selling companies that whether selling products or services, making strategic decisions, delivering solutions, or driving innovation, success is accomplished by team work. However, since the early 1990s, teams have evolved in selling activities where members were co-located, dedicated to a common mission, and directed by a single leader. As teams have become more lucrative solution to sustainable growth in sales, substantial challenges have been posed to traditional advice on team formation, leadership, roles, and process in contributing to multilevel sales or marketing organizations across countries in the world [41].

The growth of team selling is changing conventional ideas about sales and customer service. Team selling, also known as cross-functional selling, involves the efforts of different people from different areas within an organization working together to sell to and service a customer. This strategy of selling began in early 70s when sales teams were formed in high-technology organizations such as IBM to service the complex needs of their larger customers. It was observed necessary to have, along with a salesperson, technical support people and application specialists involved in inventory control, manufacturing and production planning. Besides IBM, General Motors adopted the team selling strategy and developed it as part of the culture of the company. Executives of the company felt that organizing a sales team is a major task to sell the high technology-high value products. The sales team should necessarily have a technical and finance person. Interestingly what's happened with team selling is that over the years it has filtered into less-technology-oriented businesses such as advertising, consumer packaged goods, and financial services. Several issues may emerge before they are formally addressed as the sales teams generally start to form naturally and informally. Communication among team members is difficult to establish, but is one of the keys to the effectiveness of this approach. There are many reasons for a company to choose team selling because this strategy:

- Provides strategic advantage.
- Meets the needs of demanding global customers more effectively.
- Establishes long-term, quality relationships with clients.
- Provides competitive advantage through a different kind of customer interface strategy.
- Increases flexibility in terms of how customers are served.
- Increases sales and potential profitability.
- Forces resource allocation decisions at the field level.
- Generates better input to new product and services development.

Team selling effectiveness also depends on selection and training. Training to back up the appropriate selection is so important, in fact, that IBM sends their team leaders to a customized executive development program on how to be an effective key account executive. It's a tough, rigorous program and in addition, IBM requires recertification every few years. Also at the organizational level, in order for a team selling approach to be successful, the supplier organization must have a strong market orientation as part of its organizational culture [42].

VERTICAL AND HORIZONTAL SALES FORCE DESIGNS

The vertical organizational structure possesses hierarchical reporting system and top-down control process. A top-down structure of corporate divisions gives managers tight control and allows companies to grow. Companies with vanguard management styles have developed managerial revolution with focus on horizontal processes rather than vertical structures. The management of companies engaged in sales activities should develop strategies to promote three core organizational processes that include frontline sales dynamics, enhance competence of salespeople, and sustained growth [43]. The vertical integration of corporate management acts as a catalyst for the emergence of a structural approach to strategic orientation, contractual control in the consumer goods selling and business-to-business client management. It is observed that organizational design can be restructured to take better account of the inner connection of a firm's structural properties in the vertical integration of its corporate management. The process of the vertical integration of corporate management positively influences the establishment of strategic orientation to achieve long-term development and top-down control guides the development priorities of sales oriented firms [44]. The following attributes are associated with the vertical sales organizations:

- Intra-organization communication system
- Sales managers are positioned at different levels of organizational hierarchy and salespersons constitute the bottom of the line of organizational structure,
- The span of control in the organization is vertically distributed with one to one reporting system
- Monitoring and evaluation of performance are considered as the principal tool of administrative control at various levels of employees,
- Vertical organizations emphasize on function related sales,
- Role of staff specialists and human resource managers is considered to be critical in vertical organizations in the process of recruitment and retrenchments of salespeople based on their performance evaluation by their superiors,
- Integration of sales and marketing functions is one of the key features of the vertical organization,
- Employee accountability and performance reporting are the principal determinants,
- Out sourcing salespeople in vertical integration of a common feature,
- Decision ladder in this type of organizations is top-down, and
- Vertical organizations also follow team selling and teams that are constituted with hierarchical design.

The highly competitive, turbulent business environment demands for the creation and delivery of superior customer value. In order to meet customer expectations and deliver superior value, many firms introduce significant transformations by developing a market orientation, building strategic relationships, improving processes and structures, enhancing the relationship between the marketing and sales functions, and developing appropriate new metrics. The role of vertical and horizontal designs of sales organization needs strategic alignment linkages between sales transformation strategy and sales management practice for delivering superior customer value and driving success of the firm [45].

Sales oriented companies under the increasing marketing competition are largely transforming from vertical sales force design to horizontal organizational structure in order to have the clear division of labor to enhance the outcome performance. The organization design in reference to divisions and specialization of sales force has long term advantages; however, sales force specialization may pose problem for managers due to narrow scope of their operations. When sales activities are divided and performed by individuals, it is necessary to establish effective coordination and integration with the client and organizational policies to accomplish the objectives of buying and selling. The more an organization's tasks are divided among specialists the more difficult is integrating task faced by the managers. Often this problem is aggravated when external agents such as manufacturer's representatives are inducted in the selling process limiting the managerial control over their actions.

GE Capital provides financing and services to business customers of all sizes to help them grow. The synergistic product offerings of the company include equipment leasing and lending, inventory finance, fleet services, and franchise finance in the global marketplace. GE offers business financing solutions for middle-market equipment financing and develops the financing solutions on a wide range of structured products comprising conventional loans, various lease structures and sale-leasebacks. The company is known for strategic business directives and flexible sales force. It follows the maxim of approaching the right customer in the right way by customer segmentation, sales structure alignment and sales process metrics alignment. The sales force effectiveness in the company is measured on the fundamentals of capacity that explains whether there are enough salespeople in the right marketplace or at the business accounts. GE considers quality of sales and services as the Sales DNA and believes that a good salesperson architects another good salesperson. The company ensures that sales force is spending adequate time with the clients to maximize yield in order to establish a sales lead in the competitive marketplace. The company feels that the benefits of having a scientific sales design results into the right number of sales representatives, the right customer/sales rep ratio, and the right percentage of the market covered in the given market. Such sales force design also help company in enhancing its sales performance in five customer centric ways including optimizing face time with the customer, having the right number of customer appointments, ensuring minimal response time to customers, maximizing the conversation ratio per customer, and increasing the activity per customer. One of the strengths of the sales force of GE is that its salespeople deliver value based services to the customers beyond selling products. Salespeople of the company invest time and resources with the customers in offering leadership seminar-style events to develop understanding and awareness about the products and services to be delivered by the company. Customers are also facilitated by the salespeople to clarify product or services specific issues through the customized on-site support and the company offers salespeople different levels of training based on the size and scope of the project. The Commercial Distribution Finance (CDF) has also built a reputation for understanding customers' businesses from every angle ranging from cash management and marketing to product development and risk analysis. The e-commerce capabilities allow customers to manage their accounts quickly and accurately via a secure online connection. The e-commerce tools of the virtual support speed information flow, from suppliers to vendors to distributors to resellers to end-users [46].

Sales managers should be responsible for coordination and integration of the efforts of salespeople in reference to the needs and concerns of customers, coordinating selling activities with other departments of the company such as production, product development, logistics, and finance, and tasks between generalists and specialist salespeople. Hence, the primary function of the vertical structure of a firm's sales organization is to ensure integration on the above lines. Channel-centric selling approach has a competitive advantage in coordination with the web-based technologies, such as the Internet, call centers, and database marketing. Salespeople in competitive markets also follow different routes to market to approach clients and large companies incorporate significantly more attributes related to public relations, building of user group and communities, and services to business partners to support the operations of salespeople [47].

Though companies use innovative approaches to encourage dynamism in the sales force through aligning their activities, there is no best way to dividing the selling activities among members of the sales force. Effectiveness of sales organization varies with the objectives, strategies, and tasks of the firm, its business environment, objectives, and marketing strategies. Changing economic conditions drive firms to prioritize activities that can be conducted with a client oriented perspective such as personal selling with prompt decision making. The choice between these individual and team selling tasks should encourage salespeople to develop customer-centric strategies considering the best fit with cultural factors of the clients. The division of activities among the salespeople should be considered in reference to their capabilities and competencies in managing clients and influence selling process with sustainable organizational culture that can develop positive relationship between the business culture of buyer and seller, and the selling process [48].

Most multinational companies use common bases for structuring the sales effort that include geographic sales force design, product bases sales force design, customer oriented sales force design and organizing sales force according to the different selling functions. Each of these organizational designs has unique advantages that make it appropriate for a firm under certain circumstances. However, before choosing an organizational design, it is important to decide whether the firm should hire its own salespeople or use outside agents. In many firms use of independent agents instead of company salespeople is an important option. Major computer application companies like IBM, Apple, and Hewlett Packard use both highly trained sales force of the company as well as a network of representatives of business partners around the world to help market its products and services. In such cases, firms usually rely on agents to cover geographic areas with relatively small number of customers or manage low sales potential territories that may not justify the cost of a full-time company salesperson. It is common for a firm to use both company salespeople and independent agents as horizontal sales force [49]. The horizontal sales force design in a firm is generally characterized with:

- Type of sales force of company
  - Regular sales force
  - Contracted sales force that is generally outsourced
  - Inducting casual sales force
- Category of sales force and hierarchical arrangement
- Assignment of the sales force
  - Product category-wise
  - Customer segment-wise
  - Sales function-wise
- Sales executives for major accounts management
- Cross-markets sales management.

In a horizontal sales design a company may use two common types of intermediaries or outsource agents to perform sales. An intermediary may be a manufacturer's representative who contributes in selling by achieving targets set by the firm. The intermediaries work along with the sales force of the firm. They take neither ownership nor physical possession of the goods they sell, but concentrate instead on the selling function and they are compensated solely by commissions. These representatives do not have any authority to modify their instructions concerning the prices and conditions of sale. Salespeople in the horizontal organizational design have wide opportunities for prospecting and selling products and services. They can establish contacts with potential customers across selling areas, develop familiarity with the demand of clients and understand their preferences and specializations, and they can develop cost effective selling strategies on the lines of preferences of the clients. The outsourced salespeople also can be organized in teams and administered. They have the ability to keep expenses low by spreading fixed costs over the products of several different manufacturers and work on commission basis that vary directly with the amount of goods sold. If firms are intending to develop outsourcing salespeople on a platform-based product design they should avoid three common risks including misalignment of objectives, unanticipated rivalry driven by higher commission motives among salespeople, and poor administrative control. If firms do not avoid these potential threats, entire sales force design on the horizontal organizational platform could be in jeopardy [50]. Sales organizations need to observe following approaches while designing the horizontal sales force design:

- Firms should not hire outsourced sales people who sell in unproductive ways. However, a sellerwho moves merchandise with current or potential customers in regions that are familiar to him would be the right option.
- A firm may hire an independent representative or outsource him who is alreadyselling similar products in the industry and knows about the company and customers.
- It is necessary to evaluate the credibility of the outsourced salespeople or members of sales team. The outsourced salespeople not only generate revenue generator but also represent the firm in the market.
- It is also important for the firms to check the business references before looking for outsourced salespeople so that attitudes of the salespeople can be reviewed prior to assigning the tasks.

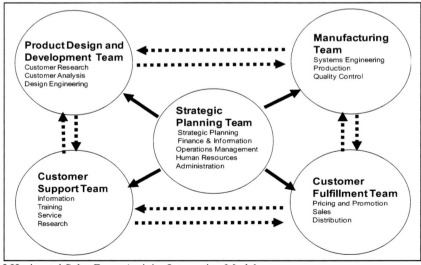

Figure 4.5.Horizontal Sales Force Activity Integration Model.

Recent business trends have received much attention on the practices of outsourcing and off-shoring. However, many cases of failure in developing outsourcing contracts suggest that the strategic use of outsourcing may not be as beneficial as believed, and hidden costs are often cited as a main source of failure. A firm can successfully develop new sourcing practices of his organization by employing strategic frameworks that will anticipate the hidden costs of outsourcing [51]. The task integration design in a horizontal sales force organization is exhibited in Figure 4.5.

Selling activities are not independent of marketing and technology led processes. Commonly in firms, the sales departments believe that marketers do not directly interact with the customer requirements and they stay out of touch with what is really happening in the marketplace. Marketing people, in turn, believe the sales force generally play myopic and act too focused on individual customer experiences. They are insufficiently aware of the larger market, and blind to the future. The marketing function takes different forms in different companies at different product life cycle stages [52]. Multi-business corporations need to develop a capability towards complex strategic integration (CSI) of inter departmental activities like sales, marketing, finance, customer relations, and technology, which involves the discovery and creation of new business opportunities by combining resources from multiple administrative units within the firm. Only some multi-business companies are currently trying to develop a CSI capability but the challenges and imperatives for all companies are the same [53].

ECONOMICS OF OUTSOURCED SALESPEOPLE

Selling agents are considered as intermediaries who do not take title or possession of the goods they sell and are compensated solely by commissions from their principals. Selling agents are usually granted broader authority by their principals to modify prices and terms of sale, and they are actively associated with the manufacturer's promotional and sales programs. The decision on inducting independent agents or a company sales force to cover a particular product or market involves careful examination of economic criteria, performance control process, transaction costs in the selling process, and strategic flexibility of the sales force design. Economic criteria are determined by measuring the productivity of sales agents at different levels of costs and sales volume in a given selling situation. While measuring the economic productivity the first step in deciding which form of sales organization is suitable to estimate and compare the costs of the two alternatives. A simple example of such a cost comparison of hiring independent sales representatives and those of the firm is illustrated in Figure 4.6.

It may be observed from the above figure that the fixed costs of using outsourced salespeople as sales representatives are lower than those of using a company sales force because of less administrative and overhead costs. The fixed cost of outsourced salespeople is also affected by zero salary or reimbursement for field selling expenses as they are recruited on commission basis. However, the costs of using outsourced salespeople tend to rise faster as sales volume increases because agents usually receive larger commissions than company salespeople. Consequently, a break-even level of sales volume exists below which the costs of outsourced salespeople are lower but on the contrary above this level a company sales force becomes more efficient.

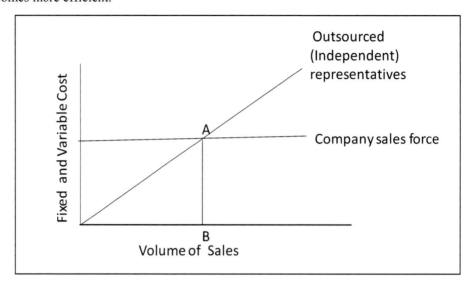

Figure 4.6.Analysis of Cost Factor in Using Sales Force.

This business phenomenon explains why outsourcing practices tend to be used by smaller firms or by larger firms in their smaller territories where sales volume is low to demand a company to work with low cost. Low fixed costs drive firms to outsource the salespeople during expansion of sales territories or product lines. Outsourcing strategy also pays to firms in selling products and services during market uncertainties as the agent does not get paid unless sales are made. Sales volume is another economic parameter for firms to look for outsourced salespeople. It is necessary to compare the productivity of salespeople in terms of volume of sales between company sales force and outsourced human recourses. Most sales and marketing managers believe sales force of the firm could achieve higher volume of sales as they can be trained on improved selling skills and driven more aggressive along with the new business policies of the firm. Regardless of organizational design, managers argue that the sales force of the firm has higher competency to produce more volume of sales in the long run as compared to the outsourced salespeople or manufacturer's agents. Such situation may occur due to lack of administrative control and performance measuring tools for outsourced salespeople. These human resources are seen as independent representatives responsible for pursuing short-run objectives. Thus they tend to be reluctant to engage in activities with a long-run strategic payoff to their principal, such as prospecting new accounts or small customers with promising sales potential. Managers can control the sales force of firm in reference to continuous supervision, establishing the product line policies, formal evaluation and reward mechanisms, and ultimately changing market and product lines of salespeople whose performance is not satisfactory.

The theory of transaction cost analysis (TCA) reveals that when adequate transaction specific assets are necessary to sell a manufacturer's product, the costs of independent agents in terms of administration and finance are likely to be higher than the costs of hiring and managing the sales force of the firm. The TCA assumes independent agents may pursue their self-interests at the expense of the firm they represent. The outsourced salespeople might provide only cursory post-sale service or invest marginal time to call on smaller accounts because they are unlikely to earn big commissions from such activities [54]. Most firms during the globalization process grew up with management practices that are suited to various modes of entry, higher transaction costs, larger capable competitors, growing and increasingly affluent markets and more hidden information. Hence in the current state of business affairs environments are less predictable, more complicated and more volatile. The result is that many core businesses are themselves becoming more uncertain and in need of renewal. Such situation is encouraging outsourcing of salespeople than having the sales force on the regular pay roll of the firm [55].

Sun Microsystems, Inc. (SMI) is a wholly owned subsidiary of Oracle Corporation, selling computers, computer components, computer software, and information technology services. Sun products include computer servers and workstations based on its own and other processors of companies and a suite of software products including the Solaris operating system, developer tools. Other major technologies of the company include the Java platform, MySQL and NFS. Sun is a proponent of open systems in general and UNIX in particular, and a major contributor to open source software. In the sales of technology oriented products service/maintenance contracts are highly profitable for many high-tech businesses. In order to reap the financial rewards companies in the informational technology sector focus on channel strategy for implementing and managing maintenance contract sales. It is observed that outsourced salespeople with high knowledge on software management can run contract sales program successfully. SMI found that without adequate outsource personnel the revenue stream of the company was eroding so they identified that as their field sales force did not have the "bandwidth" to provide cost- effective service contract sales to small-to-mid-size customers, the accounts were not adequately covered. Given the situation SMI management decided to hire third-party vendors and entrusted the task of identifying the customers who were out of warranty or whose contracts had expired. When they needed assistance, those customers would have to renew before they could get service with poor customer value. The company decided to outsource the development of a dedicated TeleServices Center - in essence creating a new distribution channel for this product. Working closely together, they designed and installed an on-site Telesales Center whose charge is to thwart third-party vendors and to comprehensively target contract renewals in small-to-mid-size accounts. Selling maintenance contracts requires TSRs who understand the service requirements of each customer. The right TSR has a background in both inside sales and customer service and identified decision makers and closed a sale. In addition, they must be "specialists" in Sun's service products. The typical sales cycle for a renewal contract is 90 days, involved two decision makers, and six-to-twelve conversations with the customer. Each TSR can contact and service 25-30 customers per day, far more than possible with field staff alone. The outsourcing program of SMI resulted into an increase in the contract renewal rate, reduced the cost to retain customers, lowered the cost of new customer contracts, liberated the tasks of company field staff to focus on major accounts, developed a renewal management database to efficiently handle the contract renewal process and designed an instructional curriculum that reduced TSR training time for salespeople [56].

Firms should select the outsourced salespeople on the basis of two principal cognitive attributes that include empathy and ego drive. Empathy, in this context, is the central ability to feel as other people do to sell them a product or service. A buyer who senses a salesperson's empathy will provide him with valuable feedback, which will in turn facilitate the sale. Another cognitive attribute, ego driver reflect in the behavior of a salesperson during the selling process not because of the money to be gained but because the salesperson feels he has to. For sales representatives with strong ego drives, every sale is a conquest that dramatically improves their self-perception [57].

BEHAVIORAL DETERMINANTS OF OUTSOURCING

Sales promoters at times fail to recognize that what influences buyers' satisfaction is not the same as what engenders store loyalty, and consequently do not effectively develop the cognitive drive to stimulate buying decisions. Hence, they need to vigilantly manage the quality of arousal by developing adequate customer involvement in the buying process. The customer values are created towards the new products through individual perceptions, and organizational and relational competence [58]. The firms need to ascertain a continuous organizational learning process with respect to the value creation chain and measure the sales performance of the innovative products introduced in the market. In the growing competitive markets the large and reputed firms are leaning on outsourcing salespeople to sell 'high-value integrated solutions' tailored to each customer's needs [59]. The sales attractiveness may be driven by the outsourced sales force through improved skills, use of advance technology, innovativeness, extended product applications, brand augmentation, perceived use value, competitive advantages, corporate image, product advertisements, and sales and services policies associated therewith which contribute in building sustainable customer values towards making buying decisions on the new products [60]. The attractiveness of new products is one of the key factors affecting the decision making of customers and in turn is related to market growth and sales. The higher the positive reactions of the customers towards the new products in view of their attractiveness, higher the growth in sales.

There is an increasing importance of building behavioral thrust among the potential buyers through effective communication and sales induction at customers' convenience. This process is largely managed by the outsourced sales people with focus on eight underlying factors of varying character important to customer satisfaction. These are awareness, selection, persuasion, convenience, trial, attention, location, promotional activities and merchandising policy [61]. Outsourced salespeople instill emotions among customers in terms of merchandise choice, visual merchandising, store environment, sales personnel attitude, pricing policies and promotional activities during the pre-purchase stage. These factors are the very foundations of consumer satisfaction and decision drivers towards buying products [62].

Systematically explored concepts in the field of customer value and market driven approach towards new products would be beneficial for a company to derive long term profit optimization strategy over the period. On a tactical level, managers need to consider the optimum spread of customers on a matrix of product attractiveness and market coverage. A Company may also need to consider emphasizing an integrated promotion strategy for new brands in reference to attributes, awareness, trial, availability and repeat (AATAR) principle. One of the challenges for the manager of a retail store is to enhance the in-store ambience to influence the young consumers for prolonged stay in the store for shopping and explore the zone of experience of new products. An augmented and sustainable customer value builds loyalty towards the product and the brand. Systematically explored customer preferences and arousal driven retailing approach towards new products would be beneficial for a company to derive long term profit optimization strategy over the period. This needs careful attention and the application of managerial judgment and experience to generate consumer arousal and develop appropriate point of sales strategies for stimulating the buying decision [63].

Visual effects associated with products often stimulate the buying decisions among young consumers. Point of sales brochures, catalogues and posters build assumption on perceived use value and motivational relevance of buying decisions of product. Emotional visuals exhibited on contextual factors such as proximity or stimulus size, drive perception and subjective reactions on utility and expected satisfaction of the products [64]. The satisfaction is the customer's perception of the value received in a transaction or relationship and it helps in making re-patronage decisions on the basis of their predictions concerning the value of a future product. Hence, many retailers develop innovative approaches to prospect new customers for new products by strengthening customer relationship and value management strategies [65].

SELECT READINGS

Bosworth, M. (1994), *Solution selling: Creating buyers in difficult selling markets*, McGraw Hill, New York, NY.

Hansen, P. H. (2006), The DNA Selling Method: Strategies for Modern-day Sales People, Brave Publishing, Alpine, UT.

Kengott, G., Downey, L. and Downey, J. (1998), *Selling from the inside out: An inspirational story for salespeople*, McGraw Hill, New York, NY.

McLeod, D. (2010), *The zero-turnover sales force: How to maximize revenue by keeping your sales force intact*, AMCOM Publications, American Marketing Association, Chicago, IL.

Zoltners, A. A., Sinha, P. and Lorimer, S. E. (2009), *Building a winning sales force: Powerful strategies for driving high performance*, AMCOM Publications, American Marketing Association, Chicago, IL.

REFERENCES

[1]   Weitz, B.A. and Bradford, K.D. (1999), Personal Selling and Sales Management: A Relationship Marketing Perspective, *Journal of the Academy of Marketing Science*, 27 (2), 241-254.

[2]   Rajagopal (2007)*Sales Management in Developing Countries: A Comparison of Managerial Control Perspectives, Journal of Asia Pacific Business, 8 (3)*, 37-61.

[3]   Zoltners, A. A., Sinha, P., Lorimer, S. E. (2006), Match Your Sales Force Structure to Your Business Life Cycle, *Harvard Business Review*, 84 (7), 80-89.

[4]   Farrell, M. A. (2005), The effect of a market-oriented organisational culture on sales-force behaviour and attitudes, *Journal of Strategic Marketing*, 13(4), 261-273.

[5]   Zoltners, A. A., Sinha, P., Lorimer, S. E. (2006), Match Your Sales Force Structure to Your Business Life Cycle, *Harvard Business Review*, 84 (7), 80-89.

[6]   Piercy, N. and Rich, N. (2009), The implications of lean operations for sales strategy: from sales-force to marketing-force, *Journal of Strategic Marketing*, 17(3), 237-255.

[7]   Ledingham, D., Kovac, M. and Simon, H. L. (2006), New science of sales force productivity, *Harvard Business Review*, 84 (9), 124-133.

[8]   Rajagopal and Rajagopal, A. (2008), *Dynamics of Buyer-Supplier Co-dependency for Optimizing Functional Efficiency, International Journal of Services and Operations Management, 4 (4),* 399-416.

[9]   MacMillan, I.C. and McGrath, R. G. (1997), Discovering new points of differentiation, *Harvard Business Review*, 74 (5), 133-145.

[10]   Ross, W. T., Dalsace, F. and Anderson, E. (2005), Should You Set Up Your Own Sales Force or Should You Outsource It: Pitfalls in the Standard Analysis, *Business Horizons*, 48 (1), 23-36.

[11]   For details see Major Pharmaceuticals and Diagnostics Providers, Case Study, West Business Services, 2008 http://www.westbusinessservices.com/

[12]   Henry, P. (1975), Manage your sales force as a system, *Harvard Business Review*, 53 (2), 85-95.

[13]   Piercy, N. F. (2006), The Strategic Sales Organization, *The Marketing Review*, 6 (1), 3-28.

[14]   Rajagopal and Rajagopal, A. (2008), Team Performance and Control Process in Sales Organizations, *Team Performance Management- An International Journal*, 14 (1), 70-85, 2008.

[15]   Tom, N. and John, H. (2008), Is your sales force a barrier to more profitable pricing or is it you?, Business *Strategy Series*, 8 (5), 365-368.

[16]   Piercy, N. F., Cravens, D. W., and Morgan, N. A. (1999): Relationships between Sales Management Control, Territory Design, Sales Force Performance and Sales Organization Effectiveness, *British Journal of Management*, 10 (2), 95-111.

[17]   Piercy, N. F., Cravens, D. W., and Morgan, N. A. (1998): Sales force performance and behavior-based management processes in business-to-business sales organizations, *European Journal of Marketing*, 32 (12), 79-100.

[18]   Bailey, E. L. (1989), Getting closer to the customer, *Research Bulletin* (Vol. 229). New York,The Conference Board, Inc.

[19]   Grant, K., Cravens, D. W., Low, G. S., and Moncrief, W. C. (2001), The role of satisfaction with territory design on the modification, attitudes, and work outcomes of salespeople, *Journal of the Academy of Marketing Science*, Spring, 165-178.

[20]   Churchill, G. A., Ford, N. M., Walker, O. C., Johnston, M. W. and Tanner, J. E. (2000), *Sales Force Management* (6th ed.), Chicago, IL: Irwin.

[21]   Bonoma, T. V. (2006), Major sales: Who really does the buying, *Harvard Business Review*, 84 (7), 172-181.

[22]   Shapior, B.P. and Posner, R. S. (2006), Making the major sale, *Harvard Business Review*, 84 (7), 140-148.

[23]   Cruickshank, J. A. (2009), A play for rurality: Modernization versus local autonomy, *Journal of Rural Studies*, 25 (1), 98–107.

[24]   Dubey, J. and Patel, R. P. (2004), Small wonders of the Indian market,*Journal of Consumer Behaviour*, 4 (2), 145-151

[25]   Blumenthal, D. (2002), Beyond 'form versus content': Simmelian theory as a framework for adaptive brand strategy,*Journal of Brand Management*, 10 (1), 9-18.

[26]   Rajagopal (2009), Branding Paradigm for Bottom of the Pyramid Markets, *Measuring Business Excellence*, 13 (4), 58-68.

[27]   Reutterer, T. and Teller, C. (2009), Store format choice and shopping trip types. *International Journal of Retail & Distribution Management*, 37 (8), 695-710.

[28]   Sehrawet, M. and Kundu, S. C. (2007), Buying behaviour of rural and urban consumers in India: the impact of packaging,*International Journal of Consumer Studies*, 31 (6), 630-638.

[29]   For details see corporate web site of Hindustan Unilever Limited, Enhancing Livelihoods, case study http://www.hul.co.in/sustainability/casestudies/enhancing-livelihoods/Shakti.aspx, also see Sabharwal, A. P., Gorman, M. E. and Werhane, P. H. (2004), Case study: Hindustan Lever Limited and marketing to the poorest of the poor,*International Journal of Entrepreneurship and Innovation Management*, 4 (5), 495-511.

[30]   Simester, D., Yu, H., Brynjolfsson, E., and Anderson, E. T. (2009), Dynamics of retail advertising: Evidence from a field experiment, *Economic Inquiry*, 47(3), 482-499.

[31]   Prahalad, C. K., and Lieberthal, K. (2003), The end of corporate imperialism, *Harvard Business Review*, 81(8), 109–117.

[32]   Prahalad, C. K. and Hammond, A. (2002), Serving the world's poor, profitably, *Harvard Business Review*, 80 (9), 48-58.

[33]   Mintzberg, H. (1981), Organization design : Fashion or Fit, *Harvard Business Review*, 59 (1), 103-116.

[34]   For similar discussion see Garvin, D. A andLevesque, L. C. (2008),The multiunit enterprise, *Harvard Business Review*, 86(6),106–117.

[35]   Allen, R. K. (2009), What is organizational design, The Center for Organizational Design, http://www.centerod.com/articles/90-what-is-organizational-design.html

[36]   Kampschroer, K., Heerwagen, J. and Powell, K. (2007), Creating and testing workplace strategy, *California Management Review*, 49(2), 119-137.

[37]   Wotruba, T., Brodie, S. and Stanworth, J. (2005), Differences in Turnover Predictors between Multilevel and Single Level Direct Selling Organizations. *The International Review of Retail, Distribution and Consumer Research*, 15(1), 91-110.

[38]   Jones, C. (2009), Direct sales offer recession-proof jobs, USA Today.com http://www.usatoday.com

[39]   Ntayi, J. M., Munene, J. C., and Eyaa, S. (2010), Sales force behavioural performance of accounts relationship managers (ARMS) in Uganda's commercial banks: A qualitative analysis, *Journal of Retail and Leisure Property,* 9(1), 5-23.

[40]   Herbig, P. and Yelkurm, R. (1997), A Review of the Multilevel Marketing Phenomenon, *Journal of Marketing Channels*, 6(1), 17-33.

[41]   Cross, R., Ehrlich, K., Dawson, R. and Helferich, J. (2008), Managing Collaboration: Improving Team Effectiveness through a Network Perspective, *California Management Review*, 50 (4), 74-98.

[42]   Rajagopal (2007), *Marketing Dynamics: Theory and Practice*, New Age International Publishers, New Delhi.

[43]   Ghoshal, S. and Bartlett, C.A. (1995), Changing the Role of Top Management: Beyond Structure to Processes, *Harvard Business Review*, 73(1), 86–96.

[44] Yan, Y., Child, J. and Chong, C. (2007), Vertical integration of corporate management in international firms: implementation of HRM and the asset specificities of firms in China, *The International Journal of Human Resource Management*, 18(5), 788-807.

[45] LaForge, R. W., Ingram, T. N. and Cravens, D. W. (2009), Strategic alignment for sales organization transformation, *Journal of Strategic Marketing*, 17(3), 199-219.

[46] For details see corporate web site of General Electric Company http://gecapsol.com

[47] Liu, S. S., Wang, C. C. L. and Chan, A. (2004), Integration of Multiple Sales Channels with Web-Based Technology–A Case of the Pharmaceutical Industry, *Journal of Business To Business Marketing*, 11(1), 131-152.

[48] Macquin, A., Rouziès, D. and Prime, N. (2001), The Influence of Culture on Personal Selling Interactions, *Journal of Euromarketing*, 9(4), 71-88.

[49] Johnston, M. and Marshall, G. (2008), *Sales force management*, McGraw Hill, New York, NY.

[50] Amaral, J. and Parker, G. (2008), Prevent disasters in design outsourcing, *Harvard Business Review*, 86 (9), 30-34.

[51] Tedelis, S. (2007), Innovative organizations-Creating value through outsourcing, *California Management Review*, 50(1), 261-277.

[52] Kotler, P., Rackham, N. andKrishnaswamy, S. (2006), Ending the war between sales and marketing, *Harvard Business Review*, 84 (7), 68-78.

[53] Burgelman, R.A. and Yves, L. D. (2001), Power of strategic integration, *Sloan Management Review*, 42 (3), 23-38.

[54] Erin, A. (1985), The salesperson as outside agent or employee: A transaction cost analysis, *Marketing Science*, 4 (3), 234-254.

[55] McGrath, R. G. and MacMillan, I. C. (2009), How to rethink your business during uncertainty, *Sloan Management Review*, 50 (3), 25-30.

[56] For details see The corporate web site of Sun Microsystems Inc. http://www.sun.com

[57] Mayer, D. and Greenberg, H. M. (2006), What makes and good salesman, *Harvard Business Review*, 84 (7), 164-171.

[58] Miranda, M., Konya, L. and Havira, I. (2005), Shopper's Satisfaction Levels are not only the Key to Store Loyalty, *Marketing Intelligence and Planning*, 23 (2), 220-232.

[59] Ganesh, J., Arnold, M.J. and Reynolds, K.E. (2000), Understanding the customer base of service providers: an examination of the difference between switchers and stayers, *Journal of Marketing*, 64, 65-87.

[60] Lafferty, B.A. and Goldsmith, R. E. (2004), How influential are corporate credibility and endorser attractiveness when innovators react to advertisement for a new high technology product? *Corporate Reputation Review*, 7 (1), 24-26.

[61] Anselmsson, J. (2006), Sources of Customer Satisfaction with Shopping Malls: A Comprehensive Study of Different Customer Segments, *The International Review of Retail, Distribution and Consumer Research*, 16 (1), 115-138.

[62] Otieno, R., Harrow, C. and Lea-Greenwood, G. (2005), The unhappy shopper, a retail experience: exploring fashion, fit and affordability, *International Journal of Retail & Distribution Management*, 33 (4), 298-309.

[63] Rajagopal (2008), Outsourcing Salespeople in Building Arousal towards Retail Buying, *Journal of Database Marketing and Customer Strategy Management*, 15 (2), 106-118.

[64] Codispoti, M. and De Cesarei, A. (2007), Arousal and attention: Picture size and emotional reactions, *Psychophysiology*, 44 (5), 680-686.

[65] Johnson, L. K. (2006), Successful business process outsourcing, *Sloan Management Review*, 47 (2), 5-6.

Chapter 5

SALES TERRITORY MANAGEMENT

Sales territory design is considered as a particularly important managerial variable, which has received little analytical attention in the traditional literature, but which appears to be an important influence on the effectiveness of the sales operation. It has been argued that organizational effectiveness is determined by sales force outcome performance and behavioral performance, as well as by the use of a behavior-based control approach. Although conventional theory has suggested that behavior performance and outcome performance result from different stimuli, behavior-based control is positively associated with both behavior performance and outcome performance [1]. The importance of designing effective sales territory is widely supported in the process of organizational restructuring [2]. Despite this importance, the impact of sales territory designs on salespeople and has not gained significant research attention. However, designing effective sales territories in the multinational sales organization is growing substantially and posing challenges towards integrating the local market differences in sales infrastructure and other competitive market attributes. Organizational commitment and sales territory design is significantly related to sales force performance. The study highlights the growing emphasis on building long-term, collaborative buyer-seller relationships that favor the use of behavior-based control systems in many sales management situations, and suggests a new agenda for management attention in improving sales force effectiveness [3].

In view of the increasing global competition, market oriented sales firms attempt to develop new sales territories to expand their clientele. However some firms overestimate the attractiveness of new sales regions in territory expansion process. By the sheer size of untapped markets, firms lose sight of the difficulties of pioneering new, often very different territories. The problem is rooted in the analytic tools including sales region portfolio analysis (SRPA), which is similar to country portfolio analysis in reference to international sales territories. In developing such strategies to develop new sales territories, managers judge the requirement of investment on building sales force infrastructure, cost of selling, account servicing cost, productivity of salespeople, and overhead costs of the company in administration of sales force in new regions. SRPA focuses on sales competition in the territory, propensity of buying, consumer income, and attributes of sales channels, calculates the costs and risks of doing business in a new market, and emphasizes potential sales. Most of the costs and risks associated in selling in new territory result from the barriers created by distance. "Sales Distance," however, does not refer only to geography bust also considers attributes including cultural distance encompassing religious beliefs, race, social norms, and language that are different for the target country and the country of the company considering expansion. Sales distance is also influenced by the administrative or political distance comprising colony-colonizer links, common currency, and trade arrangements while geographic distance in sales is observed as the physical distance between the two regions, accessibility and its size. However, the economic distance also needs to be measured in developing the sales territories by observing carefully the disparities in the two regions in reference to consumer income and variations in the cost and quality of financial and other resources. This framework can help in identifying the ways in which potential markets may be distant from existing ones [4].

The territory design is a major determinant of salespeople's opportunity to perform well, and their ability to earn incentive pay where incentives are linked directly to territory-level individual performance [5]. It has been observed that organizations may not define salespeople' territories based on mutually exclusive geographical areas. Moreover, the effective sales organizations place more emphasis on their sales territory design, and, additionally, their sales forces show significant differences in both personal characteristics and performance dimensions. Salespeople in the more effective sales units display higher levels of intrinsic and extrinsic motivation, sales support orientation, and customer orientation. Both salesperson behavior and outcome performance were rated higher by managers in the organizations with more effective sales units [6]. Experienced sales managers understand that poor territory designs will have a negative impact on salesperson morale, result in inadequate market coverage, and complicate management evaluation and control [7]. Several factors beyond control of the salesperson often guide the territory design decision including the size of the sales force, buying power of the accounts, geographical dispersion of accounts, time required to service each account, and competitive intensity. The sales territory should be designed primarily considering the following factors:

- Short-run customer breakeven point,
- Higher potential of account penetration,
- Realizing sales through extended product-mix,
- Scope of receiving volume of order,
- Consolidated orders across manufacturing plants, offices, and servicing units of large accounts,
- Possibilities of promotion technology led customer services,
- Opportunities for managing competitors and taking sales lead, and
- Likelihood of defending the sales territory by securing customers and business-to-business accounts.

Most companies aim at developing their sales force territories which have less competitive threats and can be easily cordoned by the salespeople protecting the interests of their customers and industrial accounts. Accordingly these company focus on developing sales territories with well knit technology and route to market networks to generate customer value and outperform their rivals. As a result, their strategies tend to take on both aggressive and defensive dimensions in managing sales territories. Instead of looking within the conventional boundaries that define how an industry competes, managers can look methodically across them. By doing so, they can explore territories that represent new opportunities and score for competitive advantage in the marketplace. While determining sales territories it is important that sales territory strategists should look across the substitutes in the region serving preferential customer needs and develop strategies to acquire and retain customers on personalized offers like personal finance, product training, and services. Territory selection and design process should also provide opportunities to derive insights by looking across strategic groups within an industry, across buyer groups; across complementary product and service offerings, across the functional-emotional orientation of competitors, and even across time [8].

Revival of market territories often pays advantage to the sales firms as they enter the market as leaders and pioneer sales and services activities on a very new customer relations platform. After years where all that was Portuguese was looked down on, Portuguese consumers are shifting their views and enjoying and being proud of what is Portuguese. This is most evident in wine, where the market is mainly composed of Portuguese players and some traditional drinks such as *aguardente,* liqueurs, and the former Libris, the Port wine. These products are seen as pure and traditional, and linked with the global trend towards healthy and green consumption. The Portuguese domestic players redesigned their sales territory by developing customer and distributor friendly strategies exploiting the benefit of consumer behavioremerging from global movements, the local firms are in the lead of the alcoholic drinks market. Unicer - Bebidas de Portugal SA managed to maintain its position and continued to rank first, closely followed by SCC - Sociedade Central de Cervejas SA. Indeed these two companies continued their struggle to keep their brands alive by managing customer centric sales strategies and offering consumers innovative products backed by plenty of promotions to prevent slump in the beer consumption. Providing evidence of the hard times faced by consumers, private label is the third player in the market due to its good performance in terms of beer and wine. After the recession in 2009, the alcoholic drinks market has picked up and returned to expected consumption levels, both in volume and value terms. It is because the market players kept on innovating, presenting consumers with new drinking opportunities and attracting them to the quality of Portuguese products. Nevertheless, positive growth in the on-trade is only expected to return during the latter part of the forecast period. Off-trade sales on the other side should continue to remain positive, as the home consumption trend is expected to prevail for a while [9].

Some companies incur huge losses when they enter unknown territory without proper territory management planning and overlook the possibilities of new alliances, markets, products, and technologies. Sales territories need to be formed integrating salespeople, functional technologies, partnership possibilities, and avenues for expansion of sales. Sales territories are sensitive to competitor attacks, customer defection and abrupt cost escalations. However, territorial sales failures could be prevented or their costs could be contained if managers approached innovative sales strategies with the right planning and control tools. Discovery-driven planning to design new sales territories is a practical tool that acknowledges the difference between planning for a new venture and for a more conventional business [10]. Sales territory of a firm may also be damaged by penetration of franchises which create acute local competition and affect the sales growth of the firm. In the strategy of moving franchisees the principal firm stays in remote territory but try to lead the market through local entrepreneurs who are well acquainted with the consumer behaviorand strategies of competitors. The efforts to regulate franchise encroachment suggest that competitive growth is stimulated by some cannibalization of existing system sales [11].

SALES FORCE DEPLOYMENT

The strategy of deployment of sales teams in different territories is generally embedded with 3-T power-grid which comprises a synergy of task (commitment), thrust (driving team) and time (punctuality), these indicators are weak in Mexican sales teams and affect their overall performance [12]. A sales task needs to be made clear for deriving effective sales results of salespeople in a given territory. The tasks assigned to performance in a sales territory should receive periodical feedback to make necessary modification in the selling process. A firm can deploy the sales force in a territory to manage key accounts, information systems, and prospecting new customers. The effectiveness of salespeople would be enhanced with the more specific nature of the account coverage. Hence, optimizing the number of accounts for each salesperson in a given territory is recommended. Field sales managers can use management by objectives, performance appraisal, and monthly reviews to encourage their salespeople to do their work effectively in the predetermined sales design [13]. Deployment of sales force in territories is generally based on the type of activities assigned to the salespeople in reference to making sales territories as profit centers, acquiring increasing number of customers, enhancing the volume of sales of products, and attending to various sales improvement functions such as opening new accounts, customer servicing and developing existing accounts. Decisions about deployment of sales force in territories should be based on the primary focus of gaining lead in the sales. Accordingly a coherent sales force deployment program requires clarity about the appropriate measures to use in evaluating the effectiveness of sales efforts [14].

Several factors beyond control of the salesperson often guide the territory design decision including the size of the sales team, buying power of the accounts, geographical dispersion of accounts, time required to service each account, and competitive intensity. The selling skills are determined by the degree of experience of salesperson and effectiveness of training while the adaptive selling techniques are related with increased performance [15]. Firms should be able to apply the time-based philosophy of revenue management to their sales teams. To do so, it is required to have revision in the way most sales divisions traditionally have viewed salesperson time. Hence, a different type of proposed measure, revenue per available salesperson hour, is proposed to better integrate the value of the salesperson's time as a factor in sales potential and revenue calculation [16]. Managers of sales organizations should enhance capabilities on proper deployment of sales force matching the managerial and techno-economic attributes of the selected sales territory to enable effective contribution to the company, and facilitate networking of the strategy. Salespeople in new and complex sales territories face the vital and challenging roles to play in both shaping business, connecting up with the customers and company, and executing the customer centric strategy to gain competitive advantage. However, it is necessary for the firms to discover strategies for skillful deployment of the sales force in different types of territories [17].

The hike in food and energy prices followed by the global financial crisis in 2008 eroded disposable income as well as consumer confidence in the future stability of their jobs. Consequently consumer spending propensity was reduced, which, in the case of consumer foodservice, meant fewer visits to the restaurants or switches to less expensive formats and lower-value orders. In constant value terms, sales continued declining in the global marketplace for the consecutive year, while profit margins dropped as inflation rose and costs increased. Players in this sector had scarce opportunity to transfer rising costs onto consumers in the form of higher prices. Under such circumstances many multinational firms decided to redesign sales force territories and pattern of deployment of salespeople. The financial downturn affected most retail companies in Chile and Cencosud's Costanera Center unfinished building emerged as a witness to these difficult times. Food retailers in Chile were located in the competitive sales territory and were managed by a large outsourced sales force. The multinational firms in retail food sector carefully examined the Chilean consumer preferences regarding location and offering in order to redesign the sales territories and deployment of sales force. However, lower investments in the sales in competitive territory have obliged consumer foodservice players to revisit their plans and decide if they should also postpone some openings or select less attractive locations outside shopping malls and strip centers. Among various new territorial shuffle food service sales, 17 new casinos began operating in Chile with increased competitive thrust in the food retailing sector. In addition to restaurants inside casinos, several projects include shopping centers that feature a food court. Casino openings have influenced the development of the transport infrastructure which provided access to them. Selling has been more challenging than before as managing such competitive territory demanded more outsourced salespeople and higher selling cost was estimated. Retailers, petrol stations with forecourt fast food, and independent foodservice outlets emerged alongside the competitive territory and each competing firm focused on sustainable sales than investing more resources on salespeople [18].

Many firms create cross-functional sales teams drawn from strategy, marketing, and finance for deploying in the competitive markets. Such strategies are developed on the assumptions of long-term sales plans in the context of consumer demand for products and services, and expected sales performance relative to competitors. The balance between the sales territories and the sales units constituted by the cross-cultural teams needs to be achieved during the process of territory designing. A balanced sales territory helps in building a profit center in the region. Firms may enlist various attributes on forming sales teams, develop tactics for short-run gain in the sales territory, monitor performance, track resource for sales force and develop competencies. Managers should reward sales force for upholding customer values and competitive lead, motivate and develop creativity among salespeople to narrow the strategy-to-performance gap in the sales territory [19]. However, multinational firms are engaged in designing effective sales territories by integrating the local market attributes and consumer preferences, in order to stay competitive.

Organizational commitment and sales territory design are significantly related to sales force performance and the study highlights the growing emphasis on building long-term, collaborative buyer-seller relationships that favor the use of behavior-based control systems in many sales management situations, and suggests a new agenda for management attention in improving sales force effectiveness. The taxonomy of sales territories is exhibited in Figure 5.1, which illustrates various managerial attributes of different sales regions. The deployment of sales force territory may be considered in reference to the four territorial segments including existing, competitive, potential and dormant sales areas which possess distinct operational features. Most companies deploy small number of salespeople in the existing territory as in this sales region companies have long standing and possess strong brands. As the brands are well established in the existing territory, they generally tend to generate consumer pull effect placing brand as the top of the mind decision tool. Sales and services attributes, consumer goods, brand names, logos, and corporate reputation are important attributes that are instrumental in developing pull effect among consumers. The pull effect also drives the brand value within social networks. The pull factor in the existing territories is also driven by the competitive advantage of consumers with new market players, selling systems, and underlying vulnerabilities in the marketplace [20]. Hence, companies attempt to reduce the cost of selling, increase volume of sales, increase net revenues and enhance customer value that help in building existing territory as profit center.

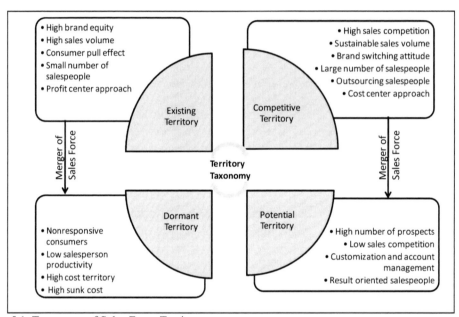

Figure 5.1. Taxonomy of Sales Force Territory.

The competitive sales territories are characterized with high competition, risk and uncertainties. Under such market situations many firms lean towards maintaining sustainable volume of sales by reducing the cost of sales. The major task of salespeople in managing such territories is to develop loyalty among consumers by preventing brand defection. In managing competitive territories firms rely on outsourcing large number of salespeople in order to reduce the operational and administrative costs. These territories can be managed as cost centers by reducing the sales cost, maintaining average volume of sales and revenue, and enhancing customer value. Salespeople face more challenging tasks in the potential territories as compared to competitive territories, which signal increasing opportunities for prospecting new accounts. However, low sales competition in these territories helps salespeople to achieve high targets. Firms need to deploy result oriented salespeople in potential territories and offer continuous skill development programs. The objective-achievement compensation system encourages the salespeople to achieve higher targets and them to provide the company with better forecasts in order to meet their quotas. Sales people in potential territories should be trained to develop their sales proposals using the objective-forecast-actual system, which is a combination of three measurements including the company objective, the forecast of the salesperson, and the actual results the salesperson achieves. Companies should also develop possible way of rewarding a salesperson according to how close his forecasts and actual results are to the company's objective [21].

Companies should not make investment on deploying salespeople and developing infrastructure in the dormant territories that show low response rate of consumers to the sales offers. In such sales territories the productivity of salespeople remains low. Some firms take the risk of investing in sales operations to enhance sales in dormant territories but due to non-responsiveness of customers fail to recover even the selling cost. Such situation leads to huge loses to the firms and increases the sunk cost. Most firms compare the cash flow trends in different sales territories to plan for future investment in the territories for improving sales in the region. In dormant sales territories firms may incur fixed cost to deploy a small number of salespeople to help building brand but do not make long-term investments in the region due to low consumer interests. In most situations, however, competitors' sustaining and disruptive investments over time result in deterioration of financial performance. The sunk-cost confers financial irrecoverable loss and shackles firms that attempt to continue investing in unproductive regions [22].

TERRITORY DESIGN AND PROCESS

Territory designing is more an engineering issue than managerial strategy. Firms should take into consideration the physical accessibility, logistics and infrastructure cost, and the factors supporting operational efficiency of the sales force while designing the territories. The sales infrastructure in a defined territory should be built around contemporary tools of telecommunication, Internet and computational services in the region, and office requirements for contributing to the working efficiency, and equipping the office infrastructure for multi-functional task administration. At the same time, innovations in technology and territory design provide salespeople with more choices to capitalize on customers and share innovative sales strategies among peers. The territory design and appropriate alignment of services help in improving the effectiveness of the salespeople by way of three different functions [23] as indicated below:

- Instrumental functions, such as improving decision making and inter-group collaboration,
- Symbolic functions that include affirming individual distinctiveness and team sales status, and
- Aesthetic functions allowing for desired sensory experiences to customers and promoting a sense of buying products and services.

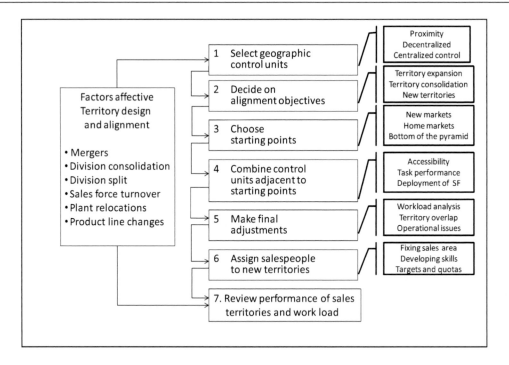

Figure 5.2.Territorial Design and Process.

Sales territory design and sales force controls are the major determinants of alignment policies and practices that govern the managers in training, supervising, motivating, and evaluating salespeople. The administrative control measures include the types of compensation offered to the salespeople and the criteria used by the sales managers to evaluate the performance of field sales force. These controls need to be set in reference to the territory design to let salespeople know which trade-offs the company would prefer them to make when the inevitable conflicts arise between spending time and money to get a sale done and what they can actually do using limited resources and still successfully sell the products and services. When sales force territories and controls are not properly aligned, functional discrepancies occur. Under such circumstances the firm may encourage salespeople to be entrepreneurial on one hand but may also ask them to file detailed call reports and check in frequently with their managers that may discourage them to take their own decision to be entrepreneurial, on the other [24]. Eventually, salespeople may develop workplace frustration and leave the company. The Figure 5.2 exhibits the sales territory design and process management steps.

Often firms need to redesign their sales territories and align them with revised corporate goals consequent upon mergers and acquisition of firms, consolidation of different sales divisions or split of large product sales divisions, turnover of sales force, relocation of manufacturing plants, and extension of product line. Upon deciding the geographical location for sales, firms identify strategic control units to supervise the operations of the salespeople. Some firms decentralize the administrative control to measure the performance of sales force while some firm are comfortable to centralize the administrative command and follow the top-down organizational policy. However the most firm indentify proximity based administrative control units to supervise the performance of salespeople in the designed territory. It is necessary for the firm to align their territory in reference to expansion or consolidation of the sales areas. The new sales territories can also be defined by the firm considering various operational and techno-economic requirements for the salespeople. Increasing numbers of sales organizations are responding to competitive challenges by improving their core competencies in existing, competitive and potential sales territories. Management of core sales processes is related to sales force design, territory alignment, and their potential to generate superior performance using the contemporary tactics [25].

The starting points can be chosen by the firms as new markets, home market or market segment lying at the bottom of the pyramid (BoP) while designing sales territories. The debate on BoP markets puts forth various perspectives including global firms which confront their own preconceptions particularly about the value of high-volume, low-margin businesses for companies to master the challenges or reap the rewards of these developing markets [26]. Successful marketing by multinational firms to the large size of consumers belonging to the bottom of the pyramid in the world can be realized through positioning global brands in the BoP market itself, managing the consumer animosity towards high profile brands, building influence of global umbrella brands and developing responsive marketing as a guiding principle for the BoP markets. Exploring the potential of the BOP markets requires positioning of global brands at radically lowered priced products and needs to enhance the consumer value [27]. Accessibility, task performance and strategies for deployment of sales force should be determined by the firm. Firms should detect any overlap in the territory and fix the problem. This would help the firm to develop strategies on deciding workload of salespeople and to assign territories to the salespeople accordingly.

Administering control measure to field sales force has become major performance guiding tool in many business firms. Industrial sales organisations largely implement direct management control and influence the activities of employees leading towards improving their efficiency [28]. The extent monitoring, directing, evaluating, and rewarding activities of salespeople in an organisation should be focused as a learning tool. Managers should guide sales force behavior through management control processes to achieve the targets [29]. Management control is thus recognized as an important performance indicator of the task performed by the salespeople. There are two control systems that are generally followed by the firms towards the administrative control of salespeople:

- Outcome control (OC) systems that measures the performance of salespeople in all quantitative terms achieved by putting the customer first, and
- Behavior control (BC) systems the system that encourages salespeople to put their managers first.

However, both systems have internal conflicts and in most cases, the right choice will be a consistent system somewhere in the middle of the OC-BC continuum [30].

Result oriented performance control and market volatility are positively related to new product selling performance. The effect of sales force adoption on selling performance is stronger where outcome based control is used and where the firm provides information on the background of the new product to salespeople through internal marketing [31]. Some studies observe that salespeople who simultaneously exhibit commitment and effort will achieve higher levels of new product selling performance. However, not much research attention is given to sales management control beyond developed countries, notwithstanding recognition of the critical role of the sales manager in international selling [32].

SALES TERRITORY MAPPING

Sales territories are built by the firms for long-term gains and sustainable growth. Hence the process of territory designing should be logical and based on scientific thinking. It has been observed that firms for short-term advantage make incorrect choices to enter a marketplace and deploy salespeople randomly in all segments anticipating the favourable response of the market to the efforts of salespeople. The process of random deployment of salespeople in different segments in the sales territory is known as mosaic, which leads to consolidation of non-responsive markets or sales areas in the long run. Non-responsiveness to the sales efforts in random deployment process occurs due to preconceived notions of consumers on the product value, resulting in slow sales. This strategy of sales force deployment may cause administrative and logistics costs. As a result, sales managers operating in high-cost and high-risk environment need to invest in sophisticated control measures to support inside sales teams for driving high performance. The cost of information to tract demand, inventory and deliveries would also increase as firms using random sales force deployment strategy may require investment on predictive analytics, data mining, and other business intelligence tools help inside sales teams to effectively manage their costs and generate sales. Firms should develop strategic logic for how to deploy the salespeople in different regions through careful sequencing of task, time and territory activities. Whether sales activities are lining up support for an internal initiative or trying to reach agreement on a sale and service, forward and backward mapping can help the to determine various business outcomes [33].

Firms may choose to begin sales operations from the key sales segments that have the potential to attract other consumer segments or neighboring markets. Sales managers should evaluate the performance of sales operations at key segments and identify the forward linkages that were developed by the key segments. The neighboring markets that are attracted towards products and services offered in the key sales segments may have the potential to drive customer attraction in further sales networks. These sales segments may be defined as strategic sales units as they possess the potential of expanding sales operations. Firms can establish their sales units in the entire territory moving gradually segment by segment. The territorial sales mapping approaches are exhibited in Figure 5.3. This process of expanding sales operations is cost effective and generates consumer pull effect that ensures high sales performance of the firm. Territory sales mapping is practiced as proximity approach.

Management thinkers within and beyond the discipline of marketing geography with focus on sales, increasingly realize that operational boundaries are not simply lines that enclose and define territories but also determine the competitive position of the firm. Sales operation boundaries, either in reference to consumer or market segments regulate the selling process and stand as principal resource, contributing to the overall revenue of the firm. Movement of salespeople, beyond and across the key territories, strategic sales units, and small segments serves to strengthen the brand equity of the products and services offered. Firms need to push salespeople performing beyond the spatially fixed activities to monitor sales competition and develop territory based strategies. It is necessary to trace the movement of competitors across, and outside the territorial boundaries and design administrative control for enhancing performance of salespeople. Sales territories have two dimensions that include inside space as a series of small segments of consumers, business partners, and service providers while the outside space represents an arena of competitors posing continuously new challenges on the inside space [34].

Hitachi High-Technologies was formed in 2001 by the combination of trading company Nissei Sangyo with the former Instruments Group and Semiconductor Manufacturing Equipment Group of Hitachi. The company makes inspection systems and other semiconductor production equipment, as well as analytical and clinical instruments, such as electron microscopes. As part of reorganization of the larger Hitachi Corporation, Hitachi High-Technologies in 2004 absorbed Hitachi Electronics Engineering, which makes inspection systems used in semiconductor and LCD production. The company also sells computers and peripherals, steel and plastics, silicon wafers, and optical components, among other materials and products. As a "global business creator" focused on nanotechnology, Hitachi High-Technologies provides systems and equipment fully capable of achieving the results that our customers demand. The company has moved in global market in stages by creating values of each product and service across the markets in Asia and Europe. In 2008, customers' earnings were affected as contraction in demand pushed down memory device prices in Asia and elsewhere. The resulting postponement or freezing of capital investments impacted heavily on the company's business, and sales slumped. The company has invested resources into exploring new sales territories and redesigning strategic sales units to enhance the customer value in the global marketplace [35].

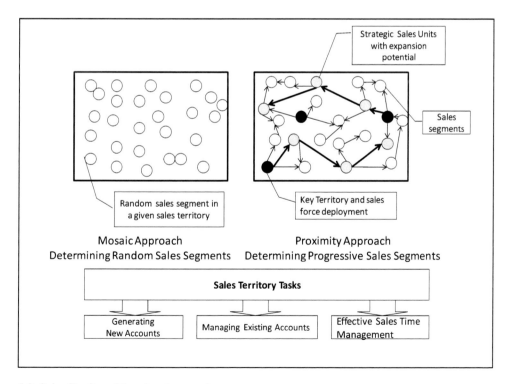

Figure 5.3. Sales Territory Mapping Approaches.

Technological development, globalization, and changes in interpersonal and inter-market relations, and their growing needs have transformed the market places making up a territory. The sustainability of a place depends on a series of factors which contribute to the quality of life, sense of place and recognition of identity. An activity such as personal selling, which in itself is obviously 'sustainable', can become effective in the context of enhancing the customer value.

DEMAND ORIENTATION IN SALES TERRITORY

In the growing competitive markets the large and reputed sales oriented firms are developing strategies to move into the provision of innovative combinations of products and services as 'high-value integrated solutions' tailored to each customer's needs than simply 'moving downstream' into services. Such firms are developing innovative combinations of service capabilities such as operations, business consultancy and finance required to provide complete solutions to each customer's needs in order to augment the customer value [36]. Demand orientation in selling increases economic satisfaction of customer as it reveals competitive advantage in making the buying decision. A strong demand oriented strategy of the firm alleviates the possibility of using coercive influence strategies by the competitors and offers advantage to the customers over competitive market forces [37]. Demand orientation is an organization-wide concept that helps explain sustained competitive advantage. Since many manufacturing firms have linked their selling strategies with services delivery attributes, the concept of demand orientation is expanding sales as a territorial system of individual customers and business-to-business accounts. The process of demand orientation contributes to continuous learning and knowledge accumulation by an organization which continuously collects information about customers and competitors and uses it to create superior customer value and competitive advantage [38].

As the new products are introduced, a firm may routinely pass these costs on to consumers resulting into high prices. However a less obvious strategy in a competitive situation may be to maintain price, in order to drive the new product in the market with more emphasis on quality, brand name, post-sales services and customer relations management as non-price factors. Studies that advocate the models of building customer value through traditional relationship marketing discuss the long term value concepts to loyal customers. Most importantly, customers are expected to rise their spending and association with the products and services of the company with increasing levels of satisfactions that attribute to their values [39]. Demand orientated sales require a different competitive mind-set and a systematic way of looking for opportunities. In the process of demand oriented selling firms can find scope for real value enhancement than looking at competitors within their own industry [40].

Adoption of a demand orientation of sales firms can help salespeople design and offer a sales-mix that provides lower perceived risk and higher customer value in the sales territories. The demand oriented sales strategies could be catalytic for dealers in driving profit and building competitive advantage for the firm in the marketplace [41]. The constituents of demand oriented sales strategyencompass various attributes exhibited below:

- Portraying customer demand
- Pricing
- Purchasing
- Customer order management
- Customer service delivery
- New product/service development
- Sales strategy development
- Managing customer lifetime value
- Multiple selling strategies
- Enhancing relationship
- Developing event-driven sales approaches.

It has been argued that the provision of integrated solutions is attracting firms traditionally based in manufacturing and services enabling them to occupy a new base in the value stream cantered on 'systems integration'. Internal or external sources of product designing, supply and customer focused promotion are used by the systems integration. The interdependent structure of a dyadic relationship refers to the extent of relative dependence that stimulates the mutual transactions between the two parties [42]. The measure of dependence may be categorized as total interdependence and asymmetry or relative dependence. Total interdependence is the sum of both parties' dependencies on each other while interdependence asymmetryrefers to the difference between each party's dependence on the other. This difference has also been referred to as the more dependent party's relative dependence or the less dependent party's relative power. Relational contracts and informal agreements between the manufacturers-suppliers and customers is sustained by the value of relationships. It has been found that the integration affects the parties' temptations to renege on a given relational contract, and hence affects the best relational contract the parties can sustain.

The recession has resulted in layoffs everywhere. But employees at small clothes may have less to worry about than those at corporate giants. This has demanded organizational re-structuring for sales and marketing in many large corporate firms as well as in Canon Mexico. The reorganization has been tuned in the company with reference to hiring process, responsive departmental network, integrated coordination and financial incentives to the sales staff. Bruce Phillips, a senior economist at the National Federation of Independent Business, USA said that "If you went industry by industry, you'd find that more small firms are hiring than big ones. That's always true in recessions. Large firms are more labour-intensive, and people don't stay as long at small firms as they do at big ones." Selling is not a competition, it's a connection. Rather than putting the competitive instinct in overdrive, devote energies to making connections with customers, their co-workers, and the people in the organization, and beyond. This strategy will not only increase sales and profits, it will speed up the sales cycle. It is necessary for the large organizations like Canon to optimize their staff strength and prune the low productive sales resources. Too many salespeople, especially those new to the profession, rush right into discussing the specifications, financing, or advantages of their wares. On the contrary the seasoned professionals know better the markets and clients. They establish personal, professional or both types of contacts well before the sales conversation moves on to the details and get to close the deal. The more effort salespeople put into analyzing the competition, the less time they will have to understand, cultivate, and maintain connections with customers. This can be a fatal mistake, since it opens the door to the very competitors salespeople were worried about in the first place! Never forget, your customers are your competitors' prospects [43]. Some companies require their salespeople to make a minimum quota of outbound phone calls every day. Unless you have an unbelievably well-targeted list to work from, chances are that telemarketing will produce a very low return on the time invested. The re-structuring of the sales organization of Canon has been proposed to maximize the per capita productivity. This has been considered simply as an arrangement of activities involving the group of people recharged with improved goals and coordinated activities. The office equipment divisionand photo and video division have been merged and the sales re-organization was brought into force. The re-organization has been brought into force as Consumer Imaging Group (CIG) with three distinctive departments including marketing, sales and service. This reorganization had aimed at inculcating the new staff, styles and shared values for better understanding of the market situations and building consensus on the marketing and sales strategies. The culture as a product has also been focused on the role of sales people for benchmarking their individual business goals as high as of the competitors [44].

A firm engaged in market-driven sales continuously observes the competitive environment in the sales territory to learn the appropriateness of its offerings to its target customers. Customers' needs and perceptions tend to change frequently and it is imperative to routinely evaluate competitive status of the firm by offering quality services to enhance the customer satisfaction and market efficiency of the firm. The demand driven sales is a function of improved customer services, quality of products. Sales and post-sales services influence the effectiveness of demand orientation and reflect the corporate culture among consumer segments. Demand orientation of the sales strategyis positively related with sales effectiveness and creation of a customer culture in the integrating sales and services in an organization. The major attributes of demand driven sales strategy to be followed by the salespeople in a defined territory should include:

- Increasing sales volume
- Efforts to outperform competitor sales
- Developing micro-segmentation based on consumer response
- Launching productive sales campaigns
- Building cost-effective selling strategies
- Multi-brand and multi-segment sales approaches
- Laying focus on transactions
- Number of relationships.

Dealers can protect customers from dysfunctional behavior and distrust by improving the relationship quality. Under situations with high total interdependence, both parties are motivated to develop, maintain, and improve the relationship. Hence, the performance of functions of dealers in selling products and services will have a stronger impact on relationship quality in a high interdependence situation. Firms with a customer-oriented business culture have shown to facilitate innovativeness in customer services to improve overall business system and develop a positive perception among the customers which is expected to yield long term loyalty. Customer focused firms which also have demand orientation, rely on developing strategies towards increasing customer satisfaction and loyalty through improved service quality. Because identified sales and service quality dimensions improve the delivery of customer perceived quality and overall sales [45]. In order to gain a customer's loyalty it is necessary to achieve his/her satisfaction. A salesperson must pay attention to customers' trust and commitment throughout all transactions. Salespeople must take care of the quality of their products and the attention given by the contact personnel, paying special attention to certain emotional aspects relating to customers' enjoyment of the product and their stay in the sales outlet during the process of deciding to purchase [46]. The benefits of customer relationships have been recently addressed and include increasing efficiency and effectiveness in maintaining current customers rather than prospecting new customers, and improved competitive advantage. The consumer benefits through consumer learning, in such situations can be stored, processed and retrieved to use in subsequent situations. This leads to an ability to manage future decisions based on simplifying problem-solving situations and reducing risk [47]. The customer-seller relationship should also be revealed in reference to the non-selling activities including convenience, comprehension to customer, cost to customer and care as major relationship management variables. It has been observed during the study that there exists a strong link between customer behavior and customer profitability, while modest links exist between repurchase intentions and referral behavior [48].

TERRITORIAL SALES PLANNING

Sales plan guides implementation and control, indicating sales objectives, and the strategy and tactics for accomplishing the objectives. The sales strategy consists of perspective plan and action plan. The perspective plan is developed with long term objectives while the action plan is formed by phasing out the activities delineated in the long term perspective plan in order to monitor the annual progress of activities and achieve the annual targets. The planning cycle is continuous. Plans are developed, implemented, evaluated, and revised to keep the sales strategy on target. Since a strategy typically extends beyond one year, the action plan is used to guide short-term sales activities. The planning process is a series of action plans guided by the sales strategy. An annual planning period is necessary since several of the activities shown require action within 12 months or less and budgets require annual planning. In the large companies the product managers are responsible for coordinating the preparation of plans. A planning workshop is conducted midyear for the kick off of the next year's plans. Top management and product, research, sales, and finance managers attend the workshop. The firm's advertising agency account manager also participates in the workshop. The current year's plans are reviewed and each product manager presents the proposed sales plan for the next year. The workshop members analyze each plan and suggest changes. Because the requested budget may exceed available funds, priorities are placed on major budget components. Each product manager must provide strong support for requested funds. The same group meets again in 90 days to review the revised plans. At this meeting, the plans are finalized and approved for implementation. Each product manager is responsible for coordinating and implementing the plan. Progress is reviewed throughout the plan year, and the plan is revised when necessary. However, sales plans can be effectively implemented provided the size of the sales forces is determined. The size of the sales force may be determined through the following equation:

$N = S / P$

Where,

    N = Size of Sales Force
    S = Forecasted Sales Volume
    P = Estimated Productivity of Sales Unit

A short-term sales plan forms the core of an overseas subsidiary's planning effort and covers sales operations usually for about a year. The complexity of planning varies among companies. In some cases, it may amount to simple preparation of sales budgets. In multinational firms, planning would involve multiple considerations to consolidate mutual interdependence between different overseas affiliates and the parent corporation. A good implementation process spells out the activities to be implemented, who is responsible for implementation, the time and location of implementation, and how implementation will be achieved. Sales representatives should target all accounts currently using a competitive product.A plan should be developed to convert 5 percent of these accounts to the company brand during the year. Account listings will be prepared and distributed by product management. The sales plan can be used to identify the organizational units and managers that are responsible for implementing the various activities in the plan. Deadlines indicate the time available for implementation. The sales manager is responsible for implementation of the plan through the sales force. Planners and implementers often have different strengths and weaknesses. An effective planner may not be good at implementing plans.

After heading towards globalization, Nestlé fairly decentralized its country business activities to provide better magnitude and direction to the regional planning and management of the company for yielding prolific sale and marketing results. The Swiss headquarters of the company offered the brand names and most of the product concepts and process information prescribed high quality standards for local managers and maintained a large energetic and business influencing staff. Simultaneously, individual country organizations of the company took the responsibility and autonomy for optimizing sales in the local markets. The company had done introspection of its business policies to know how the marketing and sales activities could be managed for higher yields and the strategies that were implemented already had resulted in augmenting the growth in these sectors. The sales force structure of the company has one of the largest treasures of human resources and shoulders greater responsibility of moving the products to the market. The case examines the influential role of sales force of Nestlé in Mexico from the point of view of responsive retail sales management. The sales force of the company has been organized in Mexico in reference to the retail performance of the superstores. Accordingly two types of sales force structures have been designed by Nestlé which were blue prints of the purchase or retail ordering network of the organizations. The mirror image strategy of Nestlé Mexico has helped it to improve the retail supply chain and just in time (JIT) strategies to get larger coverage of products in the retail market. The mirror image strategy is built on the lines of the purchase structure in the super stores. The sales structure of Nestlé in Mexico includes four major divisions of retail operation which include dry products, chilled products, ice creams and breakfast cereals. The sales directors are supported with the Sales General Managers in each product division. The operational sales teams of the company are formed of key accounts managers for the product specific categories and are in charge of all the major accounts like super stores, departmental stores and convenience stores. The key accounts managers are supervised by the national key accounts manager. The wholesales managers look after the distribution and inventory management activities in the designated territories. The activities of scheduling supply according to the orders, generating orders and reordering targets are administered by the collection manager. He also attends to the collection of payments and associated bills from the retails outlets and super stores in the country. The alternative channel manager of the company coordinates with the inter-stores inventory management, adjustment and logistics as some of the stores like Walmex has their subsidiary retail auto services stores [49]. Since the candy market is very large and needs an exclusive retail management, the company has an independent position of Candy Sales Managers to serve the product specific requirements in the auto services stores. The responsibilities of key accounts managers (KAM) include sales promotion, managing customer responses, display and retail interiors in the stores besides the ordering, stocking, inventory management and other assigned technical tasks of the company.

One way to deal with difficulty in the implementation of the sales plan is to administer the balanced scorecard methods. This process is a formalized management control system that implements a given business-unit strategy by means of activities across four areas: financial, customer, internal business process, and learning and growth. The balanced scorecard provides the framework for implementation of a common strategy that is communicated and coordinated across all major areas of the organization. The "balanced" component of the balanced scorecard reflects the need to consider how all areas of the organization function together to achieve a common goal of strategy implementation. The major benefit of the balanced scorecard is that a frequently aggregate, broadly defined strategy is translated into very specific actions. Through execution and monitoring these actions, management can assess the success of the strategy and, if necessary, modify and adjust it. Another major benefit of the balanced scorecard methodology is that it is feasible for any strategy at the business-unit level and provides a means to link performance evaluation to strategy implementation. The effective sales planning should have the following attributes:

- The ability to understand how others feel.
- Good bargaining skills.
- The strength to be tough and fair in putting people and resources where they will be most effective.
- Effectiveness in focusing on the critical aspects of performance in managing sales activities.
- The ability to create a necessary informal organization or network
- Problem solving ability when they are confronted.

	Strong	**Competitive Sales**	Weak
High	**Core Accounts** Attractive accounts High volume sales High selling resources Customer life time value	**Growth Accounts** Attractive accounts High sales potential Competitive gains Quality sales and service	
Low	**Pull Accounts** Exploring potential Need strong persuasion Investment in selling Patronizing account	**Difficult Accounts** Non-promising accounts Low response to sales Incur sunk cost Low customer value	

(Left axis label: **Sales Opportunity**)

Figure 5.4.Developing Sales Portfolios.

The corporate headquarters shares with subsidiary the perspectives of its mission and objective. This input helps the subsidiary, find its overall goals and specific sales objectives. In order to develop homogeneity among the sales plans of different subsidiaries, firms prescribe a standard procedure conducting the planning process. It is necessary for a firm to consider external factors like demands in emerging markets, regulations of the host country, international trade policies, foreign exchange regulations etc. On the basis of the above information, planning starts with a review of past sales and projections into the future. The projected forecasts are duly revised in planning inputs. The sales plan should clearly address various constituents of sales-mix (11 Ps) including product, price, place, promotion, packaging, pace (competitor analysis), people (sales front liners), performance, psychodynamics, and posture of the firm and proliferations of products, services and markets. Sales budget should be included in the sales plan for each of these constituents. The sales budget must be reviewed by the subsidiary management to add and accommodate the company-wide outlook. Strategic evaluation requires analyzing information to gauge performance and take the actions necessary to keep results on track. Managers need to monitor performance continuously and, when necessary, revise their strategies according to changing conditions. The finance function should reflect on impact of the fluctuations in local currency value, and a capital-expenditure and capital plan need to be appended. Salespeople should develop portfolio approach in reference to customer preferences and product attributes. The contemporary best sales practices suggest broadening of product portfolios to match local preferences of consumers that can increase both sales opportunities and the benefit the sales territory by making other products popular [50]. Sales portfolios can also be developed by refining the consumer segments as illustrated in Figure 5.4.

Strategic evaluation, the last stage in the sales strategyprocess, is really the starting point. Strategic sales planning requires information from ongoing monitoring and evaluation.

Evaluation consumes a high proportion of sales executives' time and energy. Evolution may be done to find new opportunities or avoid threats, keep performance in line with management's expectations, and to solve specific problems that exist. Areas of evaluation include environmental scanning, product-demand analysis, brand equity analysis, sales program evaluation, and gauging the effectiveness of specific sales-mix components such as advertising. Planning is the approach to making decisions concerning systematic allocation of resources. It is worth emphasizing that planning is a process, not an event. It is organic and ongoing and it is a key element of the overall management process.

SALES FILTERS IN TERRITORY

Strategic planning is systematic means of making the firm successful through the discipline of strategic thinking and vision used as a framework for all other decisions in the firm. Strategic planning requires an honest evaluation of the company's current situation. Sales planning, demands the commitment of the top management for its success. It requires commitment of resources, both financial and personnel, for its development. It demands complete follow up. Sales plan, which is not carried out due to lack of leadership or of the required tools needed for completion, is a total failure and a waste of time and money. Firms should consider various sales filters to increase the effectiveness of sales operations, as given below:

- Identifying unqualified opportunity
- Filtering qualified opportunity
- Selecting best few opportunities.

Most salespeople experience the routine activities like initiating sales discussions, explaining the product benefits and taking a quick look at the expressions of customers. Such tasks for the most part are neither fun nor painful; they're simply things that need to get checked off the list. However, to ensure success in the sales efforts it is necessary to develop important differentiators that could help salespeople to successfully perform their operations. This process may be defined as filtering customers to identify right customers and work with them [51]. It would be beneficial for the firms to develop powerful psychological measures to identify right customers in right territories. Nevertheless, sales organizations can overcome the uncertainties in acquiring customers by deploying a set of basic practices comprising the market preview, customer overview before prospecting, correct comparison of consumer demand and buying potential, making thorough analysis of competitive advantages or losses, filter biases of customers, competitors and markets, limit the sales options, close the deal, and facilitate the integration. Although some of these practices might seem obvious, many companies fail to follow them. Sales organizations make the mistake of commencing the predetermined results to a market situation and do not succeed in deriving the anticipated results. However, salespeople need not only to locate right customers but also manage markets, industrial clients and geographic regions [52].

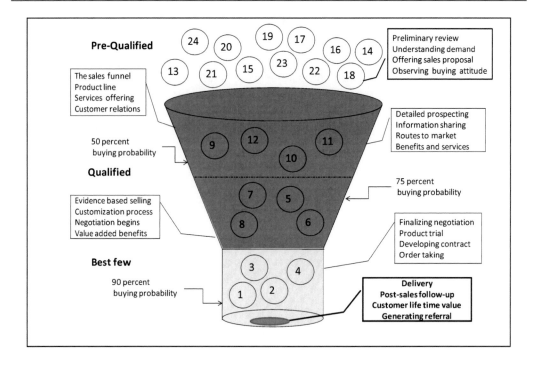

Figure 5.5. Filtering Process in Managing Effective Sales.

There are two types of sales management process in an identified sales territory for generating sales effectiveness and creation of customer value. One holds that path to sales effectiveness by random prospecting of buyers and value creation while another process illustrates prospecting customers by filtering the most committed ones. The filtering process in sales management is exhibited in Figure 5.5 that argues layered prospecting method.

It may be observed from Figure 5.5 that filtering process helps salespeople to meticulously prospect the most committed customers inclined towards buying. Accordingly, the salespeople can optimize delivering value through creativity and innovation. However, it is argued that the most successful businesses in the years to come will balance analytical mastery in selling and traditional practice of intuitive prospecting in a dynamic interplay called design-thinking in the sales operations. The design-thinking principles are associated with the funnel model that helps driving both innovation and efficiency in sales practice [53]. Such combination produces the most powerful long-term competitive edge in the sales territories.

During the filtering process salespeople may begin their exploration to determine the most committed buyers by greeting the customer as the first step to sales presentation. Salespeople should be emphatic helpers to get them to conceptual satisfaction. Salespeople may tell the customer about the features and benefits of the product or service that are of major concerns and help in stimulating the buying perception. Salespeople may also use Web 2.0 technology to research the customer buying habits on line as well as calling before the presentation to get more of a sense of what the customer is looking to purchase and for what reason. After providing comprehension about the product, services and customer relations, salespeople should always lean towards seeking the commitment of buyers. A ticket collector at the concert or a sales analyst at the food store is also salespeople working in a routine selling environment. But due to the nature of their jobs, they are asked numerous questions by customers as well as hear complaints about prices and service. A knowledgeable person with a pleasant personality is especially needed for this job because this is usually the person who is dealing with the customer during the post-sales relationship period. As noted above, more creativity is found in this job as compared to the order-handler. The counter attendant at the fast food restaurant may take the order and then suggest that the customer might also wish to buy a hot apple turnover. Pleasant personality, fast service, and suggestion selling on the part of the order-taker can result in many additional sales.

Trust is a collective behavior which emerges over a period through the personality traits of individuals in a business community. Trustworthiness in business emerges from three perspectives including structure of service providers, service quality and management. In the low trust cultures the interpersonal relationship remains obscure and business dealings are largely bureaucratized and tagged with evidences. Generally, negotiation approaches slow down the process of getting the work done and may also cause retrenchment from the business scenario over time [54]. Trust, in business relationships, is typically perceived as a result of satisfaction and experience. In business to business relations, organizational buyers may be unable to commence relationships without trusting their suppliers; thus, the stance that trust leads to commitment are interrelated. When trust is low in a cultural setting it affects the confidence of the people and so depletes their responsiveness to the given situation. If a buyer relies on trust and it is not reciprocated, the company will suffer from substantial harm [55].

During the filtering process salespeople should show two cognitive attributes, which are observation and continuous listening to customer without interrupting the conversation to derive the buying motivation and conviction of the prospect. It is not advisable to go for hard sell by trusting ideas of the salesperson. Building strong sales force and sales process are essential for business success. But when it comes to improving the selling process some companies take the right approach like filtering prospects carefully and proceeding with sales process meticulously. Many firms work on developing their sales process faster and cost effective to gain competitive advantage. It has been observed that those companies that build strong selling process are agile, adaptable, and aligned get ahead of their rivals. Great firms create both customer-centric and market oriented selling processes that respond to abrupt changes in market demand and customer preferences. Agility is a critical issue to be considered by the salespeople as customers often switch suppliers without notice. It is necessary that selling process should be resistant to these changes and typically able to cope with uncertainties. The best practices of selected firms allow managers to identify structural shifts in the sales territory early by recording the latest information, filtering right customers and accounts, and tracking key sales patterns [56].

SALES QUOTAS

Accomplishing sales objectives largely depend on the assignment of sales quotas and affect the sales performance of the firm. Failing to reach sales goals may have negative impacts on the salesperson who failed to make the goal, and the firm who relies upon his/her success. These sales goals are typically referred as sales quotas, representing the primary sales objective. Sales quotas can be assigned on the basis of individual or territory potential. Often the sales potential in a region appears optimistic and generating quotas to be achieved on regional basis should be matched with the capability and competence of salespeople. Sales managers should consider economic and behavioral factors in determining the sales quotas. There are following factors that affect fixing sales quota of the salespeople:

- Market Potential
- Sales Potential
- Capabilities of salespeople
- Network Support- Distribution, Retailing, Client Services
- Sales forecasts
- Expectations

In a territory the market potential always appears to be higher than the sales potential, which may appear deceptive for determining the sales targets. Hence, this indicator should not be taken as a guiding tool to fix the sales quotas. A high market potential may generate sales bias of selling higher volume in the region, which may lead to deception while measuring the outcome performance of salespeople. Such gap in achieving the projected sales volume may occur due to non-responsiveness of buyers to the sales proposals that were drawn on the basis of market potential. As a good quota determination process, the market potential should be considered to prospect customers and filter the highly probable buying customers.

Table 5.1. Determination of Sales Quotas

Determinants	High Sales quotas	Moderate Sales Quota
Assess general demand in the market for products and services offered	Demand tends to increase with the possibility of acquiring new customers	Demand is low but has the potential to increase over short-term
Specific demands for products and services	Product based quotas Niche sales quotas	Low product profile but high customer value based quotas
Estimated sales potential	Projected demand is high	Project demand is sustainable
Productivity of salespeople	Salespeople have high capability and competency	Skills of salespeople to be improved
Internal benchmarking	Increasing benchmarks	Following no change in the volume of sales of the previous year

The operational efficiency of distribution, retailing and client services also affects the quota of the salespeople. In the competitive marketplace, sales and operations planning is a key process for providing visibility to the enterprise, which largely depends on the market infrastructure comprising supply chain and retailing function that supports the selling process. Besides, it supports a transversal decision process, which co-ordinates different functions either in the company or between companies in a supply chain environment. An integration of supply chain, retailing and services provides better support to carry out sales activities effectively not only for gaining competitive advantages but also for augmenting sales revenue by fixing high sales quota in the region [57]. The Table 5.1 illustrates the determination of sales quotas under various conditions.

Sales quotas not only drive growth of the firm but also provide personal benefits to the salespeople. Most salespeople intend to take load of higher quotas believing that achieving quota would get quick promotions [58]. Sales quotas are considered as the initial sales objective for salespeople and sales managers, as well as one of the most important issues of the selling profession that can also be over achieved. Sales quotas are assumed to play a significant role in impacting the success of an organization. For instance, the assignment of a sales quota creates an upper management strategic direction to the marketing and sales organization about quality and quantity of the products to be sold, time and location of sales, and managing sales of various product combinations to the customers [59]. However, sales quotas are commonly used for both evaluation and control to motivate salespeople to enhance their performance. Determining the sales quota encourages outcome-based performance attitude of salespeople. It is assumed that while other products, time periods, and territories may be assigned to other salespeople, the quota participation represents one segment of a total strategic contribution of the salespeople to the organization. As a result, when assigned to a salesperson, quotas generally act as a mechanism to provide seller motivation that ultimately impacts efforts of the salespeople as part of a larger strategic managerial plan. Logically therefore, it is expected that sales quotas have a strong relationship with salesperson performance [60]. The quantitative sales quotas may be set considering the following determinants:

- Users of the product
- Rate of buying and usage
- Principal buyer of the product/service
- Market motivation for buying
- Brand image of the company
- Previous promotion records
- Success/failure of sales campaigns
- Conflict resolution and customer services support
- Demand stimulation
- Customer expectation
- Previous efforts of the sales force.

Sales quotas serve to channel salespeople's effort into the areas that management wants to prioritize. The overall performance of a business is significantly affected by the ability of an individual salesperson, and of the sales department in general, to achieve quotas. Managers who intend to lay difficult sales goal for the salespeople may not drive individual performance. Managers should not, therefore, simply impose quotas, but should communicate openly with their employees the reasons for setting a certain quota. If the salesperson highlights a logical reason for the quota being unfair, it should be reconsidered. However, setting quotas should be set to provide challenging but fair task to contribute to growth in the firm. When sales managers fail to take any action after salespeople fail to reach their quota, performance tends to be lower. Conversely, salespeople who receive "a stern verbal warning", a formal probation, or who are asked to work more closely with their manager, tend to improve their performance. Coaching appears to be a key activity for putting things right. In particular, salespeople who have failed to meet their targets, and then begin to work more closely with managers, tend to see their incomes rise. Individual performance appears to be less likely to improve when threats accompany a sales manager's help [61].

Customer-relationship management, organizational downsizing and re-engineering, quality-improvement initiatives, knowledge management, team working, and relationship marketing are the factors that help in expanding the role of selling in the firm and so broadening the role of the salesperson towards accepting challenges in selling. In particular, these factors attribute in setting high quotas to the sales people and encourage firms in placing increased emphasis on customer satisfaction and loyalty. The salesperson is the key link between the selling firm and the customer, with high levels of job involvement in managing the determined sales quotas. Sales quotas have also implications on the average number of sales calls per week and number of accounts to be serviced. The monetary incentives associated with achieving the sales quota periodically serve as an indicator of a salesperson's performance [62].

SALES QUOTA AND BEHAVIORAL PERSPECTIVES

The process of meeting the sales quota needs direct supervision of assigned salespeople, and they should be motivated to achieve quotas in team coordination by setting measurable control parameters. Generally, field salespeople are set to achieve assigned quotas in reference to their respective sales unit. In the process of behavior control, the activities of monitoring, directing, and evaluating salespersons, the efforts of sales managers are focused on the result oriented control [63]. It has been observed through empirical investigations that the method of team coordination and control system are important determinants of ethical behavior. Age of the salesperson, which is an essential factor in putting efforts to accomplish quotas, also proves to be a significant antecedent of ethical behavior. Additionally, a salesperson's ethical behavior leads to lower levels of role conflict and higher levels of job satisfaction, but not to higher performance [64]. Effective communication in the process of assigning quotas and performing task to achieve quotas is confirmed as one of the predominant critical success drivers (CSDs) in achieving sales performance in the region. Besides, top management support and commitment, clear integration objectives, quantifiable performance indexes, and qualified selling process are also identified as the CSDs of such integration [65].

Behavior of a salesperson is an important predictor of growth of sales in an organization. The study of Deshpandé & Farley (1998) evidenced that the external and internal environment of a company and personality traits affect decision making of salespeople that helps in accomplishing tasks and sales quotas faster [66]. Internal communication and the choice of a control system especially affect ethical decision making. It has been observed that informal internal communication affects the personality traits while the control system influences the ethical climate of the salespeople. Ethical climate and salespeople's personality traits also affect the process of accomplishing assigned targets. The study shows that ethical decision making can be influenced by management. Some studies have found significant relationships between behavioral and organizational variables such as job experience, closeness of supervision, performance feedback, influence in determining standards, span of control, role conflict, and ambiguity perceived by salespeople [67]. The salesperson's expectancy, instrumentality, and valence perceptions are not directly under the sales manager's control. But they can be influenced by things the sales manager does, such as how he or she supervises the salesperson or rewards the individual. Since the salesperson's motivation strongly influences performance, the sales manager must be sensitive to how various factors exert their impact [68].

The sales results emerging out of achieving quotas typically reveal performance of the salesperson and this issue is receiving increasing attention among the multinational companies. This may be evidenced from research in several countries that sales managers are concerned with the sales team building strategies towards achieving increasing quotas [69]. Both the salesperson's job behavior and psychological well being can be affected if there are perceptions of role ambiguity or conflict or if these perceptions are inaccurate. There is a good deal of evidence, for example, that high levels of both perceived ambiguity and conflict are directly related to high mental anxiety and tension and low job satisfaction. In addition, the salesperson's feelings of uncertainty and conflict and the actions taken to resolve them can have a strong impact on ultimate job performance [70]. Salespeople in a firm are influenced by sales drivers to reach high outcome performance in a given region and time. Sales drivers include territory design, compensation, scope of assigned task and cultural interaction in the market.

Effective management of sales quota is important to managers for maintaining strong organizational threshold of salespeople for continuously improving effectiveness of the sales units. The variables of economic and relational dimensions of external and internal fit have shown greater association with the performance of salespeople in managing new challenges. Sometimes it is necessary for the firm to grant greater autonomy and economic benefits for driving salespeople to enhance performance by achieving the assigned quotas. Achieving the targets largely depends on the effectiveness of the team coordination, leadership, and performance control through behavioral attributes. Effective coordination of tasks towards managing sales quotas is a consistent predictor of salespeople's performance and effectiveness of the sales units. This indicates the importance of proactive monitoring, directing, and managing the salespeople towards guiding the sales activities in right way to reach the set targets. However, some firms involve the manager above the field sales manager in the performance appraisal of salespeople. The sales managers may implement such controls effectively by establishing coordination, training, and feedback process rather than imposing command and control policy. Managers aiming to improve team task performance should recruit individuals with a strong work ethic.

Professional development of salesmen should be directed at making them aware of the advantages of adopting relational behaviors and at boosting their capabilities and skills to work in teams. The high-performance salespeople have greater commitment to their organizations and their sales managers are more satisfied with their units' sales team designs. The increasing adoption of a relational approach to customers is therefore fostering a deep-going change in the individual skills set and capabilities of the sales team and, farther upstream, a substantial rethinking of company strategies and policies of selection, training, motivation and control of the sales team. Salespeople systematically proceed towards reaching their targets by matching the right products with the right customers to enhance their performance.

WORKLOAD PLANNING IN SALES TERRITORIES

A typical salesperson's work day include making calls to prospects, manage time and beat the traffic, and count for volume of sales to report to a typical manager who monitors and evaluates his performance. The plan for a proactive workload determination for a salesperson indicates *jump* on the work to be done today despite odds. There are many contemporary issues affecting managers in determining the workload of salespeople in view of the globalization effect. It is not just buyer and supplier impact but also the pressure of new products and services in multiple markets that indicate the necessity of redefining workload of salespeople [71]. It has been observed that in several markets salespeople are deployed as standby to penetrate into new consumer segments in the same time zone to gain competitive advantage. The workload of salespeople should be determined in reference to the following attributes:

- Classifying the customer segments and build sales relationship
- Premium-mass, large-medium-small
- Determine the frequency of visits and length of each call
- Total market coverage work load
- Determine the time available per sales unit (man hours-man days)
- Appropriation of sales unit time by the tasks performed
- Unit of measurement-man hours or man days
- Preparing for sales- negotiations- selling-non selling-travelling
- Determine the size of sale force
- Total market coverage hours/ selling hours.

As firms build more international relationships, the work day often extends well beyond normal hours. In due course, as the number of relationships grows linearly, the number of communication contacts grows exponentially, and the volume of sales is expected to increase right along with as a measure of workload of a salesperson. Couple globalization with increased use of cross-functional sales within markets and it is evident that increasing workload is demanded by the firms to generate higher output with the network of salespeople [72]. Changing roles and requirements make the workload of the salespeople not only more difficult but also more difficult to work with it. The resistance of individual salespeople to these changes is another predictor in the sales performance function. Customers need accurate information fast, expecting their salespeople to return their calls immediately instead of in a day or two and manages make salespeople to cope with such demand, which affects their routine workload and drags additional tasks. Sales organizations must change to allow their salespeople to be highly connected, without succumbing to the lure of leading edge technology that is not ready for industrial use. Sales performance is the sum of a number of factors including gross sales, returns, delayed billings, delayed shipments, and credit memos; and not just the workload of a salesperson. Sales organizations use a number of different means to measure sales, and some of them can be as complicated as workload determination of salespeople [73].

SELECT READINGS

Bacon, T. R. (1999), *Selling to major accounts*, AMACOM, American Marketing Association, Chicago, IL.

Dougherty, J. and Christopher, G. (2006), *Sales and Operations Planning: Best Practices and Lessons Learned*, Trafford Publishing, Victoria, BC.

Harvard Business School (2007), *Strategic Sales Management*, Harvard Business School Press, Boston, MA.

Levitt, T. (2006), *Ted Levitt on Marketing*, Harvard Business School Press, Boston, MA

Ling, R. C and Goddard, W.E. (1995), *Orchestrating Success: Improve Control of the Business with Sales & Operations Planning*, Wiley, New York, NY.

REFERENCES

[1]   Piercy, N. F., Cravens, D. W. and Morgan, N. A. (1998), Sales force performance and behavior-based management processes in business-to-business sales organizations, *European Journal of Marketing*, 32 (12), 79-100.

[2]   Bailey, E. L. (1989), Getting closer to the customer, *Research Bulletin* (Vol. 229), New York, The Conference Board, Inc.

[3]   Rajagopal (2007), *Sales Management in Developing Countries: A Comparison of Managerial Control Perspectives, Journal of Asia Pacific Business, 8 (3)*, 37-61.

[4]   Ghemawat, P. (2001), Distance Still Matters: The Hard Reality of Global Expansion, *Harvard Business Review*, 79 (8), 137-147.

[5]   Grant, K., Cravens, D. W., Low, G. S. and Moncrief, W. C. (2001), The role of satisfaction with territory design on the modification, attitudes, and work outcomes of salespeople, *Journal of the Academy of Marketing Science*, 29(2), 165–178.

[6]   Baldauf A., Cravens D.W., Grant, K. (2002), Consequences of sales management control in field sales organizations: a cross-national perspective *International Business Review*, 11 (5), 577-609.

[7]   Churchill, G. A., Ford, N. M., Walker, O. C., Johnston, M. W. and Tanner, J. E. (2000), *Sales Force Management* (6th ed.), Chicago, IL: Irwin.

[8]   Kim, W. C. and Mauborgne, R. A. (1999), Creating new market space, *Harvard Business Review*, 77(1), 83-93.

[9]   Euromonitor International (2010), *Alcoholic drinks in Portugal, Market Report*, Euromonitor Internationa, January.

[10]   McGrath, R. G. and MacMillan, I.C. (1995), Discovery driven planning,*Harvard Business Review*, 73(4): 44-54.

[11]   Stassen, R. E. and Mittelstaedt, R. A. (1995), Territory Encroachment in Maturing Franchise Systems, *Journal of Marketing Channels*, 4(1), 27-48.

[12] Rajagopal and Rajagopal, A. (2008), *Team Performance and Control Process in Sales Organizations, Team Performance Management- An International Journal, 14 (1),* 70-85.

[13] Shapiro, B. P. and Doyle, S. X. (1983), Make the sales task clear, *Harvard Business Review,* 61 (6), 72-76.

[14] Cespedes, F.V. (2006), *Deployment, focus and measuring effectiveness,* Background teaching notes, Harvard Business School Press, Boston, MA.

[15] Johlke, M. C. (2006), Sales Presentation Skills and Salesperson Job Performance, Journal *of Business and Industrial Marketing,* 21(5), 311-319.

[16] Siguaw, J. A., Kimes, S. E. and Gassenheimer, J. B. (2003), B to B Sales Team Productivity: Applications of Revenue Management Strategies to Sales Management, *Industrial Marketing Management,* 32 (7), 539-551.

[17] Angwin, D., Paroutis, S. and Mitson, S. (2009), Connecting Up Strategy: Are Sensor Strategy Directors a Missing Link? California Management Review, 51 (3), 74-94.

[18] For details see Euromonitor International (2009), *Consumer food services in Chile,* Market Report, Euromonitor International, November.

[19] Mankins, M. C. and Steele, R. (2005), Turning great strategies into great performance, *Harvard Business Review,* 83 (7): 64–72.

[20] Micheletti, M. and Stolle, D. (2008), Fashioning social justice through political consumerism, capitalism, and the internet, *Cultural Studies,* 22(5), 749-769.

[21] Gonik, J. (1978), The salesmen's bonus to their forecasts, *Harvard Business Review,* 56(3), 116–123.

[22] Christensen, C. M., Kaufman, S. P., Shih, W. (2008), Innovation Killers: How Financial Tools Destroy Your Capacity to Do New Things, *Harvard Business Review,* 86 (1), 98-105.

[23] Elsbach, K. D. and Bechky, B. A. (2007), It's More Than a Desk: Working Smarter Through Leveraged Office Design, *California Management Review,* 49 (2), 80-101.

[24] Ledingham, D., Mark, K., and Heidi, L. S. (2006), The New Science of Sales Force Productivity, *Harvard Business Review,* 84 (9), 124-133.

[25] Hung, R., Chung, T. and Lien, B. (2007), Organizational Process Alignment and Dynamic Capabilities in High-Tech Industry, *Total Quality Management & Business Excellence,* 18(9), 1023-1034.

[26] Prahalad, C. K. and Dearlove, D. (2009), On the verge of something extraordinary, *Business Strategy Review,* 20 (1), 16-20.

[27] Wood, V. R., Pitta, D. A. and Franzak, F. J. (2008), Successful marketing by multinational firms to the bottom of the pyramid: connecting share of heart, global "umbrella brands", and responsible marketing, *Journal of Consumer Marketing,* 25 (7), 419-429.

[28] Darr, A. (2003), Control and autonomy among knowledge workers in sales: an employee perspective, *Employee Relations,* 25 (1), 31-41.

[29] Anderson, E. and Oliver, R. L. (1987),Perspectives on behaviour-based versus outcome-based sales force control systems, *Journal of Marketing,* 51 (4), 76-88.

[30] Zoltners, A. A., Sinha, P., and Lorimer, S. E. (2006), Match your sales force structure to your business life cycle, *Harvard Business Review,* 84 (7), 80-89.

[31]   Hultink, E. J. and Atuahene-Gima, K. (2000), The effect of sales force adoption on new product selling performance, *Journal of Product Innovation Management*, 17 (6). 435-450.

[32]   Money, R. B. and Graham, J. L. (1999), Salesperson performance, pay, and job satisfaction: Tests of a model using data collected in the United States and Japan, *Journal of International Business Studies*, 30 (1), 149-172.

[33]   Gary, L. (2004), *Put your moves in the right sequence*, Harvard Business Publishing Newsletter, Harvard Business School, Boston, MA.

[34]   Steinberg, P. E. (2009), Sovereignty, Territory, and the Mapping of Mobility: A View from the Outside, *Annals of the Association of American Geographers*, 99(3), 467-495.

[35]   Hitachi High Technologies (2009), Annual Reports, Hitachi High Technologies, Japan; also for details see global web site of Hitachi High Technologies Corporation http://www.hitachi-hitec.com

[36]   Rajagopal (2007), Sales Management in Developing Countries: A Comparison of Managerial Control Perspectives, *Journal of Asia Pacific Business*, 8 (3), 37-61.

[37]   Chung, J., Jin, B. and Sternquist, B. (2007), The role of market orientation in channel relationships when channel power is imbalanced,*The International Review of Retail, Distribution and Consumer Research*, 17 (2), 159-176.

[38]   Slater, S. and Narver, J. (1995), Market Orientation and the Learning Organization, *Journal of Marketing*, 59 (3), 63–74.

[39]   Reichheld, F. F. and Sasser, W. E. (1990), Zero Defections: Quality Comes to Services, *Harvard Business Review*, 68 (5), pp 105-111.

[40]   Kim, W. C. and Mauborgne, R. (1999), Creating New Market Space, *Harvard Business Review*, 27 (1), 83-93.

[41]   Chang, T. Z., Mehta. R., Chen, S. J., Polsa, P., and Mazur, J. (1999), The effects of market orientation on effectiveness and efficiency: The case of automotive distribution channels in Finland and Poland, *Journal of Services Marketing*, 13 (4-5), 407-418.

[42]   Kumar, N., Lisa, K. S., and Jan-Benedict, E. M. S. (1995), The Effects of Perceived Interdependence on Dealer Attitudes, *Journal of Marketing Research*, 32 (3), 348-356.

[43]   Michelle, N. (2003) The Seven Deadly Myths of Selling, Business Week Online, April 18

[44]   Rajagopal (2004), *Striving with Competition in Global Imaging Market : Canon in Mexican Business Environment*, Discussion Case, ITESM, Mexico, 1-22.

[45]   Choudhury, K. (2007), Service Quality Dimensionality: A Study of the Indian Banking Sector,*Journal of Asia-Pacific Business*, 8 (4), 21-38.

[46]   Moliner, M. A., Sánchez, J., Rodríguez, R. M. and Callarisa, L. (2007), Perceived relationship quality and post-purchase perceived value: An integrative framework,*European Journal of Marketing*, 41 (11), 1392-1422.

[47]   Sharma, A. and Jagdish, N.S. (1997), Relationship Marketing: An Agenda for Inquiry, *Industrial Marketing Management*, 26 (2), 87-89.

[48]   Rajagopal (2005),Measuring variability factors in consumer values for profit optimization in a firm – A framework for analysis, *Journal of Economics and Management*, 1 (1), 85-103.

[49]   Nestlé's well-known brands include: Nestlé® Toll House®, Nestlé® Nesquik™, Nestlé® Carnation®   Coffee-mate®,   Stouffer's®,Stouffer's®Lean   Cuisine®,   Nescafé®,

Libby's[®]Juice[®], Buitoni[®], Taster's Choice[®], Ortega[®], PowerBar[®], Nestlé[®] Crunch[®], Nestlé[®] Butterfinger[®], and Wonka[®].

[50] Anand, B.N. and Khanna, T. (1989), Do firms learn how to create value? Collaborate with your competitors and win, *Harvard Business Review* 67(1) 133–139.

[51] Morgan, I. and Rao, J. (2003), Making routine customer experiences fun, *Sloan Management Review*, 45 (1), 93-95.

[52] Araoz, C. F. (2005), Getting the right people at the top, *Sloan Management Review*, 46(4), 67-72.

[53] For similar discussion see Martin, R. L. (2009), *The Knowledge Funnel: How Discovery Takes Shape--How Design Thinking Produces Innovation, Efficiency, and Long-Term Competitive Advantage*, Harvard Business School Press, Boston, MA.

[54] Rajagopal and Rajagopal, A. (2006a), Trust and cross-cultural dissimilarities in corporate environment, Team Performance Management-An International Journal, 12 (7-8), 237-252.

[55] Butler, J. K. ((1991), Towards understanding and measuring conditions of trust-Evolution of trust inventory, *Journal of Management*, 17, 643-663.

[56] Lee, H. (2004), Triple-A Supply Chain, *Harvard Business Review*, 82 (10), 102-112.

[57] Affonso, R., Marcotte, F. and Grabot, B. (2008), Sales and operations planning: the supply chain pillar. *Production Planning & Control: The Management of Operations*, 19(2), 132-141.

[58] Marsh, A. (1997), A different breed of salespeople, *Forbes*, 160 (4),70.

[59] Good, D.J. and Stone, R.W. (1991), Selling and sales management in action: attitudes and applications of quotas by sales executives and sales managers, *Journal of Personal Selling & Sales Management*, 11 (3), 57-60.

[60] Schwepkar, C.H. and Good, D. J. (2004), Understanding sales quotas: an exploratory investigation of consequences of failure, *Journal of Business and Industrial Marketing*,19 (1), 39-48.

[61] Ibid

[62] Marshall, G. W., Lassk, F. G. and Moncrief, W. C. (2004), Salesperson job involvement: do demographic, job situational, and market variables matter? *Journal of Business and Industrial Marketing*, 19 (5), 337-343.

[63] Anderson, E. and Oliver, R. L. (1987), Perspectives on behavior-based versus outcome-based sales team control systems, *Journal of Marketing*, 51 (4), 76-88.

[64] Ramon, S. and Munuera, J. L. (2005), Determinants and Consequences of Ethical behavior: An Empirical Study of Sales People, *European Journal of Marketing*, 39 (5-6), 473-495.

[65] Cheng, J. M. S., Wang, E. S. T., Hsu, P. Y. and Tsai, C. C. Y. (2007), Effective communication as a critical success factor for the integration of ERP and CRM systems: the case of Taiwan, *International Journal of Technology Marketing*, 2 (2), 183-199.

[66] Deshpande, R. and Farley, J. U. (1998), Measuring market orientation: Generalization and synthesis, *Journal of Market Focused Management*, 2 (3), 213-232.

[67] Verbeke, W., Ouwerkerk, C. and Peelen, E. (1996), Exploring the Contextual and Individual Factors on Ethical Decision Making of Salespeople, *Journal of Business Ethics*, 15 (11), 1175-1187.

[68] Dubinsky, A. J., Kotabe, M., Lim, C. U. and Michaels, R. E. (1994), Differences in Motivational Perceptions among US, Japanese and Korean Sales personnel, *Journal of Business Research*, 30 (2), 175-185.

[69] Corcoran, K. L., Peterson, L. K., Baitch, D. B. and Barren, M. E. (1995), *High performance sales organizations: Creating competitive advantage in the global marketplace*, Chicago, IL: Irwin.

[70] Singh, J. (1993), Boundary Role Ambiguity: Facets, Determinants and Impacts, *Journal of Marketing*, 57 (2), 11-31.

[71] Rajagopal and Rajagopal, A. (2009), Effects of Customer Services Efficiency and Market Effectiveness on Dealer Performance, *International Journal of Services and Operations Management*, 5 (5), 575-594.

[72] Marsh, R. F. and Blau, S. (2007), Workload factors impacting managers, *Journal of Management Development*, 26 (3), 200-212.

[73] Smith, A. D. And Rupp, W. T. (2003), An examination of emerging strategy and sales performance: Motivation, chaotic change and organizational structure, *Marketing Intelligence and Planning*, 21 (3), 156-167.

Chapter 6

MANAGING SALES ACCOUNTS

Managing accounts is associated with the organizational buying process, which moves through multi-layered decision making in an organization. There are three organizational buying models that significantly influence buying and selling process. Organizational buying processes include straight buy, modified re-buy and new task buy. The buying process in organizations is influenced by buying centers that determine the behavioral aspects of decision making in business organizations. However, in many firms the buying process depends on the corporate or autonomous decisions of manufacturing units. The conflict in buyer-seller operations is resolved through joint discussion and operational process. Salespeople play a key role in the formation of long-term relationships with business partners including buyers and suppliers. As the primary link between the buying and the selling firm, they have considerable influence on buyer's perceptions of seller's reliability and the value of seller's services, and consequently the buyer's interest in continuing the relationship. This study discusses the impact of sales territory design and compensation to salespeople as predictors of performance of sales unit effectiveness. Sales management is a continuous process in a firm. The more a company learns about the sales process, the more efficient it becomes at selling, and the higher the sales yield. As the sales yield increases, sales learning process unfolds in three distinct phases including initiation, transition, and execution. Each phase requires a different size and type of sales force and represents a different stage in a company's production, marketing, and sales strategies [1]. Organizational buyers are defined as sales accounts and sales firms need to consider the following issues:

- Develop sales in an ongoing relationship
- High trust both sides
- Selling teams
- Multiple influencers on the purchase
- Strategic project work
- Don't start hard and end easily

Firms develop sales accounts by serving them over long time. A chronological growth of relationship between buying organization and selling firm makes accounts stronger and sustainable. Hence many firms Identify customer-centric business model for addressing global customers and formalize that model into a global customer management program.

Such integration of strategies is a key challenge for any firm aspiring to acquire and manage global accounts [2]. Managing sales accounts treats various operations of a buyer as one integrated account, with coherent terms for pricing, product specifications, and service delivery. Management strategies of sales accounts have proliferated over the best practices of global firms, which indicate that customer value is the principal determinant of a successful account. At the outset salespeople should determine whether products or services offered by the firm are appropriate for managing an account, whether organizations are involved in the buying process, and how managing specific accounts might affect the competitive advantage of sales firm. Managers should execute a scorecard for managing key accounts, which can help in targeting the right buying organizations. Salespeople should also develop services coordination among the accounts, develop performance measures the account satisfaction, and develop customized sales and services strategies for new business units. Sales firms should be able to map the problems in servicing accounts and should customize operations and meticulously administer strategies, given the difficulty and expense of providing multiple sales [3]. Managing accounts is a challenging task and salespeople need to take initiatives on various innovative dimensions like:

- Managing to sell vertically
- Developing strategies for horizontal sales
- Creating new business within the account
- Maintaining account relationships
- Problem solving strategies
- Facilitate high level interaction
- Monitor competition and detect problem in advance

Salespeople have the basic task of prompting frequent procurement among the accounts they are managing. The just-in-time replenishment policy should be adopted by the sales firms to keep stronger ties of buyer-seller relations. Procurement has grown up among the business-to-business clients with the increasing competition and globalization. Procurement and selling have now become the strategic business functions that increasingly recognize the importance of strategic supplier relationships driving success of key account management. Good procurement practices influence the quality of sales and services functions that would help accounts to grow continuously. There are many new techniques emerging in account relationships, including the use of psychological contracts and co-measurement and monitoring, which should be understood by the salespeople during the process of serving an account. Perceived fairness in the procurement and selling processes will also have a major impact on the customer value. Hence, account managers must recognize these changes or fall victim to sales delusion- the belief that they are performing better than they really are [4]. Procurement process of accounts needs to be supported by the sales process in order to lay functional integration among buyer and seller. The procurement process in buying firms passes through the following activity levels:

- Problem recognition
- General need description
- Product specification

- Supplier search
- Proposal solicitation
- Supplier selection
- Order-routine specification
- Performance review

Firms often consider purchasing as a complex process due to technological, economic and administrative perspectives. The problem recognition and need description stages in firms need to be supported by the salespeople during the process of prospecting. Product specification and supplier search process of procuring firms can be comprehended by the sales firms through feature presentations and benefit valuations. Such integration among buying and selling firms develops confidence in developing purchase proposal and selecting the supplier agency. At this stage, the salespeople need to be prepared to handle any product and sales related questions and develop positive impact on purchasing firm. Once the purchase proposal based on the mutual agreement emerging out of the negotiation process qualifies, the order processing can be carried out by the sales firm. However, it is necessary to provide post-sales follow-up to the account at the later stage.

Manufacturers were ready to offer products to excite consumers in the Philippines; however an economic slowdown diminished the purchasing power of many consumers. Therefore consumers were forced to focus on commodities such as rice at the expense of secondary products such as hot drinks. Under such situation it was an emergency for hot beverage manufacturers to revive their key accounts to push these products to end consumers. However among the declining trend of other products, coffee remains an important part of consumer lifestyles in the Philippines; however, consumption slowed as more people used a sachet of instant coffee twice as an economy measure. During 2008 all manufacturers supermarkets and hypermarkets had turned as key accounts and they consistently gained retail volume shares. These key accounts had increasingly become an indispensable retail format for hot drink manufacturers. Manufacturers intended to seriously prospect supermarkets and fast food chain stores to sell their hot beverage-coffee. Sales force of manufactures supported the key accounts with Tri-media campaigns, specifically TV advertisements, consolidated the leadership of big multinationals such as Nestlé Philippines Inc and Unilever Philippines Inc. By fortifying brand equity, Nescafé, Milo and Lipton generated high profits in coffee, other hot drinks and tea respectively. Sales force of manufacturers also considered new channels like convenience stores as key accounts and entered into negotiation for mutual benefits. By the end of 2008, supermarkets and other channels dominated retail volume sales as more consumers shopped for hot drinks in their outlets. However, manufacturers' profit margins had suffered as key accounts looked to exploit this growing dependency by increasing placement fees [5].

The role of procurement within global companies has changed dramatically over the past from simply buying goods and services to overseeing an integrated set of management functions. This brings new challenges and opportunities to procurement and selling functions in the global marketplace. Virtual sales teams have demonstrated increased emphasis on cost effective procurement and product specialization in sales offers to enhance the strategic character of procurement decisions of large accounts. Increasingly, procurement decisions have become intertwined with strategic management in general and sales negotiations in particular. The business transaction costs can be categorized into soft and hard costs, and internal and external factors affect these costs during the buying and selling operations. In making procurement decisions, managers need to consider the full range of decision variables including technology, use value, cost advantage, returns on investment, per unit productivity and other relational factors. In addition to traditional transaction costs such as transport costs and tariffs, managers need to recognize such elements as cultural and legal differences, government regulation, social preferences, environmental issues, political stability and risks involved in unethical sales behavior. Procurement, therefore, needs to be more closely observed by the salespeople in serving accounts successfully. However, Global sourcing creates many new opportunities for value creation, and puts salespeople at the competitive edge to be creative in managing accounts [6].

Sales accounts generally suffer with delivery pressures and the need for increased flexibility in negotiations that can help in prospecting the accounts using better communications approaches that include national account management, demonstration centers, telemarketing, and improved forms of catalog selling. Catalogs can be frequently used by the salespeople in selling the products to the accounts instead of pumping time, energy and money on straight demonstrations. There are four major steps for developing an effective sales program for key accounts, which include careful analysis of selling costs, precise demonstrations of products and services, formulation of a creative and coherent account services program, and meticulous monitoring of the total selling system [7].

BUYING CENTER AND BUYING PROCESS

Many companies undergo large and complex selling situations while managing accounts in a competitive marketplace. Sales of products and services to business-to-business accounts affect many parts of a buying company. Managers of selling firm continuously work on managing accounts that may take several years to turn as their permanent buying organizations. Sales of products and services to such principal accounts need special handling as they are more complex than smaller transactions, their potential profit is larger, and they have a long lasting effect than smaller accounts. There are sequential steps to open a contact with an account and provide services. These steps include identifying right prospects, developing profile of a company's needs and key personnel, justifying the purchase to the buyer, making the sales pitch, coordinating with company resources, closing the sale, and serving the account. Firms should be creating a senior sales force to service a multitude of major accounts, assigning a field sales manager to one or two accounts for regional sales management and having top executives take charge of the large sales [8].

Buying Center

All departments and mangers who participate in the buying process of a product or service may be referred to as a buying center. The chain of participants in the buying process depends on the number of administrative divisions, operational hierarchy and the size of the organization. Different members of the buying center may participate and exert influence of varied intensity at different stages in the decision process. Buying process in some firms also constitutes multi-cultural and multi-disciplinary teams. The most successful teams and managers need to deal with multicultural challenges by adapting to the consumer culture. Sales managers should adjust the cultural gaps for smooth working, make structural intervention in order to change the buying or selling process, set norms early in the process, and guide the teams to avoid conflicts. In general, managers who make intervention at an early stage and set norms; managers who try to engage everyone in the buying and selling process; and organizations that foresee challenges as stemming from culture may succeed in solving hierarchical problems also with good humor and creativity that interrupt the account management [9]. For example, representatives from engineering and operational areas often exert the greatest influence on the product specifications and criteria that a new product must meet, while economic, logistics and administrative indicators influence purchasing manager and he may have to play key role in choosing alternative products and alternative suppliers.

Packaging manufacturers have their sales strategies oriented to large accounts comprising food manufacturing companies. They continue to prospect industrial clients as well as government buying agencies with focus on environment friendly products due to rising concerns surrounding climate change. Bangkok Glass Co. Ltd. (BGC) invested in the promotion of its recycling campaign to support their sales focus. The company set up Glass Bottle Bank scheme to buy back glass packaging from consumers and has run a campaign on television and through other media to promote the message that used glass packaging has intrinsic value and can be sold back to packaging companies. In 2008, the company continued to expand its recycling campaigns beyond glass packaging by launching the campaign 'Change' to communicate the ideas of reduce, reuse, and recycle for materials such as folding cartons and other unused materials in industrial packaging and this strategy attracted the clients. Portable packaging has been more appealing to food manufacturer as this strategy was pro-end users. The BGC supported its selling process by influencing client's decision making at various levels using convincing technology. Widespread innovations in portable packaging have been seen in several categories including travel packs, pocket packs and single-serve packs in the Thai market. This has made food manufacturers an increasingly important target group for BCG. For food products, packaging is a key factor in terms of the speed and convenience of preparation. Microwaveable and heatproof packs are widely used for many frozen and dried processed products in Thailand, enabling them to be consumed anywhere anytime, without the necessity to transfer the products to other containers for cooking and consumption [10].

The nature of volume of purchase varies among buying centers in reference to the amount of risk perceived by the firm while buying a specific products and services. In some companies the buying center tends to be small that empowers purchasing manager to play major role in purchase decisions and the risk of buying products or services are tagged to his decision. Hence it is necessary for successful account management that salespeople should provide clear and comprehensive product or service specifications to the buying organizations and ensure that the purchase order must be large enough to provide an incentive to the buying firms. It is also required for salespeople to see that the appropriate supply market conditions exist during the process of generating purchase order and such order should be validated appropriately in the administrative hierarchy of the organization to avoid future problems in realizing payment and feed on services. Risk is a function of cost and utility measures related to the product and services. Salespeople must work on the strategies to reduce the buying risk as it affects the product's efficacy.

Sales Planning Implications

Although strategic planning has been part of the management function for as long as anyone can remember, the emergence of key account plans as a critical subset of the marketing plan in business-to-business markets has not attracted much analysis. This gap needs to be addressed, as key account plans have their own unique complexities. Moreover, the importance of key account plans is increasing. There is a need for a more widespread understanding of the benefits of key account planning, encompassing processes and outputs. Previous research demonstrates the benefits of key account planning and sets out a framework for implementing key account planning as a business process [11].Since the involvement of managers of buying and selling organizations at different stages of the planning, order raising, delivering and post-sales services process would be varying,it is important for a salesperson to know the right contact, right time of contact, and types of information and appeals each participant is likely to find most useful and persuasive. In a selling firm, hierarchical decision making in procurement means that many departments and levels are involved in the relationship, and are often located on dispersed locations.

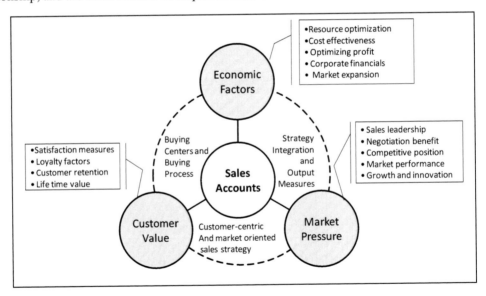

Figure 6.1: Account Management and Performance Paradigm.

The industrial clients may be very selective in choosing the buying firms and may refrain from switching to new sales firms. Sometimes a large account may ask for a national or international deal, project or contract, or harmonization of prices and conditions for their various manufacturing and operational locations. On the contrary sometimes the selling firm wants to have the same approach towards the account for all its locations. Therefore, it is necessary to coordinate between several people in hierarchy who are involved in decision making. However, by implementing customized account management strategy "one face to the customer" can be achieved, which is the secured way to proceed further with negotiations and seeking commitment of the buying organization [12]. The relationship among account management attributes and measures of sales performance is exhibited in Figure 6.1.

In many cases, however, the roles played by various members of the buying center are sufficiently consistent across similar types of firms that a company can establish policies to guide its salespeople. A firm selling products and services to an industrial account might adopt a policy of encouraging its salespeople to seek prior appointments with the top executives and to initiate contacts through the purchasing department. Firms that sell high value and high technology products often organize sales force into smaller and specialized teams or follow multilevel selling. The salesperson may be positioned at the lowest level in serving the account and this approach is the simplest for integrating the account manager in the organization, because account managers are operating from the sales department. The account manager in this approach may be a kind of senior-level salesperson [13]. It requires no special organizational structuring; however, decision making powers may not be delegated to the salespeople at this level. If the account requires coordination across organizational units, the account manager has little power to make this happen. The account manager is often no more than a senior salesperson. Account management may be considered as an advanced stage of personal selling, which requires both customer-centric and market oriented sales and services management skills. Most companies opt for exclusive deployment of account managers, separated from the sales department, because the role of account manager differs from that of salespeople, while co-ordination needs will be probably high and tasks complex [14]. Firms develop policies or organizational structures to guide their salespeople in dealing with organizational buying centers and managers perform periodical review in order to streamline the decision making.

Buying Process

Buying process includes the involvement of decision making units and their associates who influence the buying decisions. This is very common in industrial buying organizations. Salespeople need to identify the actual decision makers and determine how they view their interest in using the products and services offered by the selling firm. Most buyers show utilitarian approach to making buying decision and look for self-interest but they sometimes miscalculate. As a result, diagnosing motivation is one of the major management tasks for salespeople to be performed accurately for serving an account. Firms should invest in developing skills on applying psychological intelligence among salespeople to map the cognitive dimensions of decision making units of buying organizations. There is no formula for placing sound psychological analyses magically in the sales staff's hands. However, managers need to make sure that sales calls are highly productive and informative, listen to the buyers and influencers, and reward salespeople for rigorous fact gathering, analysis, and customization to increase sales effectiveness [15]. In order to make a decision on technologically sophisticated business solutions, a wide variety of individuals or representatives of the specialized departments may influence the decision making unit in a client firm. The participants influencing decision making units in large accounts include business analysts, customer service representatives, procurement personnel, and others. These participants in a buying process may be grouped as initiators, users, influencers, gatekeepers, buyers, deciders, and controllers. The salespeople should understand the buying process and develop job perspectives considering following approaches:

- Understand how a decision will be reached, even more clearly than the client does,
- Understand the approval process once the account has been chosen to serve,
- Analyze buyer preference, bargaining power, and the specific role in the decision-making process, and
- Initiate and stress negations on non-price factors and create the sense of urgency.

Initiators are the people who perceive a problem and look forward to the solution. Initiators influence consumers towards the purchase of a new product or service. The initiator can be almost anyone at any level in the firm. The effectiveness of the initiator is driven by his expertise and involvement in driving the buying process. Salespeople should develop contacts with the initiators, disseminate the information on product and services as well as on the firm, and draw his support in the selling process. To ensure forwarding behavior of initiator, salespeople need to develop a fit between buyer attributes and the buying emotions to enhance the opportunity of better prospecting [16]. User in the buying process may be defined as the people in the organization who are expected to use or work with the product or service being purchased. Users also play significant role in influencing the purchase decision. Users often volunteer to initiate a purchase, so it is possible that the same people may play more than one role. Firms interested in buying products and services also tap into lead users who often come up with creative ways to solve problems and influence decision makers in the organization. Besides users, influencers who are usually technical experts, alsoprovide information for evaluating alternative products and suppliers, and often play a major role in determining the specifications and criteria of making the purchase decision. They are also involved in the initial development of products and processes and set broad specifications for component and materials criteria, minimum end-product performance standards, and occasionally manufacturing techniques. The role of influencer may be seen distinctly towards:

- Recommending selling strategies
- Building grounds of interest
- Referring to alternative product, services and sales firm
- Reviewing sale presentations, and
- Gain access to decision-makers.

Purchasing agents of buying organization who are involved in controlling the flow of information to all members participating in the purchasing process may be considered as gatekeepers. They perform activities of screening and filtering in the buying process. The purchasing agent gathers proposals from various sales agencies and filters the information to circulate among the members of purchasing committee who compare the economic and operational advantages. Purchasing agents are specialists who have negotiation expertise, knowledge of buying products, and close working relationships with suppliers. They tend to become most involved in the purchasing situation in the later stages of a new buying situation. They are generally the dominant decision makers also in repetitive buying situations. The buyer is the person or firm who actually contacts, negotiates and raises the purchase order. The decider is the competent authority to make a purchase decision. In the chain of buying process the last role player is controller who determines the budget for the purchase and has the authority to critically examine the financial proposals of the selling firms. However, top management of the firm is likely to be involved when the purchase situation is unusual for the firm or when the decision is likely to have major consequences on the firm's operation.

MANAGING ACCOUNT RELATIONSHIP

Satisfaction of industrial clients is perceived to be a key driver of long-term relationships between suppliers and clients, especially when they are well acquainted with products and markets, and when industries are highly competitive. Efficiency of salespeople is one of the principal factors which influence customer satisfaction in a business-to-business context and help bridging client-seller dyadic relationship. The key services indicators, which include effective communication, cross-functional teams, and supplier integration, are followed to develop long-term relationships [17]. Client satisfaction has long been considered a milestone in the path towards company profitability. It is widely acknowledged that satisfaction leads to higher market share and stable revenues while relationship between client satisfaction levels and quality of services offered by the salespeople influence acquisition of new accounts. In fact, there seems to be little guidance for linking company costs to the key elements involved in providing client satisfaction in services, thereby diminishing the ability of a company to manage its activities accordingly [18].

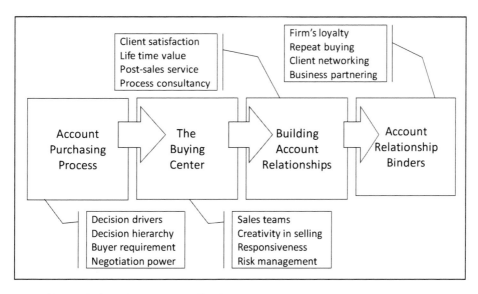

Figure 6.2. Client Relationship and Sales Effectiveness.

Professional firms providing sales and services to business-to-business accounts consider using non-technical people for client relationship managers (CRM's) rather than technical people to improve performance and customer satisfaction. Previous studies show that professional sales firms (PSF's) structure their client-relations functions around technical values while task-oriented CRM's impede market orientation because implementation of the marketing concept is dependent upon people-oriented skills. It was also found that the marketing concept does not adequately address the unique nature of PSF's, because they operate under professional and ethical obligations, and this hinders adoption and implementation of the marketing concept [19]. The effect of satisfactory client relationship on sales effectiveness has been exhibited in Figure 6.2.

Effective client relationship strategies help sales firm develop client loyalty by enhancing the lifetime value as may be seen from the Figure 6.1. The established loyalty of buying firm leads to prolonged association with sales firm by repeat buying of products and services. Sales force technology represents a variety of dynamic tools that can help in building relationship between buyer and sales force. While these tools cannot replace the salesperson nor generate deals from customers, they can enhance efficiency and effectiveness levels. Sales force technology can range from computer laptops, pagers, cellular phones, desktop personal computers and software for contact management, database management and multimedia presentations [20].

Hines is a privately owned, international real estate firm that has provided the highest level of quality, service and value to its clients and investors for more than 50 years. Hines has been active in the Mexico market since the early 1980's. Located in Mexico City, Hines Mexico office was established in 1994. The company deals with large number of clients and follows the multilevel sales strategyby negotiating the deal with the key department heads of the prospective buyer. The personnel having decision making or influencing capabilities of finance, real estate, human resources and production department are pursued by the equivalent executives of the Hines Company. The sales process includes the prospecting of the buyer, formalizing initial contract with the company, assessing the buyer needs in reference to the feasibility of infrastructure, visiting the property portfolios and developing a viable proposal for the deal. The marketing manager works on the buyer proposals to negotiate and close the deal. At the end of the contract the company approaches the client to confirm the negotiation or reset the prospecting environments for start afresh the negotiation to offer maximum satisfaction to the client. The company has the long term objective of catering to the extended relations in reopening such buyer proposal. The sales approach of the company is oriented towards delivering satisfactory services to the clients and prospecting the clients with this promise. The new buyers are paid categorical attention by providing comprehensive information and generating high level of satisfaction about the quality, brand and services of the company. The company also follows the solution approach to the problem clients and attempts to resolve the most secured solutions in real estate transactions. The salespeople target the companies listed in Fortune 100 for prospecting to sell or lease the real estate built by Hines. The buyers are also identified referring to the previous business relationships and sometimes through the real estate agents or referrals. The companies such as Coca-Cola, Amex, General Motors, and 3M are some of the clients of the Hines. The sales department verifies their need for operational area, storage and warehousing, office, parking, location, maintenance and related issues to formulate the sales proposal. The sales message is then presented by the team to the clients through electronic resources and printed information. The client is continuously pursued to keep his conviction intact on buying decision. The negotiation is reconfirmed by the marketing manager of the company and deal is closed successfully. The accounts are then offered post-sales services through a charted schedule. The client accounts services are periodically evaluated and improved accordingly. Hines has all flexibility in building relationship with new clients in order to protect the interests of the clients. The company provides some discounts to the existing clients for bringing new prospects of their subsidiary companies or referrals into the business ambience of the company. The company has focused 'attention to detail' strategy in all stages of client services [21].

Many Asian high technology consumer products manufacturers have developed customer orientation by equipping retailers and distributors to deliver quality services to augment customer acquisition, satisfaction and retention. Recent research suggests that culture of an organization builds the values, beliefs and assumptions that reflect how services need to be delivered in a competitive market which may be perceived by the customers outside the organization as well [22]. Client retention seems to be the result of a kind of repetitive decision by the customers, but their decision to cross-buy involves a more complicated process. Satisfaction of large accounts is a function of various conveniences that include sales location, delivery of products and services, one-stop shopping convenience, quality of sales and services, firm's reputation, firm's expertise in technology, and customer relations on both retention and cross-buying. Trust and satisfaction play different mediating roles in the relationships between service attributes, retaining large accounts, and inducing cross-buying [23]. Relationship value is an antecedent to relationship quality and behavioral outcomes and displays a stronger impact on satisfaction than on commitment and trust. Customer value also has direct impacts on the intention of client to expand business with the sales firm [24].

Value of relationship with the sales accounts reveals significant quality and behavioral outcomes in the sales activities. Value displays a stronger impact on satisfaction than on commitment and trust, and also directly impacts a buyer's intention to expand business with the firm. Perceived strength of the relationship with the client may be measured by salespeople in reference to technical ability, experience, pricing requirements, speed of response, frequency of client contact, degree of cooperation, trust, length of relationship, friendship and management distance barriers. The success of a sales firm is interlinked with customer value and if they serve customers by offering competitive gains, they will win. Client-sales firm partnerships are justified only if they stand to yield substantially better results than the firms could achieve on their own. Such co-dependency model elucidates the performance drivers associated with the success of sales firms. The leading business-to-business sales firms are applying new technology, new innovations, and process thinking to augment customer value, reduce costs, and enhance services their accounts [25].

Clients of hospitality industry such as high cost-high value hotels are sensitive in buying consumer durable products like high definition televisions and digital cameras in reference to the country of design (COD) and also value the country of manufacture (COM) of branded products. The transfer of the COD image to brand image affects significantly the buying decision of high value clients. Brand and COM congruity also plays an important role in buying decisions in determining the product quality and value for money [26]. As prices for durable and technology products fall over time with firms continually introducing enhanced products, clients may anticipate these prices and improvements and delay their purchases in the product category. Forward-looking business-to-business clients optimize purchase timing by trading off their utilities from buying the product and their expectations on future prices, quality levels, and brand availability. Such forward-looking behavior will result in price dynamics in the marketplace as price changes today influence future purchases [27]. The major attributes of client value enhancement for increasing sales effectiveness and developing long-term association are exhibited in Figure 6.3.

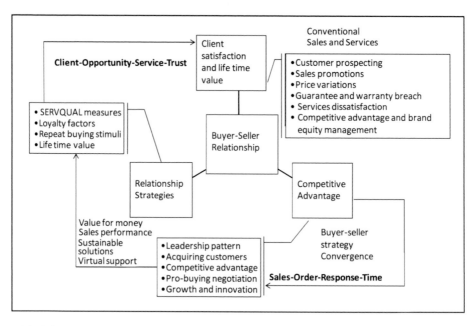

Figure 6.3 Sales, Services, and Value Enhancement Paradigm in Managing Business-to-Business Accounts.

   The effect of sales performance on relationship quality, in situations characterized by high relative dependence of the buyers on the retailers and distributors, is governed by the efficiency in delivery of services which develop high client satisfaction. Buyer-seller relationship may have significant effects on the focal firms in reference to the flexibility, responsiveness and customer relationship management which would help building capability of supplier firms towards increasing competitive advantage and gaining high client value [28]. Satisfaction plays an important role in relationships, is instrumental in increasing cooperation between channel partners, and leads to fewer terminations of relationships. Buyer-seller relationship gets closer and stronger through the information management at both the ends. Distributor information sharing helps to develop quality service oriented customer relationships. Interestingly, even if the initial level of trust in the retailer is low, the relationship quality substantially improves [29]. Customer centric approaches are practiced efficiently by the call centers to connect the customer issues with appropriate and interactive solutions. Call centers not only offer personalized attention to their problems but also help in building customer loyalty. Industrial clients rely on call centers set up by the sales firms, which exhibit high value work force towards services scheduling methods such as queuing system models to achieve optimal performance. Most of these models assume a homogeneous population of servers, or at least a static service capacity per sales agent. It is observed that a type of specialization minimizes the steady state queue size and reduces time of customer services [30].

   The key elements of buyer-seller relationships including long-term procurement plans, communication, cross-functional teams, and supplier integration are followed at different levels of transactional process. Beside, client satisfaction, reliability, and product-related performance are the major drivers that play significant role in the success of such dyadic relationship. Toyota and Honda leading Asian companies offering sales and services of gasoline based power generators in India have built taxonomy of customer-retailer-distributor relationships by consistently following six steps which include understanding functional pattern of distributors, reducing retailer-distributor rivalry and providing better opportunity, monitoring closely customer relations, enhancing supplier capabilities, sharing information intensively but in a selective way, and help customers continuously improve their association [31]. However, salespeople should understand that buyers no longer purchase products and services but lean on spending resources towards realizing sets of expectations. Thus, the relationship between buyer and seller often intensifies when the sale is made. The firm selling products and services can maintain a healthy relationship with the buyer after the purchase by regularly considering whether the relationship has been improving or deteriorating. To effectively manage relationships, managers must understand both the problems and the opportunities [32].

Through building personal relationships sales firms can develop their business with existing and potential clients. The relationship may be extended obtaining information on each contact's important dates, family events, particular interests, and life-style. The goal of nurturing such business relationship is to focus on the client satisfaction across the table. Such attention to detail requires well-trained and alert salespeople. Some direct selling companies spend enough time with salespersons and test them with long telephone conversations. These firms also reward top performers for collaborative sales efforts with partners as well as for landing big contracts [33].

The BMW data base provides a big haul of opportunities to provide the best solutions on building the existing and prospecting customers. The X Factor programs are built on data mining, which deliver highly effective marketing programs, incremental revenue opportunities, low cost per sale, and increased customer/prospect contact and satisfaction. The X factor program includes special events, referral programs, automobile certification, the scope of second BMW in member's home, lifestyle programs and felicitation to the BMW family members owning three cars. The X factor programs are drawn initially for one year which grows each quarter from prospecting to referral sessions, celebrating the car anniversary and finally launching the augmented services to the customer. Besides, dozens of new ideas are developed for entertaining the customers and building the most win-win relationship. The X1 factor programs are addressed towards the women prospects that bring them close to the brand and keep them longer by matching the corporate appeals to their unique needs. The X2 factor program is oriented towards the prime time customers who are willing to buy the second BMW. The trend of multiple BMW owner customers is analyzed and referrals are drawn to motivate the X2 factor program participants and rolled out to sales program executives for implementing the customer decisions. The X3 program is specific to the launch of series 7 BMW automobile in which these special people are invited to be among a small group who are allowed to preview the new 7 series. The X3 factor activities include appeal to their appreciation of inclusion by asking for their feedback and opinion, allowing them to be among the first to test drive, and notify them periodically of the status of the vehicle. The X4 factor is targeted for the prospects of Miniusa (MINI) model of the company and attempts to create the lifestyle concepts and messages for them to affirm their decision on buying the Mini model of BMW. These X-factor programs build a net relation with the customers and allow them to browse, post their views and requirement on net about Mini. The

X5 factor is a felicitation program on the anniversary of the BMW automobile purchased or the owner-birthday that communicates the personality of the brand and reinforces the relationship with BMW owners. This program also offers e-cards and a gift to the customers [34].

Understanding that highly accurate and timely data are not required everywhere in developing relationship, sales firms have tailored their real-time initiatives in building client relationships that can be significantly enhanced by "perfect" information. Client relationship management in other words, is coming to resemble any other valuable management tool, and the keys to successful implementation are also becoming familiar: strong executive and business unit leadership, careful strategic planning, clear performance measures, and a coordinated program that combines organizational and process changes with the application of new technology [35]. It has been observed that long-term and high-quality relationships, characterized by frequent interactions with distributors offer advantages for both retailers and customers. The quality of performance of high technology products and clients services is positively related to relationship quality. The extent of satisfaction of the organizational customers depends on the good financial conditions, a convenient assortment, good location features, clear information, and buyer friendly personnel. Hence, it may be stated that the level of performance of the sales and services delivered by the retailers and distributors is directly proportional to the level of satisfaction and loyalty of the organizational clients.

The interdependence structure of the dyadic relationship between buyer and seller as discussed in the pre-text refers to the extent of relative dependence that exists to stimulate the transactions mutually among various players. The measure of dependence may be categorized as total interdependence and asymmetry or relative dependence. Total interdependence is the sum of both parties' dependencies on each other while interdependence asymmetryrefers to the difference between each party's dependence on the other. Relational contracts and informal agreements between the manufacturers, retailers, distributors and customers are sustained by the value of length of relationships. Accordingly, there is a positive relationship between efficiency in customer services and sales effectiveness, which helps in reducing the conflicts, post-purchase dissatisfaction and complaining behavior of customers [36].

KEY ACCOUNT MANAGEMENT

Key Account Management (KAM) may be defined as the management of the customer relationships that are most important to a company. Key accounts are those held by business-to-business clients who contribute larger share in the profit for a company or have the potential to do so. The buying firms which are of strategic importance to a sales firm are also considered as key accounts. Development of customer relations and their retention is important for the sales firms to gain success in the business. In the key of accounts identification process sales firms pay particular emphasis on analyzing the type accounts that are of strategic importance to a company among other buyers, determining the needs of such distinct clients, and implementing procedures to ensure that they receive premium customer service. KAM is currently described as an important approach in creating value by implementing specific processes targeting most important customers. It has been observed that the intensity of competition as well as the intensity of sales coordination is the factor driving companies towards developing KAM programs. The companies need to pay attention towards the selection of key accounts. Many firms without KAM offer their clients special treatment without aligning their own internal organizational structures [37]. KAM is a strategic planning approach which is beyond traditional selling practices and reaches inside both seller and buyer organizations. It is a more complex and more difficult approach of selling products and services to business-to-business clients than simple sales activity directed to an individual customer. The KAM strategy helps the sales firms on following functions to:

- Understand each customer and anticipate their needs
- Appraise their potential
- Appreciate sources of risk, cost and profit
- Develop appropriate strategies for each customer
- Develop better implementation processes
- Monitor actions and improve results.

Adoption of global key account management systems typically represents a response by selling companies to two inter-related structural changes in their business environment in reference to the growing internationalization of their industry and unprecedented levels of foreign competition, and the urgency to retain customers of strategic importance *i.e.* key accounts. Global key account management is central to the ongoing and often acrimonious debate on *think global, act local*, suggesting serious implications for sales firms towards organization structure, co-ordination, control, and relationship management. The global key account management can be driven with efficacy where a *system selling* is a way of life for sales firms [38].

Key account management has emerged as a principal strategy in industrial selling. In this strategy sales managers either manage a range of major customers themselves, or develop a key accounts team within the sales organization. Sometime the need for serving the key accounts arises from the mergers and acquisition of firms and consolidation of buying requirements in industrial and retail organizations. In many developed countries a handful of sales firms dominate most product categories which provide enough strength to call for 'buyer power' of key accounts in negotiating and selecting products. Similarly, within industrial and commercial organizations buying is becoming more professional and concentrated, with concern for greater efficiency in key account management, improved margins, reducing costs, enhancing client services, and augmenting customer value. In fact the frequency of sales presentations are not high in major accounts, but serving the accounts demands more time in developing customized business, building relationships with a network of decision makers and influencers, and monitoring their performance against agreed objectives is large.

Services 800 company provided cutting-edge customer feedback and satisfaction monitoring solutions delivered via telephone interviews, e-mail, the web and other survey techniques to leading multi-national corporations since 1989. The company provides major contribution to key account management in terms of customer satisfaction measurement, as well as service level benchmarks and enterprise feedback management. The company uses eSMART web reporting tool to provide the data that helps accounts managers to drive effective management decisions in serving the client accounts. Major services of the company include customer satisfaction and internal measurement programs for medium to large companies, including multi-national companies with complex requirements. Options include everything from customer satisfaction surveys and performance measurement, to compliance and suitability monitoring or simply providing objective collection of customer's verbatim feedback. This company collaborates with The Association for Services Management International, which is the global leader in helping service professionals and service organizations, which deliver more value for their customers. The company found that in the fiercely competitive printer industry, the struggle to maintain market share gets tougher every year. So what differentiates one printer manufacturer from another? What drives customers to purchase from the same manufacturer again and again? In addition to product quality, reliability and performance features, many manufacturers are realizing the importance of service after the sale and the impact it has on driving repeat business. Services 800 Company enquired into this situation and helped the key accounts with the benchmark data that allowed matching trends to changes in programs or processes and comparing that to what the customer experiences as a result of those changes [39].

A key account manager will typically have a small portfolio of accounts to manage, but often have complex relationships to develop and manage in each of the accounts. While the key account manager's main contact may be with the principal buyer, managers will have to network within the buying organization, collecting information and analyzing feedback. Key account managers need to meticulously document the evidences in favor of the products and services offered and generate interest among buying organizations. Other tasks of managers include persuading sales amongst a range of decision influencers, carefully building contacts and relationships, and communicating effectively with all people involved in decision making [40]. Account managers have to estimate the type and quantities of goods and services to be purchased by the client organization. Such requisites are usually dictated by the demand or the firm's project outputs and operations. The criteria used in specifying the needed materials and equipment must usually be technically precise. Account relationship attributes and business development opportunities with client accounts are exhibited in Figure 6.4.

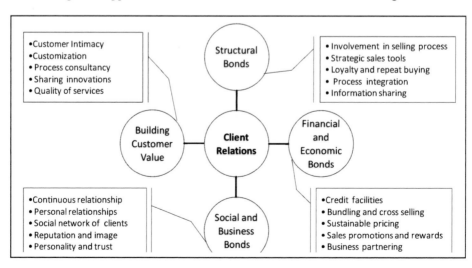

Figure 6.4.Client Relationship and Business Development Paradigm.

Top management involvement (TMI) and key account management is the necessary requirement for communicating with users in the firm. Involvement of key account manager is the major determinant of company performance and customer relationship development. The business community recognizes that top executives need to be committed to the management of the firm's most strategic customers. The TMI can also play crucial for developing new business, setting long-term agendas, and steering the firm through sticky situations, and ultimately grow business. It has been observed in a study that existence of high team spirit among salespeople drives top management involvement in internal processes. The involvement of top management helps in developing sales strategies and customer decision alignment. Firms operating in highly competitive environments should have top executives actively interacting with customers in business-related issues [41]. In the growing competitive business environment, the key accounts hold the potential to serve not only as a stabilizing force in turbulent times, but possibly as one of the better opportunities for organic growth. In some situations, these accounts [42]:

- Provide steady revenue streams
- Function as major drivers of profitability
- Offer opportunity to expand a portfolio of products and services
- Create opportunity to move up the value chain.

Key accounts managers may also develop promotional strategies to attract business-to-business clients. The sales firms may also manage inventories of activities that salespeople can carry out for their key accounts. Among these strategies there are special pricing, customization of products, provision of special services, customization of services, joint coordination of the workflow, information sharing, and customizing new requirements of buying organizations. Probably the most frequently discussed topic in key account program research is which special actors participate in key account activities. These specialized actors can be viewed as a personal coordination mode in key account management. The participation of special actors has a horizontal and a vertical component. The managers handling key accounts need support from various other departments to serve the account. The activities from such diverse functional groups include marketing and sales, logistics, manufacturing, information technology, and finance and accounting. However, even within the marketing and sales function, a key account manager may face difficulty in receiving support for implementing his tasks in serving the business accounts. One common problem is the lack of authority over regional sales executives who handle the local business with global key accounts. In many companies, KAM teams are led by a key account manager and strategies are implemented by key account coordinator who is mainly responsible for coordinating activities related to key accounts [43].

CREATIVE SELLING PROCESS

Sales managers demonstrate new concepts and convince clients about practical and profitable implementation of those ideas. However, they can be turned down by corporate decision makers who are not acquainted with the value of these selling ideas. The ability of sales managers to come up with workable ideas for convincing key accounts can quickly and permanently overshadow the buyers' feelings about worth of the creative selling idea [44]. An analysis of science of selling reveals the importance of personality traits, effective training, and close supervision in enhancing the sales effectiveness. An effective salesperson should possess good persuasion techniques, energy, self-confidence, desire for money, a habit of industry, and a state of mind which regards rejection as a challenge. Sales manager serving both consumer and industrial accounts should be provided training on creative sales in reference to orientation, the product or service, procedures for booking orders, time management, and sales techniques. An aggressive, imaginative, disciplined account manager is necessary in a firm to drive strong and dynamic buyer-seller relationship between salespeople and customers [45].The creative selling process consists of eight steps, none of which is less important than any other if the process is to be effective. It should be emphasized to all employees that all steps are vital to the achievement of all effective selling.

Pre-Customer Contact

Customer-contact is the principal requirement to manage the interface between the customer and the organization and is probably considered as one of the most critical tasks for salespeople in terms of the building a strong and prolific interface with the client organization. It is reasonable to expect managers with responsibility for organizing, developing and monitoring salespeople to ensure effective sales. However, a major problem concerning the performance of salespeople in developing customer contact as it is a outcome of the quality of management training they receive [46].Customer contact centers managed by the salespeople to serve accounts provide different types of services including call transfers, e-mail management and instant messaging. Clients calling might sometime turn impatient; hence salespeople need to show quick response time. E-mails that are not responded within a specific time limit need to be prioritized. It has been observed in a research study that limited cross-training with two skills per salesperson results in considerable performance improvements [47].

Tools of Selling

Increasing transportation costs and unprecedented travel difficulties are driving up the expenses and uncertainties associated with use of an outside sales team. As a result, sales managers operating in today's high-cost and high-risk environment need to invest in sophisticated data analytics to support inside sales teams that do not travel [48]. There are many tools of selling, however, they cannot replace the salesperson and generate deals from customers. Tools of selling can enhance efficiency and effectiveness levels. Sales force technology can range from computer laptops, pagers, cellular phones, desktop personal computers and software for contact management, database management and multimedia presentations [49].

Target Market Knowledge

There is a growing belief among the sales firms that effectively managing market knowledge can enhance effectiveness of salespeople. However, it has been observed that in a limited way resource-based and market oriented knowledge develop strategic sense making in order to drive competitive sales strategies by the salespeople. Firms introduce the concept of 'knowledge management orientation', and examine the relationships among knowledge management orientation, market orientation, and sales performance. The best salesperson knows something of the likes and dislikes of the firm's primary customers. The firms should tell all about client accounts and their life-styles to enable the salespeople to develop integration of persuasion and selling skills. It is observed that knowledge sharing, knowledge absorption, and knowledge receptivity serve as first-order indicators of sales effectiveness, which, in turn, has a positive link with enhancing the customer value, sales turnover, and revenue of the firm. Importantly, knowledge on current market conditions and sales requirements mediates the relationship between knowledge management orientation on one hand and subjective and objective sales performance on the other [50].

Product Knowledge

Organizational culture's influence on relationship development increases through the customer-oriented behavior of salespeople. But, considering those attributes of salespeople implies internal processes such as reflection, intuition, or interpretation, something that makes the 'individual's knowledge' indispensable. A salesperson gains confidence by knowing about the products and services being offered [51]. If a person sells shoes, it helps to know the merchandise as well as how to fit them. If a person sells building materials, the selling job is probably more effective if the salesperson can also answer questions about home repairs. It helps the person who sells clothes to know something about fabrics and current fashions. If the person is in the lawn service business, he should know about lawn care. Most will not take the initiative to acquire product knowledge. Management should make such knowledge available to them.

Prospecting

Prospecting is the process of finding customers who are ready to buy and can generate high net income for an organization. Leads for prospects come from three categories of sources including organization-initiated, acquired leads, and marketing activity-initiated leads.Too often, salespeople take easy prospecting the customer and analyze their afterthoughts. Given all the expense and effort involved, they should put their efforts towards winning the customer and maximizing productivity [52]. Prospecting can be performed with new and regular customers. Although it is not appropriate to every selling situation, prospecting should be customer-centric. Essentially, prospecting involves not waiting for the customer to show up at a store or to phone about a service. It is concerned with taking the initiative by going to the customer with a product or service idea.

A salesperson sees that a customer is getting married. Action is taken on this knowledge by contacting the person and telling about products and services being offered, which might be of assistance to a new bride. By using newspapers and personal contacts, a salesperson can take the initiative to contact and create new customers. A firm's best prospects are its current customers. A salesperson should make a practice of calling regular customers on a periodic basis to tell them about products or services. All salespeople should be encouraged to prospect by phone and in-person whenever they see regular customers. Prospecting with the same regular customer on a frequent basis can make prospecting loose the special feeling that it can create in customers. However, salespeople should not overuse this approach. Every salesperson should be challenged to treat each customer as an individual by responding differently to each customer. Initial contact also means responding to customers when they enter the sales areas even when they cannot be waited on immediately. Salespeople should be responsive to the waiting customers and such action will reduce the number of customers who leave without being served.

Presentation of Merchandise

Personal selling and personal sales professionals use a number of tools to develop customer relationships. Among these, the sales presentation offers a number of advantages to the sales representative, including strategic advantages. Yet many personal selling tools tends to view the sales presentation as a tactical tool but not strategic. However, salespeople should be given training on the special ways of conveying sales by making effective sales presentations that can add strategic value to the selling process.In presenting products or services sales proposal to the customer, the salesperson should use product knowledge to take the best advantage [53].

Buyer Benefits

Product knowledge is important but the salesperson must remember what makes the customer buy. Clothes may be made of durable fabrics but it is also important to stress on the implied benefit that they will also appeal to the opposite sex. Salespeople should learn to sell benefits to attract clients and develop long-term relationship. Selling well-designed consumer benefit would appreciate business relations of both sales firm and its client. The benefit mapping drives clients towards simplifying decision making, enhancing customer satisfaction, reducing risk, and driving profitable purchases. On the contrary, deceptive benefits to the buyers can lose negotiations on the table, fuel client disappointment, put buyers at risk and trigger lawsuits costing heavy tangible and intangible damage to the companies. Setting defaults benefits requires sales firms to balance a complex array of interests, including customers' preferences and the desire of sales firms to maximize profits and minimize risk. At a basic level, default benefits of products and services being offered can catalyze the deal and help aligning with preferences of buying organizations [54].

Customer Involvement

Product knowledge can be used to get customer involvement. Show the customer several features of the fuel saving automobile device and then have the customer put it on and work it. If they are involved in the product, it will be hard for the customer to take off the device so that the salesperson can put it back into the case. The best way to present many products is to get involvement. When buyers can locate themselves in the story, their sense of commitment and involvement is enhanced. By conveying a powerful impression of the process of winning, narrative plans can mobilize an entire organization.

The Bosch Group today comprises a manufacturing, sales, and after-sales service network of approximately 280 subsidiaries and more than 12,000 service centers in over 140 countries. Smart electronic stop-start system has been produced by the company for BMW since 2007, with first variants of the BMW 1-series featuring the start-stop system as standard equipment. The start-stop system switches off a vehicle's engine in stationary situations, such as traffic jams or stop lights. Bosch supplies the key component for this system: a starter that has been developed specifically for stop-start applications. The starter switches the engine off when the vehicle is stationary and starts it again as soon as the driver wishes to accelerate. During each trip, the Bosch system reduces fuel consumption and carbon dioxide (CO_2) emissions by as much as eight percent, depending on the vehicle. If the stops last longer, the actual saving of CO_2 emissions and fuel can be significantly higher. For Bosch, increasing fuel prices and stricter emissions standards call for innovative solutions. The Bosch Smart Electronic start-stop system is a cost-effective way of conserving resources and protecting the environment. This technology significantly reduces fuel consumption, especially in city driving and will help to reduce CO_2 emissions further in the future. The company already produces the battery sensor that detects the battery's current charge and communicates this information via the energy management system [55].

Sales organizations have been realized that involving customer in new product procurement has the potential for significant results. The customer involvement in buying process include the product experience *through do it yourself, take home* and *simulation* sales programs. Numerous studies have shown that buyer participation in sales have helped reduce perceived risk, increase concept-to-customer development process, and improve use value. Sales process integration is most successful when driven by a formalized process that considers buyer preferences, level of complexity of the technology, and degree of risk. Leading sales companies conduct a formal in-depth buyer preference evaluation and risk assessment prior to buyer involvement [56].

Routes to Market

The emergence of e-commerce is having a substantial impact on purchasing behavior. Yet, the rate of consumers' acceptance of e-commerce has been slower than many predicted. Using the view of consumption as an institution, this study examines buyer's preferences for shopping on-line. It is observed that consumers perceive relative advantages and disadvantages of shopping on-line vs. shopping in traditional stores. Some studies suggest that consumers have substantive reservations about shopping on-line which may be adversely affecting the acceptance rate of this new channel [57]. Buyers should be given options of sales channels illustrating the buyer benefits.

> Sears, the leading retailer for energy-efficient and Energy Star qualified appliances, has made it even more convenient for consumers to save significant amount of money on their electric bills and reduce household carbon emissions by making available the GE Hybrid Water Heater. Providing significant savings for consumers, the GE Hybrid Water Heater is up to 62 percent more efficient than a traditional electric water heater.As part of Sears' commitment to providing energy efficient products, Sears is the first retailers to offer the new GE Hybrid Water Heater to customers. Sears Blue Appliance Crew is dedicated to providing the best products on the market that combine innovation and conservation at affordable prices. Offering hybrid products like the GE Hybrid Water Heater that marry advanced eco-friendly technology with traditional elements is just one of many ways to remain customer-centric and make it easier for customers to go green [58].

Choice among channels, satisfaction with those individual channels, product satisfaction and payment equity are also some factors that influence overall satisfaction. Direct and indirect customers served by the salespeople derive the perceived benefits of the channels. Indirect customers are prospected by outsourced sales promoters or customers manage to buy products or services through Internet. In the indirect selling process it is important for the firms to ensure consistency between the information of intermediary's sales force and the supplier's website. Managers should therefore track and optimize both the individual channel experience and multi-channel integration [59].

Use Showmanship

Successful public speaking and showmanship consists of three key elements to a persuasive presentation: a clear structure, strong body language and physical presence, and the creation of powerful visual images in the minds of the audience. Salespeople should combine attributes such as good listening skills, audience focus, and simple optimism in presentations for driving effectiveness among audience.

Handling Objections

Many companies have only one response to handle the customer objection, which is by way of remaking the situation to provide better satisfaction with a product orientation based on attribute quality and a short-term internal perspective. In contrast, an extrinsic customer satisfaction focus emphasizes finding new ways to increase the positive, emotional aspects of the customer experience over time and in following such approach customer also gets involved in finding solution to the objections raised. This supports salespeople to handle objections in a customer centric way and find suitable solution to satisfy the customer. Sales managers who wish to climb out of their customer dissatisfaction groove must move beyond the mere measurement of quality, refocus their practices on the customer's actual experience, and formulate a comprehensive strategy for using that knowledge throughout the organization. Objections of customers during the selling process should be handled using the following tools:

Counter-questioning:	Salespeople should be inquisitive to know why an individual customer or client is raising objection. The counter-question drives the buyer to introspect and put forth the right argument revealing dissatisfaction on specific issues.
Restate Objection:	By restating the objection, buyer may clarify the situation to facilitate the salespeople to find the right ways to deals with the situation. This approach tends to reduce the magnitude of the objection in the eyes of the customer.
Direct Response:	Although offensive to some, this approach may be necessary if the customer is not going to buy unless the untruth can be corrected. Self-efficacy theory suggests that people will perform better when they believe that they have the skills necessary for success. It also suggests long-term success in reference to response-outcome expectations between buyers and sellers [60].

Consumer satisfaction with the complaint-resolution process can be a critical factor in customer retention. The consumer satisfaction and dissatisfaction with the complaint-resolution process has various critical dimensions. The two consumer-input dimensions are consumer time and effort expended in an attempt to resolve the problem while the customer-outcome dimensions are compensation received and attitudes of the sales representative. In handling customer objections certain demographics like age, gender, social class, and occupation should be examined when selecting complaint-resolution strategies [61].

Closing the Sale

In various ways, the salesperson can assist the customer by helping him or her to make the buying decision. Closing techniques that can aid in this effort include offering post-sales services, product choice and incentives for future association. Sales incentives may drive commitment and relationship to retain customers and meet their future aspiration to stay with the sales firm. In this way salespeople can outperform competition and lead the market.

Consultative Sales

Consultative sales skills, strategy and knowledge of industry conditions are the major tools for salespeople to solve client problems and provide value based solutions. A consultative sales approach process permits the salespeople to partner with the prospective client emphasizing on benefit to gain advantage in the deal and avoid probable risk factors. Consultative skills techniques focus on enhancing the relationship by leveraging the credibility gained by the supplier through knowledge and performance. The key element in successful execution of the consultative sales approach technique is to focus on cost justifying the solution. In consultative sales process salespeople should understand that customer value is an important element in developing solutions to optimize satisfaction of clients.

MONITORING SALES PERFORMANCE

Salespeople play a key role in the formation of long-term relationships with business partners including buyers and suppliers. As the primary link between the buying and the selling firm, they have considerable influence on buyer's perceptions of seller's reliability and the value of seller's services, and consequently the buyer's interest in continuing the relationship. Buyers often have a greater loyalty to salespeople than they have to the firms employing the salespeople, as salespeople develop buyers' perceptions [62]. Sales management is a continuous process in a firm. The more a company learns about the sales process, the more efficient it becomes at selling, and the higher the sales yield. As the sales yield increases, sales learning process unfolds in three distinct phases including initiation, transition, and execution. Each phase requires a different size and type of sales force and represents a different stage in a company's production, marketing, and sales strategies [63].

Effective management of sales force typically requires both planning and action to deliver powerful aspiration, establishing performance measuring, rewarding systems to reinforce specific goals of sales force, and developing mind-sets of salespeople by creating a sense of shared responsibility in performing tasks. Although companies devote considerable time and money to administer their sales forces, it is necessary to focus management strategies on how the structure of the sales force needs to be closely associated with the performance indicators of a firm's business. Firms must consider the relationship between the differing roles of internal salespeople and external selling partners, the size of the sales force, its degree of specialization, and how salespeople share their efforts among different customers, products, and activities. These variables are critical because they determine how quickly sales force of a firm responds to market opportunities, influences performance of sales people, and affects revenues, costs, and profitability of the firm [64]. The task of a salesperson changes over the course of the selling process. Different abilities are required in each stage of the sale including identifying prospects, gaining approval from potential customers, creating solutions, and closing the deal. Success in the first stage, for instance, depends on the salesperson acquiring precise and timely information about opportunities from contacts in the marketplace. Managers often view sales as a set of stereotype activities which begins with prospecting and terminates with closing the deal. But in practice it is a diligent task which involves variety of conflicts and high risk. Hence, sales performance requires effective administration in a phased manner to resolves conflicts and ambiguities during the negotiation and transactional process.

Administering control measure to field sales force has become major performance guiding tool in many business firms. Industrial sales organizations largely implement direct management control and influence the activities of employees leading towards improving their efficiency [65]. The extent monitoring, directing, evaluating, and rewarding activities of sales managers in an organization intends to guide sales force behavior through management control processes in order to achieve favorable results to the organization and the employees.

Management control is thus recognized as an important performance indicator of the task performed by the salespeople [66].

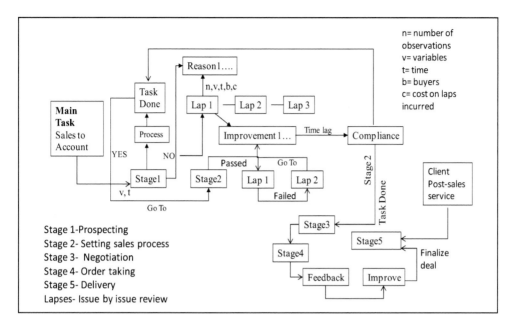

Figure 6.5.Sales Process Audit and Supervision Model.

Result oriented performance control and market volatility are positively related to new product selling performance. The effect of sales force adoption on selling performance is stronger where outcome based control is used and where the firm provides information on the background of the new product to salespeople through internal marketing [67]. It has been observed that salespeople who simultaneously exhibit commitment and effort will achieve higher levels of new product selling performance. The sales performance in an international context has been emerged of significance managerial consideration in view of growing globalization and the need to sustain competitive advantage. However, not much research attention has been given to sales management control beyond developed countries, notwithstanding recognition of the critical role of the sales manager in international selling [68]. An effective control system in sales process can enable firm to develop long term relationship and enhance sales effectiveness. The sales process audit and supervision model is exhibited in Figure 6.5. Managers need to supervise the sales activities by client accounts at various levels, review the performance at each level, identify reasons, suggest improvements, evaluate compliance of salespeople, and advise them to move to second stage as illustrated in the Figure.

On a tactical level, managers need to consider the optimum spread of customers on a matrix of sales attractiveness and market coverage. Besides market demand, managers also need to consider planning differentiation with large product portfolios operating in volatile markets which are governed with asymmetric behavior of retail organizations. In addition, the complexity of market, channel and supply networks makes supply chain planning more intricate [69]. The strength of supplier/customer relationships is again measured by applying a mix of objective, judgmental or subjective factors that include:

- Length of relationship,
- Importance of the customer,
- Friendship,
- Co-operation in product development,
- Social distance

Benefits are calculated by taking the revenue from the sales to buying organization(gross value of sales minus the commission paid) and subtracting from it direct costs, pseudo-direct costs (the costs that could be attributed to groups of similar customers and therefore apportioned accordingly) and indirect costs [70]. The liberal environment of the self-service stores for merchandise decisions, service quality and learning about competitive brands are the major attributes of retail self-service stores [71]. Managerially, multi-brand testing process allows the firm to forecast the impact of the new product introduction on the market shares of competing brands (including those marketed by the firm) at both the aggregate and segment levels. Hence, the firm can use the results to measure segment-specific cannibalization and switching effects; in addition, it can identify segment-specific adoption patterns for achieving organizational sales. Importantly, the method allows the sales people to choose customized marketing mix strategies for different segments after allowing for the effects of competitive retaliation [72].

A few studies have found significant relationships between personal and organizational variables such as job experience, closeness of supervision, performance feedback, influence in determining standards, span of control, and the amount of role conflict and ambiguity perceived by salespeople. It was found that idealized influence behavior of salespersons positively influences customer trust which, together with individualized consideration of salespersons, in turn influences customer commitment. It was also found that the joint effect of both customers' trust and individualized considerate behavior of the salesperson is greater than each alone on customer's relationship commitment [73]. Other studies relate personal characteristics to variations in motivation by showing salespeople's desires for different job-related rewards (*e.g.,* pay, promotion) differ with such demographic characteristics as age, education, family size, career stage, or organizational climate. The salesperson's expectancy, instrumentality, and valence perceptions are not directly under the sales manager's control. But they can be influenced by things the sales manager does, such as how he or she supervises the salesperson or rewards the individual. Since the salesperson's motivation strongly influences performance, the sales manager must be sensitive to how various factors exert their impact [74]. The sales results typically emerge from the performance of the salesperson and this issue is receiving increasing attention among the multinational companies. This may be evidenced from research in several countries that indicates that sales managers are concerned with the team and customer relationship building activities of salespeople as well as their sales results. Outcomes are certainly important, but managers also need to consider the behavior (inputs) of the salesperson that is instrumental in guiding the sales results.

The effective management of salespeople is important to managers of international marketing operations spanning multiple countries and simultaneously maintaining strong behavioral threshold of salespeople for continuously improving effectives of the sales units. The degree of salesperson input and involvement does, however, appear to vary across firms.

The multinational pharmaceutical companies generally assign weights to different performance objectives and incorporate territory data when establishing these objectives. Most salesperson performance evaluations are conducted by the field sales manager who supervises the salesperson. However, some firms involve the manager above the field sales manager in the performance appraisal of salespeople. The behavior control is a consistent predictor of salespeople's performance and effectiveness of the sales units. It indicates the importance of proactive monitoring, directing, and evaluating salespeople by the managers. Sales managers may implement such controls effectively by establishing coordination, training, and feedback process rather than imposing command and control policy.

Managers may define clear sales tasks to ensure significant results from the salespeople and evaluate timely performance feedback. A firm may focus on administering sales activities in four major processes including territory-target based deployment of sales force, efficiency in managing client account, working with improved information systems, and minimizing field level task conflicts. Hence, limiting the number of accounts for each salesperson may work out as a good option to improve the efficiency of sales activities in the market. Field sales managers may be encouraged to adapt management by objectives, performance appraisal, and monthly reviews to encourage their salespeople to do their work effectively. However, some companies fail to coordinate salespeople's efforts with other customer-facing teams and while operating at cross-purposes, these functions conflict over roles and resources jeopardizing the relationship with customers. To correct such misalignments, it is necessary to match sales management practices with strategic priorities of the firm and allow effective coordination among customer-facing teams to enable seamless service for customers.

SELECT READINGS

Capon, N. (2001), *Key account management and planning: The comprehensive handbook for managing your company's most important strategic asset*, Free Press, New York, NY.

Corcoran, K. L., Peterson, L. K., Baitch, D. B. and Barren, M. E. (1995), *High performance sales organizations: Creating competitive advantage in the global marketplace*, Chicago, IL, Irwin.

Page, R. (2008), *Make winning a habit: Five keys to making more sales and beating your competition*, McGraw-Hill, New York.

Stefan, W. (2006), *Key account management in business-to-business markets*, DUV Publishers, Frankfurt, Germany.

Sallie, S., Sperry, J., and Reese, S. (2003), *The seven keys to managing strategic accounts*, McGraw-Hill, New York.

REFERENCES

[1]   Ledingham, D., Mark, K., and Heidi, L. S. (2006), The New Science of Sales Force Productivity, *Harvard Business Review*, Vol. 84 No. 9, pp. 124-133.

[2]   Capon, N. and Senn, C. (2010), Global Customer Management Programs: How To Make Them Really Work, *California Management Review*, 52 (2), 32-55.

[3]   Yip, G. S. andBink, A.J. M. (2007), Managing global accounts, *Harvard Business Review*, 85 (9), 102-111.

[4]   Ryals, L. J., Rogers, B. and Rogers, B. (2006), Holding Up the Mirror: The Impact of Strategic Procurement Practices on Account Management, *Business Horizons*, 49 (6), 41-50.

[5]   For details see Euromonitor International (2009), *Hot drinks in Philippines*, Market Report, Euromonitor International, June.

[6]   den Butter, F.A.G. and Linse, K. A. (2008), Rethinking Procurement in the Era of Globalization, *Sloan Management Review*, 50 (1), 76-80.

[7]   Shapiro, B.P. and Wyman, J.(1981), New ways to reach your customers, *Harvard Business Review*, 59 (4), 103-110.

[8]   Shapiro, B. P. and Posner, R. S. (2006), Making the major sales, *Harvard Business Review*, 84 (7), 140-148.

[9]   Brett, J., Behfar, K. and Kern, M. C. (2006), Managing cultural teams, *Harvard Business Review*, 84(11), 84-91.

[10]   For details see Euromonitor International (2009), *Packaging industry in Thailand*, Market Report, Euromonitor International, August.

[11]   Ryals, L. and Rogers, B. (2007), Key account planning: benefits, barriers and best practice. *Journal of Strategic Marketing*, 15(2), 209-222.

[12]   Kempeners, M. A. and van der Hart, H. W. (1999), Designing account management organizations, *Journal of Business and Industrial Marketing*, 14 (4), 310-335.

[13]   Colletti, J.A. and Tubridy, G.S. (1987), Effective major account sales management, *Journal of Personal Selling & Sales Management*, 7 (2), 1-10.

[14]   Wotruba, T.R. and Castleberry, S.B. (1993), Job analysis and hiring practices for national account marketing positions, *Journal of Personal Selling & Sales Management*, 13 (3), 49-65.

[15]   Bonoma, T. V. (2006), Major Sales: How really does the buying? *Harvard Business Review*, 84 (7), 172-181.

[16]   Dobele, A., Lindgreen, A., Beverland, M., Vanhamme, J., and Wijk, R.V. (2007), Why Pass on Viral Messages? Because They Connect Emotionally, *Business Horizons*, 50 (4), 291-304.

[17]   Rajagopal (2010), Bridging Sales and Services Quality Function in Retailing of High Technology Consumer Products, *International Journal of Services and Operations Management*, 6 (5).

[18]   Cugini, A., Carù, A., and Zerbini, F. (2007), The Cost of Customer Satisfaction: A Framework for Strategic Cost Management in Service Industries,*European Accounting Review*, 16 (3), 499-530.

[19]   Simon, G. L. (2005), The Case for Non-Technical Client Relationship Managers in B2B Professional Services Firms. *Services Marketing Quarterly*, 26(4), 1-18.

[20]   Carter, T. (2002), Technology and the Sales Force -Amach, Inc., *Journal of Hospital Marketing and Public Relations*, 14(1), 79-91.

[21]   For details see corporate web site of Hines Internatinal Inc. http://www.hines.com, also see Rajagopal (2003), *Analysis of sales audit pattern in Mexico: A case of Hines International Estate Developers*, Discussion Case, ITESM, Mexico City Campus, Mexico.

[22] Macintosh, E. and Doherty, A. (2007), Reframing the service environment in the fitness industry,*Managing Leisure*, 12 (4), 273-289.

[23] Liu, T. C. and Wu, L.W. (2007), Customer retention and cross-buying in the banking industry: An integration of service attributes, satisfaction and trust,*Journal of Financial Services Marketing*, 12 (2), 132-145.

[24] Ulaga, W. and Eggert, A. (2006), Relationship value and relationship quality: Broadening the nomological network of business-to-business relationships, *European Journal of Marketing*, 40 (3-4), 311-327.

[25] Rajagopal and Rajagopal, A. (2009), Buyer-supplier Relationship and Operational Dynamics, *Journal of Operations Research Society*, 60 (3), 313-320.

[26] Essoussi, L. H. and Merunka, D. (2007), Consumers' product evaluations in emerging markets: Does country of design, country of manufacture, or brand image matter?*International Marketing Review*, 24 (4), 409-426.

[27] Song, I. and Chintagunta, P. K. (2003), A Micromodel of New Product Adoption with Heterogeneous and Forward-Looking Consumers: Application to the Digital Camera Category,*Quantitative Marketing and Economics*, 1 (4), 371-407.

[28] Squire, B., Cousins, P. D. and Brown, S. (2005), Collaborating for customization: an extended resource-based view of the firm, *International Journal of Productivity and Quality Management*, 1 (1-2), 8-25.

[29] Ganesan, S. (1994), Determinants of long term orientation in buyer-seller relationships, *Journal of Marketing*, 58 (2), 1-19.

[30] Ryder, S. G., Ross, K. G. and Musacchio, J. T. (2008), Optimal services policies under learning effects, *International Journal of Services and Operations Management*, 4 (6), 631-651.

[31] Rajagopal (2008), Consumer Response and Cyclicality in New Product Management, *Journal of Customer Behaviour*, 7 (2), 165-180.

[32] Levitt, T. (1983), After the sales is over, *Harvard Business Review*, 61 (5), 87-93.

[33] Mackay, H. B. (1988), Humanize your selling strategy, *Harvard Business Review*, 66 (2), 36-38.

[34] For details on the X factor program see BMW Mexico web site www.bmw.com.mx, also see Rajagopal (2003), *Building customer loyalty through relationship networking: A case of BMW Mexico*, Discussion Case, ITESM, Mexico City, Mexico.

[35] Ledingham, D. and Rigby, D. K. (2004), CRM done right, *Harvard Business Review*, 82 (11), 118-128.

[36] Frazier, G. L. and Rody, R. C. (1991), The use of influence strategies in inter-firm relationships in industrial product channels, *Journal of Marketing*, 55 (1), 52-69.

[37] Wengler, S., Ehret, M., Saab, S. (2006), Implementation of Key Account Management: Who, why, and how?: An exploratory study on the current implementation of Key Account Management programs, *Industrial Marketing Management*, 35 (1), 103-112.

[38] Millman, T. F. (1996), Global key account management and systems selling, International Business Review, 5 (6), *Project Marketing and Systems Selling*, 631-645.

[39] For details see corporate web site of Service 800 at http://www.service800.com

[40] Noonan, C. J. (1998), *Key account management: Sales Management*, Butterworth-Heinemann, Oxford, 370-401.

[41]   Guesalaga, R. (2007), *Top management involvement with key accounts*, Working Paper, Institute for the Study of Business Markets, Smeal College of Business Administration, Penn State University, PA.

[42]   Shea, L. G., Frankel, W. B., and Libb, A. (2009), *Driving organic growth through key account expansion*, Web Discussion, Opinion Research Center, June,09.

[43]   Homburg, C., Workman, J.P. Jr., and Jensen, O. (2002), A Configurational Perspective on Key Account Management, *Journal of Marketing*, 66 (2), 38-60.

[44]   Elsbach, K. D. (2003), How to pitch a brilliant idea, *Harvard Business Review*, 81 (9), 117-123.

[45]   McMurry, R. N. (1961), Mystique of super salesmanship, *Harvard Business Review*, 39 (2), 113-122.

[46]   Strong, C. A. (2006), Is managerial behaviour a key to effective customer orientation? *Total Quality Management & Business Excellence*, 17(1), 97-115.

[47]   Ahghari, M. and Balcoğlu, B. (2009), Benefits of cross-training in a skill-based routing contact center with priority queues and impatient customers, *IIE Transactions*, 41(6), 524-536.

[48]   Gessner, G. and Scott, R. A. (2009), Using Business Intelligence Tools to Help Manage Costs and Effectiveness of Business-to-Business Inside-Sales Programs. *Information Systems Management*, 26(2), 199-208.

[49]   Carter, T. (2002).,Technology and the Sales Force-Amach, Inc., *Journal of Hospital Marketing & Public Relations*, 14(1), 79-91.

[50]   Wang, C. L., Hult, G., Ketchen, D. J. and Ahmed, P. K. (2009), Knowledge management orientation, market orientation, and firm performance: an integration and empirical examination, *Journal of Strategic Marketing*, 17(2), 99-122.

[51]   Cegarra-Navarro, J. and Rodrigo-Moya, B. (2007), Learning Culture as a Mediator of the Influence of an Individual's Knowledge on Market Orientation, *The Service Industries Journal*, 27(5), 653-669.

[52]   Wreden, N. (2002), *Making your proposal come out on top*, Harvard Business Publishing Newsletter, Harvard Business School, Boston, MA.

[53]   Hershey, L. (2005), The Role of Sales Presentations in Developing Customer Relationships, *Services Marketing Quarterly*, 26(3), 41-53.

[54]   Goldstein, D.G., Johnson, E. J., Herrmann, A., and Heitmann, M.(2008), Nudge your customers towards better choice, *Harvard Business Review*, 86(12), 99-105.

[55]   For details on the Bosch electronic stop-start system for automobiles, see corporate web site of the company at http://www.bosch.com and http://www.bosch-press.com

[56]   Slater, S. F. and Olson, E. M. (1996), Value based management system, *Business Horizons*, 39(5), 48-52.

[57]   Inks, S. A. and Mayo, D. T. (2002), Consumer Attitudes and Preferences Concerning Shopping On-Line, *Journal of Internet Commerce*, 1(4), 89-109.

[58]   For details see GE Press Room http://www.geconsumerproducts.com/pressroom/press_releases/

[59]   Madaleno, R., Wilson, H. and Palmer, R. (2007), Determinants of Customer Satisfaction in a Multi-Channel B2B Environment, *Total Quality Management & Business Excellence*, 18(8), 915-925.

[60]   Barling, J. and Beattie, R. (1983), Self-Efficacy Beliefs and Sales Performance, *Journal of Organizational Behavior Management*, 5(1), 41-51.

[61] Richard, M. D. and Adrian, C. (1995), A segmentation model of consumer satisfaction/dissatisfaction with the complaint-resolution process, *The International Review of Retail, Distribution and Consumer Research*, 5(1), 79-98.

[62] Weitz, B.A., Bradford, K.D. (1999), Personal Selling and Sales Management: A Relationship Marketing Perspective, *Journal of the Academy of Marketing Science*, 27 (2), 241-254.

[63] Money, R. B. and Graham, J. L. (1999), Salesperson performance, pay, and job satisfaction: Tests of a model using data collected in the United States and Japan, *Journal of International Business Studies*, 30 (l), 149-172.

[64] Zoltners, A.A., Sinha, P., and Lorimer, S. E. (2006), Match Your Sales Force Structure to Your Business Life Cycle, *Harvard Business Review*, 84 (7-8), 80-89.

[65] Darr, A. (2003), Control and autonomy among knowledge workers in sales: an employee perspective, *Employee Relations*,25 (1), 31-41.

[66] Oliver, R. L. and Anderson, E. (1994), An empirical test of the consequences of behavior- and outcome-based control systems, *Journal of Marketing*, 58(4), 53-57.

[67] Hultink, E. J. and Atuahene-Gima, K. (2000), The effect of sales force adoption on new product selling performance, *Journal of Product Innovation Management*, 17 (6), 435-450.

[68] Honeycutt, E. D. and Ford, J. G. (1995), Guidelines for managing an international sales force, *Industrial Marketing Management*, 24 (2), 135-144.

[69] Rajagopal (2007), Stimulating Retail Sales and Upholding Customer Value, *Journal of Retail and Leisure Property*, 6 (2), 117-135.

[70] Rajagopal and Sanchez, R. (2005), Analysis of Customer Portfolio and Relationship Management Models: Bridging Managerial Gaps, *Journal of Business and Industrial Marketing*, 20 (6), 307-316.

[71] Babakus, E., Bienstock, C. C. & Van Scotter, J. R. (2004), Linking perceived quality and customer satisfaction to store traffic and revenue growth,*Decision Sciences*, 35 (4), 713-737.

[72] Sharan, J., Jedidi, K., and Jamil, M. (2007), A Multi-brand Concept-Testing Methodology for New Product Strategy, *Journal of Product Innovation Management*, 24 (1), 34-51.

[73] Jayakody, J. A. S. K. and Sanjeewani, W. M. A. (2006), The impact of salesperson transformational leadership behavior on customer relationship marketing behavior: A study of the Sri Lankan corporate banking sector, *The International Journal of Bank Marketing*, 24 (7), 461-474.

[74] Dubinsky, A. J., Kotabe, M., Lim, C. U., and Michaels, Ronald E (1994), Differences in Motivational Perceptions among US, Japanese and Korean Sales personnel, *Journal of Business Research*, 30 (2), 175-185.

Chapter 7

SALES FORCE RECRUITMENT AND TRAINING

Effective recruitment process of salespeople is a vital element in ensuring convergence of an organization with its goals, human resources and growth in business. Recruitment process can be expensive for an organization to get an employee who is inadequately qualified, fails to perform his assigned tasks and leaves the organization without making significant contribution. The time and effort invested carefully in planning the process of recruitment can help to get the right person for the job, reduce labor turnover and enhance competitive advantage. A Firm's reputation as an employer can be stronger if it has efficient managers contributing significantly to its growth. The profitability of firms improves when potential jobseekers consider them an employer of choice. Hence, many firms have devoted their energy in recent years to developing not only the recruitment practices that will enable them to find the top talent, but also the retention practices that will keep current employees' hearts and minds engaged. However, a firm cannot become an employer of choice unless it has a manager of choice [1]. Firms can be frequently chosen by the potential employees provided they rightly identify strategies for attracting talent in global markets where brand, opportunity, job challenges, and organizational culture play important role in determining the employment preferences. A desirable brand affiliation in conjunction with inspirational leadership appeals to attract young high potentials. Opportunity should imply an accelerated career track or at least a drive acquisition of skills and experience. The culture of the firm should be meritocratic, value both individual and team accomplishments, and follow over time promises implied during recruitment of people [2].

Globalization in business environment has caused a series of changes in the process of recruitment, global staffing, organizational management, and rewards. The changes are affected by globalization highlighting the difference between standardized and localized human resource processes. Sales firms need to respond fast to staffing demand, changing organizational structure, and economic indicators to stay ahead in competition [3]. Firms make continuous efforts to attract the talents of competing companies to make sustainable business growth. Open competition for the talents of competitive firms had been rare before the globalization, but is now an accepted fact. Fast-moving organizations continuously require new talent because when a good employee walks away, the business takes a hit. The recruitment strategies need to be set around the socio-economic benefits of employee, career paths, and training efforts. However, a market-driven approach allows employees to stay long with the firm and develop loyalty so that the firm can use highly talented people for competitive growth in the long run. Compensation is only one of many useful retention mechanisms. Firms can also redesign jobs to reduce employee turnover. Firms can promote loyalty of salespeople to particular client accounts or work teams. Generally sales firms do not count on talent and hire people who are not in high demand, and place key players in locations where they are not constantly tempted by job offers. Such approaches result into high turnover of salespeople affecting the business of the firm. Firms can outsource, strengthen recruitment, standardize jobs, cross-train employees, and organize work around short-term projects to derive advantage of talented human resources [4]. Hence firms should plan for recruiting best salespeople considering the following organizational requirements:

- Identifying requirement
- Pooling staff needs
- Classifying job specific criteria for recruiting
- Specialist, Generalist, Auxiliary sales force
- Defining job perspectives
- Delineating probationary criteria
- Scheduling announcement.

In driving high-value talent in a company, managers need to market the workplace environment, creating job offerings and employment relationships that provide mutual value for firms and their employees. Firms can use 9Ps comprising product, price, place, promotion, people, performance, psychodynamics, posture and proliferation to create compelling job offerings that attract high-value employees. Many firms use relationship marketing concepts to build long-lasting employment relationships that create mutual value for the firm and its employees. Firms should also generate high-value employment relationships, which include greater employee commitment and satisfaction and lead to greater employee advocacy of the firm and lower levels of employee turnover [5].

Recruiting salespeople forms a major part of overall resourcing strategies of sales oriented firms, which seek to identify and secure the salespeople needed for the organization to stand competitive and achieve its goals at different temporal spans. Recruitment activities are responsive to the increasingly competitive market where human resource managers make efforts to secure suitably qualified and capable recruits at all levels. To be effective these initiatives need to include how and when to source the best recruits internally or externally. Common factors associated with the success of either human resources strategies may be as stated below:

- Well-defined organizational structures with sound job design,
- Robust task and person specification and versatile selection processes,
- Reward, employment relations and human resource policies,
- Commitment for strong employer branding, and
- Employee engagement strategies.

Job designs and recruitment process should be based on performance control, accountability, influence, and support determinants. It is necessary for a firm to have efficient salespeople, managers should meticulously set the span of control to reflect the resourcesallocated to each position and unit that plays an important role in delivering customer value. This setting, like the others, is determined by how the business creates value for customers and differentiates its products and services. Mangers should also examine the entrepreneurial behavior of the recruits and measure the creative tension by fixing accountability to handle tasks [6].

Human resources managers should consider the job attraction parameters to drive good candidates. Some studies have identified that compensation is one of the important factors attracting the candidates for jobs followed by fringe benefits, career advancement prospects, category of employment, job autonomy, and supervision method. The organizational culture comprising the task, thrust, time, target and territory of work are also considered as vital decision variables for the prospects to choose jobs. Another behavioral dimension that appears to be critical in determining the employer and employee relationship is the opportunity of unlearning knowledge and skills of the employees. Often firms succeed in convincing highly talented candidates to accept lower positions assuring that they will be promoted to the position that has a close match with their qualification and experience, but employees under such situations succumb to frustration as they do not find an appropriate platform to share their knowledge and implement skills. There are four factors that make higher impact on the recruits including intangible job context, employee development and skill utilization, remuneration packages and workplace, culture and rewards [7].It has been observed that in many industrialized countries there is a concern among growing firms about skills shortages, and mismatches are appearing in the labor market and that human resources managers are aware of recruitment difficulties and skill shortages that affect the market competitiveness of both small and large firms. However, there is a growing awareness among recruiters that education and training systems can influence the skill and occupational mix of a locality and local economic wellbeing. In services industry like tourism and hospitality, with their reliance on the secondary job market and high rates of employee turnover, there is a strong tendency to have high levels of recruitment and low levels of training [8].

Manpower Inc. (NYSE: MAN) is one of thelargest providers of temporary employees next to Adecco and Randstad, which has a long standing in USA, providing human resource services to manufacturing and marketing industry. The company has capability of placing about 4 million people in office, industrial, and professional positions every year. It has some 4,000 owned or franchised offices in 80 countries and territories including France, Italy, UK, and USA. Manpower Company offers employers a range of services and solutions for the entire employment and business cycle including permanent, temporary and contract recruitment; employee assessment and selection; training; outplacement; outsourcing and consulting. The company also provides employee testing, training, and other contract services. Its Global Learning Center give employees access to training materials over the Internet. The focus of Manpower's work is on raising productivity through improved quality, efficiency and cost-reduction across their total workforce, enabling clients to concentrate on their core business activities. Manpower Inc. operates under five brands: Manpower, Manpower Professional, Elan, Jefferson Wells and Right Management. The company provides business solutions to the client organizations through three main services: Task-Based Outsourcing (TBO), Recruitment Process Outsourcing (RPO) and Managed Service Programs (MSP). The digital strategy of the company is a key focus area as technology is ubiquitous and along with this ubiquity is the rise of social networks, Web 2.0 and mobile connectivity. As a result, clients of the company and candidates expect an interaction in a different way. This is a perfect opportunity for the company to inject the Manpower Experience into the digital world.

The introduction of MyPath, combined with core offerings like Direct Source, Direct Time, Direct Office and Direct Talent, enhances the position of the company as the bricks and clicks premier brand [9].

Many firms rely on external recruitment methods to secure the best candidates. Rapid changing firms demand high selling skills and experience which is not an easy combination to get a candidate particularly in sales that has monotony in carrying the tasks. It would be unusual for an organization to undertake all aspects of the recruitment process without support from third-party dedicated recruitment firms. This may involve a range of support services, such as; provision of resumes, identifying recruitment media, advertisement design and media placement for job vacancies, candidate response handling, and short listing, conducting aptitude testing, preliminary interviews or reference andqualificationverification. Internal recruitment can provide the most cost-effective source for recruits if the potential of the existing pool of employees has been enhanced through training, development and other performance-enhancing   activities   such   as performance   appraisal   and succession planning development centers review performance and assess employee development needs and their promotional potential. Typically, small organizations may not have in-house resources or, in common with larger organizations, may not possess the particular skill-set required to undertake a specific recruitment assignment. The cycle of recruitment and stages of human resource movement process to be followed by the sales firms is exhibited in Figure 7.1.

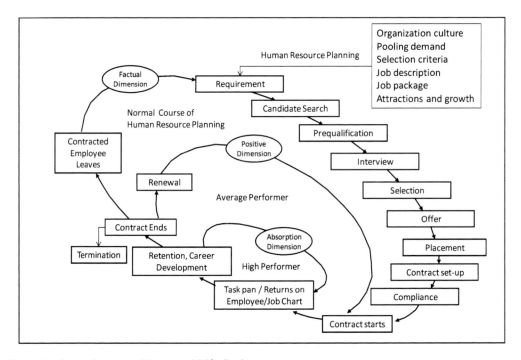

Figure 7.1.Rectruitment and Personnel Life Cycle.

Most firms intend to retain high performers by providing economic and relational incentives like determining the career path and organizational patronage. It may be seen from the above Figure that high performers have larger task pan (list of tasks to be performed), higher returns on their position (cost to firm), and transparency in the job description. On the contrary the average performers struggle at the edge of revalidation of contract as firms may look into hiring new people in lieu of average performers [10].It has been observed that good salespeople possess a balance of cognitive and normative behavior. A study upon comparing behavior with performance revealed that top sales performers focus the conversation with clients and manager simultaneously to draw a balanced judgment on selling the products and services while weak sales performers allow customers to control the dialogue. One of the attributes of a good salesperson is to be cost effective by performing the tasks on time so that productivity and returns to time ratio may increase. Persuasiveness is another important cognitive attribute of the salespersons that attracts customers and helps them stay with the firm for long term.Sales managers should not only try to hire only those who will become top performers, but also nurture their salespeople with potential [11].

Except in some sectors where high-volume recruitment is the norm, organizations may face unexpected requirement situation for an unusually large number of new recruits at short notice. Such human resources situations are generally managed by the outsourcing specialist within the firm or an external recruiter is contacted to manage the point-to-point resourcing program. Recruiting people in executive and senior management levels demands acquisition of 'high-potential' recruits. In order to search highly potential recruits, firms need to look for candidates havinglong standing successful track record. Human resources department of such firms may start their search in association with wide range of 'search and selection' or 'headhunting' consultancies. Such recruits may be identified typically form long-standing relationships with client organizations. Certain organizations with sophisticated recruitment practices have identified that there is a strategic advantage in outsourcing people. In this process complete responsibility for all workforce recruitment may be delegated to external agencies. In such arrangements the external recruitment services providers may not only physically locate talents or 'embed' their resourcing team(s) within the client organization's offices but also collaborate with the senior human resource management team in developing the longer-term resourcing strategy and human resources plan.

RECRUITMENT AGENCIES

The manpower supply industry operates within five types of agencies. The recruiting agencies in the industry channel candidates into the hiring organizations through formal application process and layered screening. Categories of manpower supply agencies play a significant role in the coordination of human activities. Providing manpower by the private agencies develop recruitment strategiesaccording to institutional priorities It is argued that multi-functionality of the employment agency attracts talentsin reference to the derived manpower planning, employment benefits, training opportunities and workplace environment. This functional value of employment agencies plays a significant role in the services provided to clients and placement of candidates. As a general rule, the agencies are paid by the companies, not the candidates [12]. The recruitment agencies provide the following services to their clients in processing the applications of prospecting candidates:

- Classifying the prospects
- Ice Breaking/De-freezing
- Educational issues
- Referrals and recommendations
- Experience
- Strengths and weaknesses
- Relationships.

Firms seeking the services of private agencies in supplying suitable manpower should categorically filter the prospective candidates to fit into specific areas of sales. They may be pooled in two broad categories including line salespeople, who are generalists and product specific salespeople who demonstrate specific skills. It is necessary for the firms to develop exercises to defreeze the perceptional blocks, preconceived notions and external influences against products and services, clients and the recruiting and serving firms. It is necessary to check self-enhancing bias in personality, subjective dissatisfaction, and negative perception among the recruits as they are risk averse in performing tasks. Highly skilled candidates with higher subjective happiness as compared to those with indifference or negative perceptions show higher potential of accomplishing tasks [13]. Educational issues concerning qualifications, quality of education, and job experience of prospecting candidates need to be managed carefully by weighing their value against experience. Among many, the main challenge for companies is to rethink and restructure their recruiting activities in order to attract competent employees as well as balancing between educational credentials and functional skills.Such a challenge broadens the scope of recruiting and requires developing multifaceted recruitment strategies. The employment agencies should also aim to reveal how nontraditional (such as corporate involvement in public education) and improved traditional recruiting strategies can enable firms to continue to attract and retain capable employees from different genders, marital statuses and ethnic backgrounds. However, the strategies that proved effective in the past will continue to work in the future as long as companies are willing to adapt their messages and their workplaces to the needs and demands of this changing workforce [14].

Effectiveness of recruiting agencies has to be primarily assessed by the firms hiring these agencies examining the turnover of employees provided by recruiting agencies, job survival rates and job performance. As reported in most narrative reviews and all five quantitative reviews, referrals by current personnel, in-house job postings, and the re-hiring of former employees are the most effective tools followed by the recruiting agencies. Walk-in recruitments have been slightly less effective, and the least effective sources are newspaper advertisements, school placement services, and low profile employment agencies [15].

Traditional Manpower Suppliers

Manpower suppliers operating with traditional values are known as employment agencies, which function historically from a fixed physical location unlike Internet services providing agencies. In such agencies candidate visit a local branch of the agency for a short interview and an assessment before being taken onto the candidate pool of the agency.

Recruitment consultants then work to match their pool of candidates to their clients' open positions. Suitable candidates are short-listed and put forward for an interview with potential employers on a temporary or permanent basis. Settling compensation to agencies takes several forms, the most popular:

- A contingency fee paid by the company when a recommended candidate accepts a job with the client company which ranges typically 20-30 percent calculated on the basis of first-year base salary of candidates;
- An advance payment to recruiting agencies is made by the firms towards compensation for high level executive search/headhunters, and
- A pre-negotiated hourly or daily fee, in which the agency is paid by the firm and in turn recruiting agencies pay the applicant as a consultant for services as a third party. Such recruitment contracts can be terminated without any notice by either party. In recruiting salespeople, sales consultants in direct selling, sales supervisors and sales service providers are preferred by the firms to be recruited in this manner.However, some contracts allow a consultant to transition to a full-time status upon completion of a certain number of hours with or without a conversion fee.

Talent Search Organizations

The Talent Search Organizations (TSO) function as third-party recruiters that seek out candidates on demand of client firms. These organizations are generally more aggressive in searching best candidates than in-house recruiters. They use advanced sales techniques, such as initially posing as clients to gather employee contacts, as well as visiting candidate offices. In order to attract good candidates TSOs need strong brand equity of client firms. A desirable brand affiliation in conjunction with inspirational leadership appeals to talented young candidate who show potentials and interest inan accelerated career trackandacquisition of skills and experience at a fast pace. The TSO can successfully search talent if the client firm has a meritocratic culture and value. For instance, talent strategies that work at the home company may probably need extensive tailoring to gain competitive advantage. In some job markets proficiency in English may impede spotting talent. It's critical to develop a core of local talent and to embrace and leverage diversity. In the talent race, a local company that creates genuine opportunities and exhibits the desired cultural conditions will often win out over a Western multinational offering higher pay [16].TSOs generate their own lists of candidates and sometimes also buy databases of prospects. They may prepare a candidate for the interview, help negotiate the salary, and conduct closure to the search. TSOs hold memberships in industries and trade associations. They operate on small scale with high margins on the candidate placements.

ABC Consultants Private limited is a placement company based in India, which is engaged in providing recruitment services for the client firm in solving the middle and senior management talent needs. The company recruits across various industry segments for multinational corporations as well as leading Indian business houses. In the past four decades the company has helped in shaping the careers of over 85,000 professionals and have partnered with more than 350 multinational companies to set up and staff their operations in

India. ABC Consultants has a large organization network including 9 offices spread across 7 major Indian cities and an international presence in the United Kingdom and Dubai. The company operates through domain-specialist teams spread across the country, providing high quality permanent hire services. The company attends to generic corporate functions like finance, human resources and legal cut across all industries and is increasingly holding strategic importance. The company has developed specialized competencies in these functional areas. Its effort is to bring together both industry and functional expertise to successfully close critical assignments. The human resources selection process of the ABC Consultants is based on a structured and systematic approach, which is proactive, detailed and target oriented. This process begins with understanding the client's needs and goes all the way to post recruitment follow-up with both client and candidate. The key principles of our search methodology focus on understanding the client organization with regards to existing business activities, organization structure, culture and future growth plans. The placement consultants map position specific competencies including job title, deliverables, principal accountabilities, and other relevant details. Developing a sourcing strategy the company draws up a target list of appropriate industries and companies from which to recruit and utilizing the corporate network, along with the database to develop a list of most promising candidates within these companies. This is presented to the client as a comprehensive list of people the company intends to approach. Upon establishing the mutual interest between ABC Consultants and Client Company, the process of pursuing specific candidates is drawn. Candidates are contacted and apprised of the opportunity for eliciting their interest in the position. Consultants of the company carefully evaluate the candidates with focus on determining a profile match with the job, their career aspirations and suitability in the context and culture of the client organization. In addition to personal interviews, the candidates are subjected to psychometric assessment tools such as Thomas Profiling and SHL Testing. Detailed Reference Checks on candidates selected for hiring is administered before the offer is formally extended. In the mean time negotiations for the compensation and pattern of requirements are carried out with the client firms. Accordingly, the candidates are briefed to join the position ensuring smooth transition [17].

Firms can fill the talent gap in their company through various lost cost and high benefit strategies instead of engaging external TSOs in the recruitment process. It is suggested that by talking to employees before they make their retirement plans, firms can optimize chances of finding mutually beneficial work arrangements, such as flexible hours, telecommuting, part-time or temporary work, and job sharing. In this way smart companies build loyal workers look around for new challenges to contribute their time and talent [18]. According to a survey conducted, employees of age 55 years and above have the highest rates of job satisfaction and pride in their work. Many retirement-age workers also feel a desire to serve others and give back to the community.

Niche Placement Agencies

Niche Placement Agencies (NPA) are the specialized firms in talent search for professional and scientific sectors like health care, marine engineering, energy production, petroleum, and aviation industry. Because of their focus, these firms produce superior results and drive resources for developing specific skills. The specialization of NPAs in staffing allows them to search for better candidates within predefined demographic segments, develop large proprietary databases and cater to the demand of client firms in short notice. These Niche firms are focused on developing long term relationships with their candidates.

In-house Recruitment

A larger number of employers tends to undertake their own in-house recruitment, using human resources department of the firm. These operations are supported by the front-line hiring managers and recruitment personnel who handle targeted functions and prospecting candidates. In addition to coordinating with the external agencies mentioned above, in-house recruiters may advertise job vacancies on their own websites, coordinate internal employee referrals, work with external associations, trade groups and focus on campus graduate recruitment of reputed institutes. While job postings are common, networking is by far the most significant approach when reaching out to fill positions.

Hybrid Placement Firms

These firms are the new hybrid firms in the manpower placement industry attempting to combine the research aspects (discovering passive candidates) of recruiting, training, and developing a strategic fit between the candidates and job designs to hire them for their clients. These firms provide competitive passive candidate intelligence to support company's recruiting efforts. However, recruitment of people in sales organizations should be taken as an asset building exercise in the firms. Companies should balance develop-versus-buy decisions on human resources by using internal development programs to produce optimum needed talent. However, resource gaps can be filled in with outside hiring. Firms can reduce the risks in forecasting the demand for new positions by sending smaller batches of candidates through better modularized training systems. This process may be similar to the practices of manufacturers, deploying people in just-in-time production and operation lines. Sales organizations can also improve their returns on investment in development efforts by adopting innovative cost-sharing programs and seek to protect their investments by generating internal opportunities to encourage newly trained managers to stick with the firm. Taken together, these principles form the foundation for a new paradigm in human resource management and a talent-on-demand system for the competitive and fast growth business firms [19].

It has been argued over time that more than hiring human resources, retaining them in the firm is a difficult proposition. This appears to be true for the sales firms where every recruit builds his own contribution in terms of market share, client relationship, prospecting strategies and contributing revenue to the firm. If a salesperson leaves the firm, his walking away is considered not only as a gap in the human resources but also loss of market share and of his personalized efforts that contributed to the firm.Fast-moving markets require fast-moving organizations that are continuously refreshed with new talent. But none of the firms would like to experience their talents to leave but as said earlier, when a good employee quits, the business is adversely affected. However, a market-driven approach to retention is based on the assumption that long-term, across-the-board loyalty is risk averse. By considering the long-term association with employees, firms need to develop useful retention mechanisms. It is possible to redesign jobs to reduce employee turnover and promote loyalty through particular work teams. Firms can team up with other companies to offer cross-company career paths and when there is no effective way to prevent attrition, firms may resort to the alternative of outsourcing, strengthening recruitment, standardizing jobs, cross-training employees, and organizing work around short-term generalized projects [20].

RECRUITMENT PROCESS

Firms apply diverse stages in recruiting salespeople. Commonly firms rely on interview process, however some firms also measure attitude of prospects through group discussion process. Large companies like pharmaceuticals, computer electronics, and heavy engineering machinery manufacturing companies propose a group decision support system (GDSS), with multiple criteria to assist in recruitment and selection processes of human resources. In a two-phase decision-making procedure various techniques involving multiple criteria and group participation are then defined corresponding to each step in the procedure. The purpose of GDSS is to determine the wide scope of personnel characteristics and to measure the degree of consensus in the prospect to work in sales teams. Firms use both traditional and network-based PC system with web interfaces to support the group discussion activities [21]. Firms should lean towards evaluating the individual and group behavior of prospecting candidates during the selection process in reference to the following attributes:

- Individual and group behavior
- Leadership and motivation levels
- Commitment attribute
- Work culture management traits
- Team spirit, individualism, indecisiveness, subordination, creativity, diplomatic skills in managing conflicts, and
- Problem solving capabilities.

With very high number of job application in sales area the competition for the selection of suitable employees is extremely keen. Firms that strive to maximize candidates' job offer acceptance should realize increased competitive advantage through more effective recruitment programs. It is clear that applicants are interested in job attributes andpay schemes, but in the current seller's market, pay and job enrichment levels may be considerably depreciated. The recruitment interview process should emphasize organizations' strategic efforts to maximize the acquisition of top talent in competitive markets through interviewer persuasion and active promotion of organizational reputation [22]. There are various challenges for the firms in developing effective recruitment process as discussed in the Figure 7.2.

The cultural differences among firms and employees have become increasingly critical with growing effects of globalization. In the area of recruitment, a cross-cultural knowledge base is vital as the demand for result oriented employees, poses escalating challenges to effectively attracting desirable applicants. However, the effectiveness of recruitment practices across cultures has been a subject of continuous research in reference to developing a common framework of recruitment exploring how organizational values influence the effectiveness of recruitment practices in different cultural contexts. It has been argued that cultural values may moderate relationships between recruitment practices and recruitment outcomes across all phases of the recruitment process [23]. The total quality management (TQM) oriented firms identify the recruitment process in an "explicit TQM" (where applicants were specifically told that the firm was a TQM employer) way and "inferred TQM" (where TQM was inferred on the basis of specific reported practices) pattern. It has been observed in a study that there were significant differences between TQM firms (regardless of classification) and non-TQM firms, and between explicit TQM firms and inferred TQM firms [24].

User needs	Client, candidates, contractors, other involved parties
Security	Data encryption, external to internal, integrity
Legal issues	Data protection, employment laws, discrimination laws,
Process	Flexibility, accommodating direct contact, specific processes
Language	Translation and meaning, maintaining a single interface
Interfaces	Search / retrieval, usability, point and click, on screen help
Geography	Time zones, user support, training
Compensation	Competitive, isolated, subjective
Data types	Back office systems, client systems
E-recruiting	Browser based environment, utilising existing systems, data management

Figure 7.2.Major Challenges in Developing Human Resources.

The recruitment process should begin with job analysis to document the actual or intended tasks to be performed by the incumbents. This information should be detailed in a job description, which provides the transparency on the recruitment process. Often firms develop unclear job descriptions presenting a historical collection of tasks performed in the past. These job descriptions need to be reviewed or updated prior to a recruitment effort to reflect present day requirements. The recruitment process should be initiated with an accurate job analysis and job description to successfully converge the goals of the firm and prospecting candidates. Many job descriptions are too narrow in their scope and do not allow for future changes in programs or work assignments. A good job description should be broad enough to allow for occasional as well as routine duties, and flexibility in the program. Job descriptions, unlike job analyses, should also be written so as to apply to many varied job locations or situations. Job description should comprisethe rationale of tasks assigned, job descriptions as guiding manual to perform tasks in the workplace, working process, job analysis, and monitoring and evaluation strategies [25]. It has been observed in a study that some firms develop provocative job description to attract high talents and such practice is strongly related to organizational performance. This relationship appears to be strongest to the extent that companies maintain human resources information systems and involve in the recruitment process and strategic manpower planning. It is also common that the firms relying on competency-based attributes of employees attempt to develop performance linked job analysis. However, a company-wide policy of job analysis is an important source of competitive advantage for the attention of HR professionals, line managers and top management [26]. The factor influencing the job description of salespeople is exhibited in Table 7.1.

Table 7.1. Factors Influencing Job Description for Salespeople

Sales related tasks	Administrative tasks	Career Path and Performance
New vs. established account selling Selling through distributors Entertaining customers Decision making power Physical activities Technical knowledge Territory management Developing sales proposals Individual vs. team selling One time vs. systems selling Type of prospects/customers multilevel vs. group selling Customer relations and concept selling	Reports to management Customer service and training Sales promotion Educational seminars Collecting receivables Marketing plans Negotiations Travel	Compensation plan Promotion timing Earnings potential Promotion leaders Sales quotas Performance levels Key performance indicators

In the recruitment process sourcing involves advertising and employing personalized candidate search. Advertising for finding potential salespeople or sales managers is a common tool to drive the recruitment process. Firms advertise the positionsoften on multi-media platform such as the Internet, general newspapers, job advertising newspapers, professional publications, window advertisements, job centers, and campus graduate recruitment programs. Candidate search efforts induce relevant talent who may not respond to job postings and other recruitment advertising methods. The personalized search efforts for potential candidates is known as passive prospects approach that results into enlisting the prospecting candidate and soliciting their interest for the announced positions.Online recruitment websites can be very helpful to find candidates that are very actively looking for work and post their resumes online, but they will not attract the "passive" candidates who might respond favorably to an opportunity that is presented to them through other means. Also, some candidates who are actively looking to change jobs are hesitant to put their resumes on the job boards, for fear that their current companies, co-workers, customers or others might see their resumes. The emergence of meta-search engines allows job-seekers to search across multiple websites. Some of these new search engines index and list the advertisements of traditional job boards. These sites tend to aim for providing a "one-stop shop" for job-seekers. However, there are many other job search engines which index pages solely from employers' websites, choosing to bypass traditional job boards entirely. These vertical search engines allow job-seekers to find new positions that may not be advertised on traditional job boards, and online recruitment websites. Various stages of recruitment process that is commonly used by the sales oriented firms are exhibited in Figure 7.3.

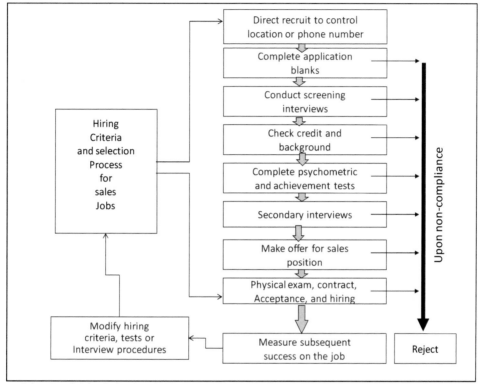

Figure 7.3.A Model for Selecting Salespeople.

Suitability for a job is typically assessed by looking at the skills of the prospecting candidates besides acquired qualifications. These attributes of the prospecting candidate can be examined during the process of conducting personal interviews. However, recruiters make some common mistakes as detailed below while managing the personal interviews:

- Failure to establish rapport with the candidate
- Lack of interview plan
- Insufficient time
- Poor listening and interruption
- Personal bias
- Direct questions
- First impressions

Business management software is used by many recruitment agencies to automate the testing process. Many recruiters and agencies are using an applicant tracking system to perform many of the filtering tasks, along with software tools for psychometric testing. Recruitment should be followed by a well-planned induction process for the new employees to become fully operational and get integrated with the workplace culture. Induction process for salespeople is to be programmed with a practical approach to familiarize them with the work culture of the firm, client attributes, logistics and operational process, financial boundaries and working in teams.

In recent years, many concepts for organizational excellence have put greater emphasis on employees making the recruitment process more goal oriented. This refers to the need for increased participation human resource personnel in the recruitment process for change on one hand and drives more intensive approaches on choosing appropriate methods of selecting potential employees to augment the productivity and organizational growth. In view of low budgetary provisions, firms focus on single aspects, such as recruitment process design and evaluating various human resources concepts by short-term cost factors, which does not yield encouraging results in many cases. Therefore, more holistic concepts are needed to realize a sustainable success in the organization. Among others factors, demands for the development of organizational culture are crucial to acquire and retain potential employees [27].It has been argued that to build a highly competent organization of highly skilled employees, firms may not simply engage human resources in hiring the best and brightest candidates and then turn them loose into a Darwinian competition of survival of the fittest. Organizations need to provide and maintain the right environment for these employees to flourish, which indicates avoiding a number of common pitfalls, such as often mistakenly believing that one individual can single-handedly turn around an entire department or organization, and overestimating the importance of pay as compared to the talent. Firms often commit mistakes by providing employee autonomy to achieve goals, which makes them carry overload of tasks. Such tendency of achieving goals in short span may cause various functional risks. However, achievements under such practices mayappear by chance and not by design. Many firms may also be biased assuming that high performers need efficient colleagues throughout the organization in order to continuously improve talent in the organization [28].

The aim of the human resources function in reference to management and development is to ensure the availability of competent, motivated and learning employees to facilitate the achievement of its business objectives in the organizational context. The human resource function has been quite successful in performing its role in terms of developing capacity, knowledge, attitude, and skills of employees besides performing recruitment activities. However, in the recent past a shift has been taking place in the expected model role from both the human resource function and the human resource manager. This new role of human resources department is assumed to serve as active partner along with other functional departments like production, finance, marketing, and sales. This has implications for the human resource managers and practitioners including developing human capital through head-hunting, recruitment, training and career advancement plans. The implications will also be towards investment in human resources to enhance contributions to business results [29].

OUTSOURCING SALESPEOPLE AND ROLE SPECIFICATIONS

Outsourcing salespeople has become a necessity for most companies in today's competitive environment, which has resulted in reducing the cost of human resource management and implementing higher control measures. Growing competition and choice in the outsourcing salespeople has gradually altered the way organizations select and contract human resources. In the conditions of economic recession, organizations are even more concerned than before about the value proposition and risk involved in signing over their information systems and technology activities to third party organizations. A study suggests that although outsourcing human resourcesoffers new business opportunities for salespeople, much of the outsourced work continues to be in areas of low risk and low value to client businesses. The main challenge for outsourcing salespeople is therefore to develop strategies that facilitate low cost, high value contracts, without incurring additional risk. From the perspective of the sales firm, being able to identify and assess the value and risks of the different outsourced salespeople business models helps them to make better choices during the decision making process when selecting salespeople [30].

Outsourced salespeople identify appropriate customer from different markets, facilitate the dialogue with the firm and bring customers together with the firm to process the sales. Sales promoters support interactive learning processes and bridge the gap between the customers and firm [31]. The bargaining power of firms increases with outsourced salespeople who stimulate the demand for products and contribute to the enhanced sales at retail outlets. It has been observed that pull effect for the brands supported by the sales promoters increases at the retail stores as customers gather the pre-buying information from outsourced salespeople [32]. There are three common outsourcing contracts, which include fixed-price, cost-plus, and gain-sharing. The firms should address issues of when to outsource and which contracts to select for developing effectiveness in sales functions.Outsourced contracts generate organizational effectiveness and assists managers in developing cost effective sales programs, time bound achievements and managing customer-centric negotiations [33].

Firms that introduce new sales territories frequently or sell less familiar brands in the marketplace engage in pushing sales by working with customers and outsourced salespeople. A broad set of process standards makes it easy to determine the selling strategies for the outsourced teams in a given market. Such standards help firms evaluate the costs versus the benefits of outsourcing. Eventually these costs and benefits will be visible to buyers in terms of satisfaction and develop demand pull effects [34]. Firms must consider and alter four factors over time that include the differing roles that internal salespeople and external selling partners should play, the size of the sales force, its degree of specialization, and the way salespeople apportion their efforts among different customers, products, and activities. These variables are critical because they determine how quickly sales forces respond to market opportunities, influence salespersons' performance, and affect companies' revenues, costs, and profitability [35].

A new strategy among the multinational companies engaged in competition of identical products like carbonated drinks and purified bottled water has emerged which can be explained by two-dimensional framework in reference to manufacturer's capability levels (low, medium, and high) and outsourcing sales force to enhance market coverage. A match between distributor capability levels and outsourcing sales force needs is posited to be the key to a sustainable relationship between salespeople and their customers [36].However, some studies reveal that relationship of customers with the outsourced salespeople may not be sustainable in the long run. There is a common set of key contractual elements that exist between most outsourcing contracts. The nature of client-salespeople relationship functions as the key in the selling process for new products. In this process the perceptional problems with customers can greatly devalue the customer-salespeople relationship and brand as a whole [37].

Outsourcing has become a critical strategic decision that can allow organizations to develop and leverage the capabilities required to compete in today's global business environment. In particular, the outsourcing framework provides a mechanism for understanding which services should be kept internal and which should be outsourced based on both organizational capability and opportunism considerations [38]. It has been observed in some studies that clients, who intend to raise purchase order at a short notice, generally lean towards the advice of the salespeople. However, outsourced salespeople at times fail to recognize that what influences buyers' satisfaction and consequently do not effectively develop the cognitive drive to stimulate buying decisions. Hence, they need to vigilantly manage the quality of negotiations by developing adequate customer involvement in the buying process [39]. The outsourced salespeople should be driven towards continuous sales learning process with respect to the value creation chain and measure sales performance in the marketplace. In the growingcompetitive markets the large and reputedfirms are developing strategies to move into the provision of innovative combinations of products and services as 'high-value integrated solutions' tailored to each customer's needsthan simply 'moving downstream' into services [40]. The sales attractiveness of outsourced salespeople may comprise personal selling skills, use of advanced technology, innovativeness, and knowledge of post-sales services associated with effective selling process, which contribute in building sustainable customer values towards making buying decisions [41].

The growing pressure to reduce costs and improve efficiency induces many organizations to undertake shared services initiatives. This consolidation and streamlining of common business functions is also known as in-sourcing, in-house services, business services, or staff services. While adoption of a shared service structure is viewed by many as an appropriate strategy to pursue, most companies still struggle to devise optimal strategies and to generate adequate returns on sales for their projects, because none of the approaches that are commonly adopted is recognized as universally effective [43]. Practically clients react favorably to leisure sales campaigns of goods and services. However, customer value plays a decisive role in the buying process.It may be argued that individual consumer behaviorto a buying option is a function of perceived product value, value for money and competitive advantage. The consumer response is affected by search cost associated with making a buying decision [43].

Creating value through outsourcing has emerged as a popular competitive strategy for firms in various industries. In order to survive in the domestic and international marketplaces, firms, especially in developed countries, are seeking opportunities offshore. Offshore outsourcing has emerged as a popular competitive strategy and emerging markets have become increasingly attractive locations. As firms in developed countries continue to face enormous challenges to sustain competitive advantage, outsourcing to emerging markets is becoming an increasingly important source of business renewal and corporate transformation [44]. The driving forces behind outsourcing can be a combination of financial problems, a need for clear focus on core competence, and cost efficiency considerations. Firms opt for outsourcing human resources because the outcome is more successful while hiring people from an outside organization is based on strategic decisions including core competence and cost efficiency considerations, rather than being emergency action prompted by financial problems. This strategy contributes to the performance of the company as it develops strong resource base and a customer network in the long-run [45].

DEFENSIVE DOWNSIZING OF SALES FORCE

Salespeople are assets to the organizations. Firms generally go for mass recruitment of salespeople when the products and services of the company have rising demand or company is planning for rapid business expansion. On the contrary, firms tend to downsize their human resources during lean sales period or economic recession. The downsizing practice has a strong relationship with survivor attitudes and the practices survivors use in managing their careers. Specifically, indirect downsizing methods tend to avoid negative attitudes and promote career management strategies that are more beneficial to the organization [46]. Strategic downsizing without affecting the morale of employees should be considered by the firms. Such strategy includes deciding to downsize, planning the program, making the announcement and implementing the program. In this sequel, it is argued that downsizing programs are not just about 'doing more with less' but also to provide opportunities to build a sense of trust and empowerment between managers and employees, which can provide significant benefits going forward. If the firm has downsized recently, it is import for the firm to pay attention towards managing a bunch of survivors. Those employees who are not affected should be taken care of in boosting their morale for sustainable performance [47].

During the period of economic recession it is necessary for the business organization to make broad and striking changes in the organizational structure to reduce the cost of hiring people.As market recovery and economic growth improves, human resources productivity should aim at a higher level of market-oriented flexibility. Despite the connection between the seniority-based salary system (SSS) and the permanent employment regime, more effective results can be expected in the former than the latter. The SSS measure of human resources would encourage existing staff of the organization to take on additional responsibilities, and career path of employees during the crisis period remains sluggish. Under such staffing conditions organizational and individual benefits on professional development and growth need to be established in reference to lifestyle, mental capacity and emotional growth [48]. In addition, such organizational streamlining would also suspend the hiring of new staff. These measures can be considered as proactive strategies to combat with the anticipated slump in the financial organization. Alternatively the financial organizations may strategically implement the downsizing of personnel in anticipation of projected recession in the business. Human resources division may plan the manpower downsizing process in the following stages:

- Downsizing should be considered as the need of the hour and should never be made a practice to curtail the expenditure. Also downsizing should be part of a clearly defined, long-term vision that fits into the company's overall strategic plan.
- Before implementing the downsizing plan, the marketing and sales organizations should form a cross-functional team to represent all stakeholders' interests. At the announcement stage, senior managers should explain the necessity of the downsizing and how it helps the firm in the long term. Management should communicate frequently and be open and honest.

When looking for ways to cut costs, most human resources managers reach for the head-count hatchet, and usually get approval of the top management. Human resource managers should reconsider costs during the period of economic recession. They need to keep a long-term view and strive to maintain the loyalty of employees and customers.

Another approach to manage the human resource during the period of economic crisis in a financial institution is to pursue *defensive downsizing* rather than strategic restructuring, as a result of both internal and external constraints on restructuring strategies. Defensive focus advocates short-term efficiency with increased emphasis on training and restructuring job descriptions of existing staff in order to augment productivity [49].The process of defensive downsizing envisages two complementary core values that include personal initiative and the capacity for collaboration that are essential for organizational survival. Some additional skill sets need to be developed in reference to prevent mental agility, reinforce personal visibility, and team facilitation. The work culture in the organization should be streamlined in reference to 3Ts- task, thrust and time [50]. The concept of defensive downsizing emphasizes that organizational downsizing process should not be governed by retrenching employees in abundance, but should be guided by retaining those most valuable to the organization. However, downsizing strategy aimed towards organizational survival has both merits and demerits. It is perceivedby the employees ofsome downsized organizations that retrenchingexcess employees and offering outplacement to those who lost their jobs goes some way to improving the financial performance of downsized firms. This may suggest that if downsizing is necessary then attention needs to be given to how the process is implemented in order to maximize the financial return [51].

The 3-T determinants lay the groundwork for success inside the new organization. The typical work culture of 3-T power-grid may be described as synergy of task (commitment), thrust (driving force) and time (punctuality). These attributes of the Japanese distinguish them from most of the existing work cultures across the countries in the world. This has reflected into the material culture (technology and economy) of Japan toward continues improvement (Kaizen). In fact, Kaizen is a social culture, which has been later adopted by the Japanese organizations.

Customer centered companies should be engaged in outsourcing people to promote bankable products during the situation of financial siege. Though there are some perceived risks, managers in a widening range of financial institutions are exploring the scope of sales outsourcing. This strategy of promoting sales has helped financial institutions, which have undergone downsizing process, cut costs on salaries and other overheads. Outsourcing for promoting bankable products and services sharpens the corporate core competencies.In outsourcing human resources for augmenting the overall returns to the convergence of sales promotion, investors' perceptions and confidence of investors need to be considered as critical tools.

TRAINING SALESPEOPLE

In the real market conditions performing sales is more challenging than it appears during the stage of planning. Firms need to place a strategic emphasis on disseminating knowledge among employees and become a learning organization. Numerous training programs of organizational change, client management, competitor analysis, and profit center development and renewal must be undertaken to keep firms competitive. This requires ongoing training of all employees to be successful in selling in today's global marketplace. The training units of the sales firms should be well equipped with human resources, applied learning tools and computer based learning programs. External consultants and professional staff are often called upon to provide training to the employees; however, many organizations are turning to their own employees as an effective, lower cost alternative [52]. The common objectives of training salespeople laid by the firms include:

- Increase in productivity
- Improve morale
- Lower rate of employment attrition
- Improved customer relations
- Better management of time and territory
- Confidence in decision making
- Capabilities for situation handling and strategy building, and
- Satisfaction in performing job tasks.

There are several benefits in using this strategy, but significant support must be provided to realize these benefits. To help employees adopt new work culture, organizations can develop various orientation training programs to provide continuous learning. One of the most needed areas of training for salespeople is towards developing their negotiation skills. Salespeople are advised by their superiors to read selected books and newsletters to improve their negotiating skills. Many also take time out of their busy work schedule to attend classes and training programs focused on negotiation. Trainers pass on interesting concepts and war stories about great negotiations. However, when the executives return to their offices, the ideas and stories learn in various training modules rarely affect their actual negotiation behaviors. Some companies like Motorola have considered training as mandatory for salespeople and is delivered to them as a part of their job design. Motorola began a program of skills training in 1980 to teach its work force statistical process control and other quality improvements techniques. After many problems, Motorola changed the program from specific techniques to a comprehensive graduate work in computer-integrated manufacturing. The company developed partnerships with local schools and colleges to provide elementary education as well as business courses. In view of the quality of training programs offered and advanced management pedagogy followed by the company, academic associations have conferred the company the status of Motorola University. This pattern was followed by other companies like AXA University which is committed to improve the quality of services by imparting training to the employees through continuity programs. Similarly 3M University has focused on imparting training to its employees associating various universities [53]. These companies aim at developing a salesperson with following attributes by imparting quality training:

- Phases of business development
- Core competencies
- Quality of salesperson
- Account management
- Consultative skills
- Sales leadership.

Competitive success depends on learning and a good training enhances the scope of learning. However, most people, including professionals in leadership positions, are not very good at it. Learning is a function of how people reason about their own behavior. Yet most people engage in defensive reasoning upon confronted with problems and avoid examining critically the cause and effect issues to fix the problem. Companies need to analyze various reasoning patterns of managers and employees and develop the focus of continuous improvement efforts [54]. There are various types of sales force training programs are organized by firms covering the following aspects:

- Territory management
- Corporate sales policies
- Managing people in the sales play
- Sales decision making
- Strategic selling
- Interactiveselling skills.

An effective technique, which is used for years for training salespeople in a natural and friendly manner, is role playing. Employees like role playing and it has dramatic appeal in learning. In role playing based training, salespeople retain more through these sessions than they would from manuals or lectures.Managers of organizations, often in coordination with personnel departments, create training programs for their staff. The emphasis in sales organizations is to train staff to use behavior modification techniques, especially to reinforce appropriate salespeople-client behavior [55]. With the rapid development of the Information Communication and Technology (ICT) there is an increasing demand for distance training programs for salespeople on virtual platforms to learn without affecting their routine field work. Consequently, the web-based applications course is designed by many sales firms and imparted to the salespeople through their regional training centers. Some online virtual training programs like Negotiate and Deal Environment (NADE) system connect the learning place between the technical and managerial skills. NADE is a role-playing virtual environment for teaching the art of creating secured web-based systems for information processing and backend applications. The system provides an environment where group interactivity and strategic planning are key success factors [56]. It is observed that the multinational firms encourage designing and delivering continuity sales training programs in order to improve the knowledge of salespeople on competitive best practices and policies, sales presentation and communications skills, sales objectives, product information and technical selling skills, and customer relation skills [57]. Different types of sales force training programs organized by firms include:

- Individual training
- In-house programs-interactive
- External sessions-short duration
- Home assignments
- Applied training
- Team training
- Situational training
- Sales orientation programs-repeat courses on basic sales themes.

Some recent studies indicate that there is a wide divergence between global and local practices followed by the firms in offering training to the salespeople. It appears that global firms perceive that sales training leads to greater sales force performance and view sales training as a part of their strategic marketing plan. Global firms also employ a more formal training curriculum, focus on soft competencies of salespeople, and build an established training culture. Conversely, local firms rely upon On-The-Job training (OJT) and appear not to understand that sales training programs can be a source of competitive advantage [58].

Firm that leverage sales training processes attain a distinct advantage over their competitors by developing customized training programs for their sales force. It is increasingly critical that training programs are optimized to ensure that new skills translate into immediate performance. Large and experienced firms spend a substantial proportion of their time training and supervising salespeople, delivering knowledge on mapping consumer preferences and sales learning skills. This knowledge not only enhances the effectiveness of salespeople, but also benefits salespeople in driving new ideas on acquiring new customers and retaining the existing customers. In this process it is necessary for the firms to make an assessment of trainees needs to cover many different domains, including; knowledge, clinical competence, communication skills (written and verbal), procedural skills, teamwork, and professionalism. Implementing multiple methods of delivering training would increase the scope of learning [59]. In designing and conducting sales training programs firms should consider the strategic fit for the salespeople on the following lines:

- Need assessment
- Job description and actual tasks, previous programs and gaps thereof, specific issues
- Setting objectives of training
- Review needs, duration of the program, overall learning and corporate gains
- Developing and implementation of program
- Training methods, applied learning
- Evaluation and review
- Training kit, follow-up system, gains
- Cost of training.

Designing sales training programs should include customer insights and competitor learning. For example, customer research indicates that training to salespeople should be developed in reference to buying economics, competitive advantages, buying trends, decision factors, mystery shopper strategies in leaning competitor, and customer preferences. Most customer centric firms develop sales excellence with focus on the customer demands and customer satisfaction. Salespeople need to be prepared to develop strategies for excellence in integrated sales and services, quick sales processes and customer relationships. In a competitive marketplace salespeople needto be trained in creating excellence in customer experiences, as it is the most significant way to create competitive advantage in the market. Sales performance should be driven by a broader focus on customers' expectations and multiple ways of measuring customers' satisfaction should be delivered in the sales training program. The sales training programs should support the sales strategydevelopment process with a broad focus on direct-to-customer sales strategy. Customer satisfaction measurement based on surveys and deriving results to continuous sales improvement are also important issues to be covered in the sales training programs [60].

Global Knowledge Inc. is the worldwide leader in IT and business skills training. The training is delivered to the employees of the client organizations via training centers, private facilities, and the Internet, enabling customers to choose when, where, and how they want to receive training programs and learning services. The company provides technical training services to the employees of clients such as travel reservations firm Sabre, ACTS Retirement-Life Communities, and the Netherlands Ministry of Defense. Specializing in networking systems and telecommunications, Global Knowledge offers more than 700 vendor-authorized and proprietary courses covering the products of specific manufacturers, including Microsoft and Cisco Systems. The company also provides custom integration and training services and outsourced education management services. Global Knowledge operates in about two-dozen countries around the world. Customers rely on the company to deliver superior IT and business courses and certification preparation with exceptional curriculum, hands-on labs, and experienced instructors. Courseware is based on applied instructional design methodology that promotes rapid skills development. The training programs delivered by the company are designed with focus to improve capabilities, resources, and commitment [61].

Additionally, training programs must incorporate employee research to better understand both value-creating activities and obstacles to delivering them. In delivering applied training programs activity focus is a critical differentiator between top performing sales representatives and those with average performance. Only by understanding how sales people spend their days and what non-sales activities are imposed upon them can sales leaders begin to better inform, align and promote the right sales behaviors. Training programs built around customer and employee research hold the power to transform sales teams by creating effective business allies. Consultative selling is an educational activity that is part of the process of interpretation of products and services being offered to the clients. Most companies recruit untrained salespeople to handle consultative sales for short term gains with low investment on training. However, training to develop consultative sales for high value and high technology products is now common in most emerging firms. Consultative sales training is an adult education activity, but much training is competency-based with an emphasis on knowledge transmission and skill acquisition. Delivering a good training program should lead to change, not only in terms of knowledge and skills, but also in attitudes and behavior. It is argued that good applied sales training should alter how salespeople perceive the problems of the customers, offer solutions, stay proactive to customers, share knowledge and experience, and develop customer-centric attitudes and behavior [62].

Sales firms that want to maximize training effectiveness should understand that individuals learn differently and one size does not fit all training methods. Training programs need to cater to a broader mix of learning styles based on learning profiles of their unique sales force, and develop those trainings to get the most out of their investment. By providing variety of programs, firms can make an impact across a much wider audience and create a more positive sales training experience that translates into sales growth. Training for individual preferences requires an in-depth evaluation of current programs coupled with an inherent understanding of how the sales force learns. Many salespeople have found success by modifying existing curriculums to incorporate classroom training, self-directed learning and role-playing opportunities, which should be based on customers' needs. By better aligning curriculums to sales force preferences, firms can improve return on investment made in sales training. Some multinational companies including GE, 3M, AXA are committed in developing an integrated quality management policy applied throughout the training development cycle, using a consistent approach across all training media. This strategy is based on all-round commitment to quality and on sensitively defined standards. Quality management techniques and independent quality reviews and audits successfully complement established validation techniques [63].

The training programs offered by the outsourced agencies should design training programs on the basis of core requirements of the client firm and profile of the trainees. In case of a client's top request for help in generating more demand among its customers, sales representatives should be learning through various training techniques how to address that request. Active reinforcement of training is usually accomplished by technology links to both the training itself and the formal coaching process. By adding a post-work agenda to training initiatives, it is possible to communicate reminders at 24 hours, one week and one month intervals to both the trainee and their training manager. This process promotes tasks and discussions about the new material.The use of participant observation provides a detailed picture of training content and firms can delineate the contents of the training programs that contribute to learning [64].

TRAINING PEDAGOGY

In the contemporary sphere of designing and delivering training programs, there are frequent pedagogical shifts in the learning contents. The most significant changes include increased manifestation on current training approaches, introduction of developing competitive strategies, increased focus on the design and delivery of training programs on applied learning platform, organization of sessions, meticulous contents and coverage, more self-learning space in teams, increase in confidence about learning and sharing, and a more work-centered approach towards learning [65]. The relationship between the applied training skills and training experiences of participants demonstrates the effectiveness of applied learning. It is observed that self-determined learning motivation along with trainer autonomy provides trainees greater satisfaction, which leads to comprehensive learning, better course grades and higher post-training evaluations [66]. Learning effectiveness is developed through careful selection of learning activities, application of best practices in both delivery and content of programs, and particularly effective learning relationships between learning approaches and moderations of learning stages by the trainers. Management training courses effectively requires learning autonomy wherein trainees can develop logical path of problem solving through preferred self-learning techniques. This may lead to the learning insight to 'think global' for students and 'teach local' strategy for professors familiarizing employees of the firm to deal with local market situations [67]. Sales firms can develop training programs through common collaboration tools including existing intranets, shared drives and email groups, which are just a few reasonable and low cost alternatives to connect sales people. Irrespective of pedagogical approach to training programs, the focus should be laid on creating an informal network that promotes viable, on-going connectivity to contributed content essentially building a virtual database that can solve future needs.

Moderator Supported Learning Approaches (MSLA)

In this learning approach trainers serve as bridge inintegration of information, technology, and participant teamsinto a unified leadership frameworkstimulating trainees to analyze the given subject, develop content interactivity, identify problems and frame logical solutions in teams instead of learning core concepts conventionally 'across the training curriculum'. Following learning tools can be practiced by the trainers in delivering the program:

- Learning Appraisal Seminars (LAS) are designed to be conducted as learners' forum at the end of each phase of the curriculum of the course. Trainees are divided into groups and a concurrent topic is assigned to each group for developing a paper and presenting it in the seminar. Chairperson and discussants for each presentation are chosen from different groups. Participants of all groups are expected to use major concepts discussed in the previous sessions to develop papers on the given theme. Breakdown of the analysis includes introduction to the topic, brief resume of relevant concepts discussed in the previous sessions, thematic convergence of the major issues of the seminar topic and theoretical issues, situational reflections on the topic, and take-home lessons. All trainees should be encouraged to participate in each presentation. LAS has been considered as a tool to revise the topics discussed in previous sessions in a liberal environment by trainees consulting books, periodicals, professional journals and trainer for expert opinion. Key elements for the success of classroom seminars include an upfront needs assessment to prioritize topics, interactive sessions promoting skill development through actual practice of various strategies, open discussions to identify challenges and solutions, and a convenient and customary time slot [68].
- Case debates are organized in the course and administered to teams after three different cases on the given topic have been discussed in the class. Issues emerging from all cases are steered to develop analytical insights on the theme and are debated using case breakdown method which is analogous to ZOPP [69] method used by a

German consulting company. In this process insights of participants are invited in the four analytical categories including management focus (customer centric, service orientation, profit centric, competitive etc.), strategy construct (advanced marketing mix- 11Ps which include product, price, place, promotion, packaging, pace, people, performance, psychodynamics, posture and proliferation), causes and effects, and lessons learned. This process is termed as Data Board Preparation (DBP).Upon listing the insights of all participants, the common attributes from each category of DBP are integrated. Thus final attributes appear in each analytical category. This process is termed as Data Integration Process (DIP). In conclusion of the case debate a strategy statement is constructed based on the DIP attributes which serves as case learning experience.

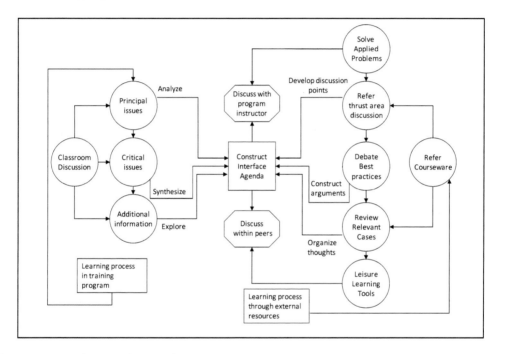

Figure 7.4.Managing Learning interface.

Training can also be effectively delivered during the leisure time such as luncheon meetings and coffee breaks. One of the effective learning tools can be the administration of crosswords on various topics related to sales. This exercise can be administered to the students in teams as an open book leisure exercise. Working teams enjoy full autonomy in solving the crosswords and learn new concepts while referring to various resources. Later unsolved points of crosswords are discussed in the classroom and the solution is presented by the trainer. Many trainees found that solving academic crosswords is a unique experience and it helps promoting generic learning skills. Recent popularity of crossword puzzles in management education has

to be viewed in the context of the sweeping changes in learning perspectives of trainees. Crossword solving supplements the efficacy of new learning processthrough self-learning in management education. Learning through crossword puzzle   isalsoconsidered as an approach to develop *learning entertainment* of marketing and sales concepts [70]. The effective learning map in executive training programs is exhibited in Figure 7.4.

The contemporary practices in designing and delivering the training programs are focused on Learning and Skills Sector (LSS). Recent training initiatives relating to this concept have insisted on the development of subject specialist pedagogy. Powerful social, political and economic considerations, together with practical concerns, have been combined in recent decades in a way that has propelled the LSS curriculum. The momentum for change at both the macro and micro levels within the LSS has been towards the provision of multidisciplinary curricula and generic training. Resistance to this trend has largely been seen as reactionary and conservative [71].

Self-learning Tools (SLTs)

Self-learning perspectives enable students to work remotely using open-learning techniques, as self-managed learners in a task-centered program. The legitimacy of self-learning in a higher education setting is crucial in understanding the differences and similarities both within the classroom learning and liberal learning [72].Self-directed learning skills lean towards evidence based problem solving, improving interpersonal and group skills, self-assessment, and knowledge building. The competence and aptitude in self-learning of management issues include knowledge comprehension, competitive communication skills, leadership and self-concept [73].

Poster sessions are one of the pedagogies in self-learning tools where topics for poster sessions within all major areas of sales are identified in a brainstorming session. Creating and presenting posters requires students to develop and use the vital skills of enquiry, critical analysis, and dissemination of findings. Poster presentation would appear to have potential as a learning strategy and as a method of assessment. It is observed that despite the intensity of the assessment procedure, observing the emergence of analytical debate and creative achievements of students, learning in poster sessions is an exhilarating and worthwhile experience. The *Break and Build* is another learning tool in this category. This is an interesting self-learning game in teams of trainees. A set of sales cases is given to the different teams of trainees to read them thoroughly and build the case with the document available with them. Teams enjoy the liberty to consult Internet, books, periodicals and professional journals. Finally, the case built by the trainees has to be presented in the open house session. In this self-learning method teams are relatively large to work on information gathering.This method of task management develops ability to learn continuously and the learning process is directed towards understanding the problems and obtaining the results [74].

All training pedagogies should be subject to evaluation to measure the effectiveness of training. Evaluation of MSLA and SLT training tools are aimed at measuring abilities for concept development, information search, information utilization, problem solving, creative thinking, logical presentation of ideas and matching decisions with the contextual requirements. This process helps in building context-bound approaches in improving the pedagogical quality of various learning tools that may be productively used in conjunction with students' preferences in effective learning [75].

SELECT READINGS

Cantrell, S. (2010), *Workforce of one: Revolutionizing talent management through customization*, Harvard Business School Press, Harvard Business School, Boston, MA

Gardner, H. (2007), *Five minds for the future*, Harvard Business School Press, Harvard Business School, Boston, MA.

Pike, R. W. (2003), *Creative training techniques handbook: Tips, tactics, and how-to's for delivering effective training*, Human Resources Development Press, Amherst, MA

Magee, J. (2001), *Sales training handbook*, McGraw Hill, New York, NY.

REFERENCES

[1]   Gary, L. (2004), *Do people want to work for you*, Harvard Business Publishing Newsletter, Harvard Busoness School, Botson, MA.

[2]   Ready, D. A., Hill, L. A., and Conger, J. A. (2008), Winning the Race for Talent in Emerging Markets, *Harvard Business Review*, 86(11), 62-70.

[3]   Sparrow, P. (2007), Globalization of HR at function level: four UK-based case studies of the international recruitment and selection process, *The International Journal of Human Resource Management*, 18(5), 845-867.

[4]   Cappelli, P. (2001), Market-Driven Approach to Retaining Talent, *Harvard Business Review*, 83(1): 25-32.

[5]   Schweitzer, L., and Lyons, S. (2008), The market within: A marketing approach to creating and developing high-value employment relationships, *Business Horizons*, 51 (6), 555-565.

[6]   Simons, R. L. (2005), Designing high performance jobs, *Harvard Business Review*, 83(7), 54-62.

[7]   Qu, H., Ryan, B. and Chu, R. (2001), The Perceived Importance of Job Attributes Among Foodservice Employees in Hong Kong Hotel Industry, *International Journal of Hospitality and Tourism Administration*, 2(2), 57-76.

[8]   Bradley, S. and Taylor, J. (1996), Human capital formation and local economic performance, *Regional Studies*, 30 (1), 1-14.

[9]   Manpower Inc. (2009), Annual Report. For details on Manpower Inc. see corporate website http://www.manpower.com

[10] Rajagopal and Rajagopal, A .(2010), Measuring Performance of Sale Force: Analysis of Administrative and Behavioral Parameters, *International Journal of Economics and Business Research*, 2 (5), 399-413.

[11] Gellerman, S. W. (1990), Tastes of a good salesperson, *Harvard Business Review*, 68(3), 64–65.

[12] Mäkitalo, A.,and Säljö, R. (2002), Invisible People: Institutional Reasoning and Reflexivity in the Production of Services and"Social Facts" in Public Employment Agencies, *Mind, Culture, and Activity*, 9(3), 160-178.

[13] Lee, J. Y. and Im, G. S. (2007), Self-enhancing bias in personality, subjective happiness, and perception of life-events: A replication in a Korean aged sample, *Aging & Mental Health*, 11(1), 57-

[14] Wiley, C. (1992), Recruitment strategies for changing times, *International Journal of Manpower*, 13 (9), 13-22.

[15] Zottoli, M. A. and Wanous, J.P.(2000), Recruitment Source Research: Current Status and Future Directions, *Human Resource Management Review*, 10 (4),353-382.

[16] Ready, D. A., Hill, L. A., and Conger, J. A. (2008), Winning the Race for Talent in Emerging Markets, *Harvard Business Review*, 86(11), 62-70.

[17] For details on the functional pattern of the ABC Consultants Private Limited, see its corporate web site http://www.abcconsultants.net

[18] Cynthia, M. P. (2008), *How to close the talent gap*, Harvard Business Publishing Newsletter, Harvard Business School, Boston, MA.

[19] Capelli, P. (2008), Talent management for the twenty first century, *Harvard Business Review*, 86 (3), 74-81.

[20] Capelli, P. (2000), Market driven approach to retaining talent, *Harvard Business Review*, 78 (1), 103-111.

[21] Shih, H. S., Huang, L. C., and Shyur, H. J.(2005), Recruitment and selection processes through an effective GDSS, *Computers & Mathematics with Applications*, 50 (10), 1543-1558.

[22] Ferris, G. R., Berkson, H. M., and Harris, M. M. (2002), The recruitment interview process: Persuasion and organization reputation promotion in competitive labor markets, *Human Resource Management Review*, 12 (3), 359-375.

[23] Ma, R. and Allen, D.G. (2009), Recruiting across cultures: A value-based model of recruitment, *Human Resource Management Review*, 19 (4), 334-346.

[24] Connerley, M.L. and Rynes, S.L. (1996), Does total quality management affect applicant perceptions of recruitment and selection processes?,*Journal of Quality Management*, 1 (2), 207-225.

[25] Jerabek, A. (2003), Job Descriptions: Don't Hire Without Them, *Journal of Interlibrary Loan, Document Delivery & Information Supply*, 13(3), 113-126.

[26] Siddique, C. M. (2004), Job analysis: a strategic human resource management practice, *The International Journal of Human Resource Management*,15(1), 219-244.

[27] Zink, K. J. (2008), Human resources and organizational excellence, *Total Quality Management & Business Excellence*, 19(7), 793-805.

[28] Groysberg, B., Lee, L. E., and Abrahams, R.(2010), What It Takes to Make 'Star' Hires Pay Off, *Sloan Management Review*, 51 (2), 57-61.

[29] Singh, K. (2003), The effect of human resources practices on firm performance in India, *Human Resource Development International*, 6(1), 101-116.

[30] Weerakkody, V. and Irani, Z. (2010), A value and risk analysis of offshore outsourcing business models: an exploratory study, *International Journal of Production Research*, 48(2), 613-634.

[31] Walter, A. and Gemnden, H. G. (2000), Bridging the gap between suppliers and customers through relationship promoters: theoretical considerations and empirical results,*Journal of Business and Industrial Marketing*, 15 (2), 86-105.

[32] Gómez, M. I., Maratou, L. M. and Just, D. R. (2007), Factors Affecting the Allocation of Trade Promotions in the U.S. Food Distribution System,*Review of Agricultural Economics*, 29 (1), 119-140.

[33] Yao, T., Jiang, B., Young, S. T. and Talluri, S. (2010), Outsourcing timing, contract selection, and negotiation, *International Journal of Production Research*, 48(2), 305-326.

[34] Davenport, T. H. (2005), The Coming Commoditization of Processes, *Harvard Business Review*, 83 (6), 100-108.

[35] Zoltners, A. A., Sinha, P., and Lorimer, S. E. (2006), Match Your Sales Force Structure to Your Business Life Cycle, *Harvard Business Review*, 84 (7-8), 80-89.

[36] Kim, K. I.,Syamil, A., and Bhatt, B. J. (2007), A resource-based theory of supplier strategy, *International Journal of Logistics Systems and Management*, 3 (1), 20-33.

[37] Platz, L. A. and Temponi, C. (2007), Defining the most desirable outsourcing contract between customer and vendor, *Management Decision*, 45 (10), 1656-1666.

[38] McIvor, R. (2008), What is the right outsourcing strategy for your process, *European Management Journal*, 26 (1), 24-34.

[39] Miranda, M., Konya, L., and Havira, I. (2005), Shopper's Satisfaction Levels are not only the Key to Store Loyalty, *Marketing Intelligence and Planning*, 23 (2), 220-232.

[40] Davies, A. (2004), Moving Base into High-value Integrated Solutions: A Value Stream Approach, *Industrial and Corporate Change*, 13 (5), October, 727-756.

[41] Lafferty, B.A. and Goldsmith, R. E. (2004), How Influential are Corporate Credibility and Endorser Attractiveness when Innovators React to Advertisement for a New High Technology Product? *Corporate Reputation Review*, 7 (1), 24-26.

[42] Aksin, O. Z. and Masini, A. (2008), Effective strategies for internal outsourcing and offshoring of business services: An empirical investigation, *Journal of Operations Management*, 26 (2), 239-256.

[43] Rajagopal (2006), Leisure shopping behavior and recreational retailing: a symbiotic analysis of marketplace strategy and consumer response, *Journal of Hospitality and Leisure Marketing*, 15 (2), 5-31.

[44] Javalgi, R. G., Dixit, A., and Scherer, R. F. (2009), Outsourcing to emerging markets: Theoretical perspectives and policy implications, *Journal of International Management*, 25 (2), 156-168.

[45] Brandes, H., Lilliecreutz, J.,and Brege, S.(1997), Outsourcing--success or failure?: Findings from five case studies, *European Journal of Purchasing & Supply Management*, 3 (2) 63-75.

[46] Jiang, J. J. and Klein, G.(2000), Effects of downsizing policies on IS survivors' attitude and career management, *Information & Management*, 38 (1), 35-45.

[47] Nyberg, A. J. and Trevor, C. O. (2009), After layoffs, help survivors be more effective, *Harvard Business Review*, 87 (6), 14-16.

[48] Gandolfi, F. (2006), Personal development and growth in a downsized banking organization: summary of methodology and findings,*Human Resource Development International*, 9 (2), 207-226.

[49] Klaus, M. E. and Enese, L. D. (2003), Post-Acquisition Restructuring as Evolutionary Process,*Journal of Management Studies*, 40 (2), 459-482.

[50] Rajagopal and Rajagopal, A. (2006), *Trust and Cross-Cultural Dissimilarities in Corporate Environment*, Team Performance Management-An International Journal, 12(7-8), 237-252.

[51] Carswell, P. (2005), The Financial Impact of Organizational Downsizing Practices: The New Zealand Experience,*Asia Pacific Journal of Management*, 22 (1), 41-63.

[52] Martin, H. J. and Hrivnak, M. W. (2009), Creating disciples: The transformation of employees into trainers, *Buisness Horizons*, 52 (4), 357-365.

[53] Wiggenhorn, W. (1990), Motorola U: When Training Becomes an Education, *Harvard Business Review*, 68 (4), 71-83.

[54] Argyris, C. (1997), Teaching smart people how to learn, *Harvard Business Review*, 69 (3), 99-109

[55] Adams, G. L., Tallon, R. J. and Rimell, P. (1980), A comparison of lecture versus role-playing in the training of the use of positive reinforcement, *Journal of Organizational Behavior Management*, 2(3), 205-212.

[56] Depradine, C. (2007), A role-playing virtual world for web-based application courses, *Computers and Education*, 49 (4), 1081-1096.

[57] Jantan, M. A., Honeycutt, E. D., Thelen, S.T. and Attia, A. M. (2004), Managerial perceptions of sales training and performance, *Industrial Marketing Management*, 33 (7), 667-673.

[58] Honeycutt, E. D., Mottner, S. and Ahmed, Z. U. (2005), Sales training in a dynamic market: The Singapore service industry, *Services Marketing Quarterly*, 26(3), 55-69.

[59] Taylor, P. J., Russ-Eft, D. F.and Taylor, H. (2009), Transfer of Management Training From Alternative Perspectives, *Journal of Applied Psychology*, 94 (1) 104-121.

[60] Van der Wiele, T., Hesselink, M. and Van Iwaarden, J. (2005), Mystery shopping: A tool to develop insight into customer service provision. *Total Quality Management & Business Excellence*, 16(4), 529-541.

[61] For details on the Global Knowledge Inc. see corporate web site http://www. globalknowledge.com

[62] Christie, M. F. and Mason, P. A. (2003), Transformative Tour Guiding: Training Tour Guides to be Critically Reflective Practitioners. *Journal of Ecotourism*, 2(1), 1-16.

[63] Williams, M. (1994), Quality Management and Training Design, *Innovations in Education& Training International*, 31(2), 121-125.

[64] Kakavelakis, K., Felstead, A., Fuller, A., Jewson, N. and Unwin, L. (2008), Making a sales advisor: the limits of training 'instrumental empathy', *Journal of Vocational Education and Training*, 60(3), 209-221.

[65] Donnelly, R. (2008), Lecturers' self-perception of change in their teaching approaches: reflections on a qualitative study,*Educational Research*, 50 (3), 207-222.

[66] Filak, V. and Sheldon, K. (2008), Teacher support, student motivation, student need satisfaction, and college teacher course evaluations: testing a sequential path model, *Educational Psychology*, 28 (6), 711-724.

[67] Rothwell, A. (2008), Think global, teach local,*Innovations in Education& Teaching International*, 42 (4), 313-323.

[68] Pinheiro, S. O. and Heflin, M. T. (2008), The Geriatrics Excellence in Teaching Series: An Integrated Educational Skills Curriculum for Faculty and Fellows Development,*Journal of the American Geriatrics Society*, 56 (4), 750-756.

[69] ZOPP, from the German term "Zielorientierte Projektplanung" translates in English to "Objectives-Oriented Project Planning." ZOPP is a project planning and management method that encourages participatory planning and analysis throughout the project cycle with a series of stakeholder workshops.

[70] Cinnéide, B Ó (2006), Developing and testing student oriented case studies: The production process and classroom/examination experiences with "entertaining" topics, *Journal of European Industrial Training*, 30 (5), 349-364.

[71] Fisher, R. and Webb, K. (2006), Subject specialist pedagogy and initial teacher training for the learning and skills sector in England: the context, a response and some critical issues. *Journal of Further and Higher Education*, 30(4), 337-349.

[72] Taousanidis, N. I. and Antoniadou, M. A. (2008), The Greek challenge in work-based learning,*Industry and Higher Education*, 22 (3), 177-182.

[73] Pearson, V., Wong, D. K. P., Ho, K. and Wong, Y. (2007), Problem Based Learning in an MSW Program: A Study of Learning Outcomes,*Social Work Education*, 26 (6), 616-631.

[74] O'Keeffe, T. (2005), Towards zero management learning organizations: A honey-bee perspective, *Journal of European Industrial Training*, 29 (9), 764-778.

[75] Hosie, P., Schibeci, R. and Backhaus, A. (2005), A framework and checklists for evaluating online learning in higher education, *Assessment & Evaluation in Higher Education*, 30 (5), 539 – 553.

Chapter 8

SALES PERFORMANCE AND
COMPENSATION PLANNING

Effective management of sales force typically requires both planning and action to deliver powerful aspiration, establish performance measuring and rewarding systems to reinforce specific goals of sales force, and develop mind-sets of salespeople by creating a sense of shared responsibility in performing tasks. Although companies devote considerable time and money to administer their sales forces, it is necessary to focus management strategies on how the structure of the sales force can be closely associated with the performance indicators of a firm's business. Firms must consider the relationship between the differing roles of internal salespeople and external selling partners, the size of the sales force, its degree of specialization, and how salespeople share their efforts among different customers, products, and activities. These variables are critical because they determine how quickly sales force of a firm responds to market opportunities, influences performance of salespeople, and affects revenues, costs, and profitability of the firm [1]. The task of a salesperson changes over the course of the selling process. Different abilities are required in each stage of sale including identifying prospects, gaining approval from potential customers, creating solutions, and closing the deal. Success in the first stage, for instance, depends on the salesperson acquiring precise and timely information about opportunities from contacts in the marketplace. Managers often view sales as a set of stereotype activities which begins with prospecting and terminates with closing the deal. But in practice it is a diligent task which involves variety of conflicts and high risk. Hence, sales performance requires effective administration in a phased manner to resolves conflicts and ambiguities during the negotiation and transaction process.

Mangers should find qualitative and future business enhancement measures while evaluating a company's performance. It is necessary to measure the sales performance of firms without major biases that include overdependence on self-appraisals, external data to develop benchmarks, relying heavily on employment motivational practices to boost performance, considering the results of the tasks in absolute way than deriving relative interpretations, and looking backward to find faults in the performance. Managers can develop a service quality index to measure customers' repeat purchase intentions based on firm specific conditions to help measures that lead rather than lag the profits of the firm. A health insurance company found that the 10 percent of its sick patients accounted for about 80 percent of its costs, and after successful performance evaluation the company is now offering customers incentives for early screening. This strategy on one hand supported the performance and helped retaining potential customer by developing the loyalty factor. A really good assessment will combine managers' relative independence with their expertise [2].

Administering control measures to field sales force has become major performance guiding tool in many business firms. Industrial sales organizations largely implement direct management control and influence the activities of employees and towards improving their efficiency. The extent monitoring of sales managers, directing, evaluating, and rewarding activities in an organization intend to guide sales force behavior through management control processes to achieve results favorable to the organization and the employees. Management control is thus recognized as an important performance indicator of the task performed by the salespeople [3].

Sales performance is affected by internal and external factors of the organization. Internal factors of the organization include, work culture, guidance by the managers, and administrative support. Of all the things sales managers do to develop salespeople, many people believe the most important one is coaching. Coaching is defined as extending guidance towards using skills, experience, and direction to help someone improve their performance. Coaching consists primarily of giving people feedback to reinforce what they do well while suggesting ways and means to improve. Coaching implies that everybody can improve and hopefully everybody wants to get better at what they do. Many sales managers erroneously believe that they have to be better than the person they are coaching at the skill set they are discussing in order to be a useful coaching resource to that person. To be a good coach requires an understanding of the skills being discussed and a desire to help someone improve one's performance. The sales manager who knows how to provide that feedback will have more productive, effective, satisfied, and motivated salespeople. Effective Sales Management Coaching can provide sales managers with the skills to help salespeople grow and develop professionally [4].

Result oriented performance control and market volatility are positively related to new product selling performance. The effect of sales force adoption on selling performance is stronger where outcome based control is used and where the firm provides information on the background of the new product through internal marketing [5]. It has been observed that salespeople who simultaneously exhibit commitment and effort achieve higher levels of new product selling performance. In view o the growing competition, the measurement of sales performance has become critical to sustain competitive advantage [6]. Customer and the competitive environment in a sales organization are the external factors that affect the work process and the type of selling skills required. The central focus of external factors affecting sales performance is to push management to increase the skills and knowledge of salespeople or to encourage the use of a more rigid and less autonomous form of work organization regulations. It is argued that firms should select performance improvement strategies distinctively without putting rigid administrative controls. Firms should empower and improve skills of salespeople, through effective knowledge transfer and recruitment of qualified individuals. In order to improve the performance of salespeople, managers should aim at increasing autonomy and routinize the work process. The routinization approach is more pervasive despite many salespeople believing that it is inappropriate for successful selling within the particular customer environment [7].

The basic objective of salesperson performance evaluations is to determine how well individual salespeople have performed. However, the results of salespersons' performance evaluations are used by the foreign firms for various administrative purposes as summarized under:

- To ensure that compensation and other reward disbursements are consistent actual salesperson performance
- To identify salespeople who might be promoted
- To identify salespeople whose employment should be terminated and collect evidence to support the need for termination
- To determine the specific training and counseling needs of individual salespersons and the entire sales force
- To provide information for effective human resource planning
- To identify criteria that can be used to recruit and select salespeople in future
- To advise salespeople of work expectations
- To motivate salespeople
- To help salespeople set career goals, and
- To improve salesperson performance.

The above mentioned diverse purposes affect all aspects of the performance evaluation process. Performance evaluations for identifying salespeople for promotion into sales management positions should focus on criteria related to potential effectiveness as a sales manager and not just current performance as a salesperson. One of the methods that is workable and effective, discusses the development sales benchmarking to perform competitively in the market. Some managers find it difficult to measure the performance of sales representatives because of the variations in the attitudes of sales representatives, customers and business conditions. However, a few questions as stated below may be put forth to the managers before getting started with the evaluation process:

- Volume of sales in national currency/US Dollars
- Amount of time spent in office
- Personal appearance: for example, clothes, hair, cleanliness and neatness
- Number of calls made on existing accounts
- Number of new accounts opened
- Completeness and accuracy of sales orders
- Promptness in submitting reports
- Amount spent in entertaining customers
- Extent to which sales representative sells the company
- Accuracy in quoting prices and deliveries to customers
- Knowledge of the business
- Planning and routing of calls.

Usually managers evaluate their sales representatives primarily on the basis of sales volume, they rely too much on the number of sales calls made by each of their sales representatives, compare each sales representative's present sales results with past sales for a corresponding period, expect their sales representatives to follow explicitly the selling methods that worked for them when they were selling or they give their sales representatives too much freedom. It may not be possible to determine the accuracy all the performance evaluation approaches used by sales organizations. Several studies have revealed some general conclusions on the quality of performance evaluation methods employed by multinational companies. Most sales organizations evaluate salespersons performance annually, while some firms conduct evaluations semi-annually or quarterly. It has also been observed that a large number of multinational companies engaged in selling their consumer goods use combinations of input and output criteria that are evaluated by quantitative and qualitative measures. However, emphasis seems to be placed on outputs, with evaluations of sales volume results being most popular. Sales organizations, which set performance standards or quotas, also provide information to the sales people on achieving the targets. The multinational pharmaceutical companies generally assign weights to different performance objectives and incorporate territory data when establishing these objectives while most salesperson's performance evaluations are conducted by the field sales manager who supervises the salesperson. However, some firms involve managers who are above the field sales manager in evaluating the performance appraisal salespersons. Also some foreign firms use more than one source of information in evaluating performance of salespeople. Computer printouts, call reports, supervisory calls, sales itineraries, prospect and customer files, and client and peer feedback are some of the common sources of information. Although performance appraisal continues to be primarily a top-down process, in some companies it leads to the implementation of a broader assessment process. An increasingly popular assessment technique of 360° feedback, involves performance assessment from multiple evaluators.

A good sales strategycan be made operational using multidimensional construct consisting of four dimensions, namely customer segmentation, customer prioritization/targeting, developing relationship objectives/selling models, and use of multiplesaleschannels. Moreover, by drawing on the resource-based view of the firm, as well as on industrial organization theory, moderating effects of transformational leadership, customer solution orientation, and demand uncertainty on the sales strategy–performance relationship can be measured. It is observed that sales strategy is consistently and positively associated with performance while transformational leadership, customer solution orientation, and demand uncertainty are found to exert significant moderating effects on this relationship. A well-crafted sales strategy contributes in improving the performance of salespeople. Hence, managers should adopt a transformational leadership approach, assess changes in demand, and assess the degree to which their customers are demanding solutions in order to improve the sales performance of the firm [8].

An effective win-win situation can be acquired by the sales firms and customers by sharing services and brand value through long term equity. This would allow the sale managers to improve the performance by focusing fully on targeted sales, salesperson productivity, cash flow, volume of sales and profitability. Indicators to measure the sales performance should include:

- Organizational standards and systems design
- Responsiveness, professional and value-added sales management strategy

- Long-term customer relationship building strategies
- Consistently higher volume of sales levels
- High tenant satisfaction ratings
- Customer retention rates that outscore competitors
- Efficient and cost-conscious operations with lower operating costs in different markets at the same time to augment the brand equity of the sales firm
- Financial stability and conservative fiscal policies, with a debt to asset value ratio of less than 40 percent
- Commitment to optimize sales in prime locations, large accounts and influential customers
- Maintaining high standards of quality and service
- Strict adherence to best practices
- Leveraging autonomy of salespeople to make quick and appropriate decisions
- Help developing social networks with positive buyer attitude.

Performance of salespeople is also a function of managerial decision making process. Sales training plays a significant role in enhancing performance. Hence, sales executives should take aggressive action to train and retain sales talent, manage the sales process, and use sales support technologies to meet the challenges of this new environment. Consumer satisfaction is one of the major drivers of sales performance towards prospecting customers for buying time sharing proposals at leisure facility centers. Hence, sales teams need to be oriented towards providing high convergence of tangible and intangible factors to the customers during sales process. In order to improve performance, salespeople should always standby innovative strategies. The individual innovation in planning and delivering sales and services should be encouraged by the firms. Salespeople should be driven to represent quality of services, promote consumer experience and develop loyalty among customers. Salespeople should be trained to exhibit company's innovation portfolio and clearly express the sales intentions as an "innovation effectiveness curve. This curve lets companies plot annual spending on innovation projects against the financial returns from those projects and solve for growth [9]. A strategy paradigm for augmenting such effect is exhibited in Figure 8.1.

There is an increasing competition in developing countries in consumer goods sales sector and many regional sales firms are emerging as low investment high profit ventures. Hence, salespeople should focus on customer value emphasising the design, technology and services innovation. It is observed that the higher the substitutioneffect of products and service, the lower the competitive advantage. Sales drivers including territory design, compensation, task performance pattern and cultural interface need to be effectively planned and administered at managerial level to inculcate higher perceived values on task management among the salespeople. Higher perceived values on the drivers among salespeople help in maximizing the outcome performance in leisure facility centre developer firms. The highly competitive, business environment calls for the creation and delivery of superior customer value. In order to meet customer expectations and deliver superior value, many firms are undergoing significant transformations by developing market orientation strategies, building strategic relationships between salespeople and customer, improving sales processes and services, enhancing the relationship between the marketing and sales functions, and developing appropriate new metrics. Key alignment linkages include transformationstrategy of sales firms, sales performance, and sales management practice towards improving delivery of superior customer value [10].

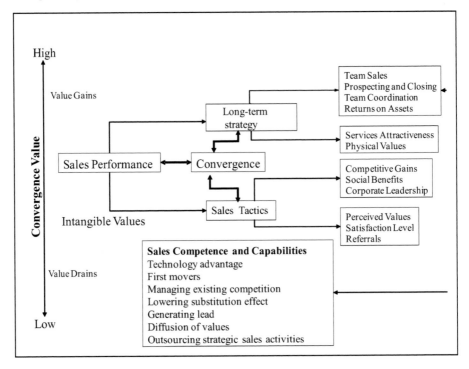

Figure 8.1.Attributes and Process of Sales Performance.

The high-performance salespeople have greater commitment to their organizations and their sales managers are more satisfied with their units' sales territory designs. The mangers may need to re-conceptualize the fact that the salespeople should shift from a "hard selling" to a "smart selling" approach. The SMART variables may be considered to administer sales teams which include-strategy orientation, measurability, approach, reality and time frame. The strategy orientation would drive the brainstorming discussion to result orientation and the measurability of sales performance and would count on the success of the deliberations [11]. Teams which need to work within an organization and across functional activities such as sales, marketing, procurement, personnel and finance find that team working fosters collaborative sales approach rather than a competitive approach. It is important that terms of reference of teams must be capable of doing the job for which they have been selected and this clearly implies that the membership should include people that are able to contribute towards completion of the task [12]. Advanced selling strategies are based on the lean operating principles of Toyota Japan. Companies around the world, in all sectors of the economy, now embrace these approaches to improve quality, cost and productivity. Different purchasing and sourcing practices in lean organizations mean that winning sales requires major changes in sales strategy and practice. Serving lean customers, who demand complete transparency across the supplying organization and focus on capability not cost, has proven problematic for traditional sales departments. Further, as supplying organizations themselves become lean, organizational redesign extends from operations into sales [13].

Many companies integrate sales performance management with the customer relationship management (CRM) application to automate and manage the pre-sales process, spreadsheets, and working through manual procedures take care of all performance indication from incentive compensation to post-sales analytics. CRM application drives sales, enhances sales effectiveness by increasing productivity of a salesperson, profit contribution, and strengthens the buyer-seller relationship. Thus managers consider CRM applications in a firm in reference to economic factors that affect overall sales performance.

Salespeople should be motivated not only to make further sales, but also to consistently make profit centric sales. Managers should use incentive compensation to align desired payment of salespeople with strategic corporate objectives. Most companies consider that incentive compensation is a serious indicator of sales performance, as they tend to motivate salespeople towards achieving and over achieving the given targets. Most of the compensation plans do not support real-time visibility, so salespeople cannot readily see how they are being paid. The compensation plans are difficult to audit, elevate the risk of non-compliance, and provide a poor basis for meaningful post-sales analytics, which is a valuable tool for improving pre-sales processes. However, managers should understand that there are no 'one-size-fits-all' compensation plans processes. Hence sales firms should motivate and reward performance with contests, sales promotion incentive fund, and non-cash incentives. It is observed that sales contests and special performance incentive funds can promote desired short-term sales behaviors and allow salespeople to accelerate their sales efforts. Automating the compensation management process makes it easy to create, launch and manage these kinds of opportunistic sales performers. Alternatively offering non-cash incentives may be another rewards option that can drive sales performance, particularly when these reward points can be redeemed for instant gratification.

Mary Kay Cosmetics motivates the sales force of 200,000 independent agents who comprise the firm's only distribution channel. There is a powerful effect on sales-force behavior that results when creative types of employee recognition are combined with financial incentives. In sales, motivation plays an important role in maintaining happy employees and workers. Happy workers "produce more product and better service" (Hahn, 2007). Mary Kay's philosophy is that every person associated with the company lives by the Golden Rule (Case). By treating customers, business associates, and all others with such respect, Mary Kay improves not only the morale of its consultants, but also its public image. Mary Kay's business depends on its consultants to not only purchase and sell its products, but also recruit new consultants (Case). To retain its consultants, Mary Kay must recognize their accomplishments. Mary Kay provides prizes to consultants based on the increase in their sales, as well as recognizes its consultants for their progress in recruiting new consultants. Career Car program offers rewards to the very important performer (VIP) of the company. Mary Kay Career Car qualifiers have been stopping traffic with their prestigious Career Cars since Mary Kay Ash awarded the first one in 1969. Since then, more than 100,000 independent sales force members worldwide have earned the use of a Career Car.

Managers should develop a strategic model and forecast to formulate compensation plans prior to deployment. This is a crucial performance in management practice. Managers should align the sales team to assess the perceptions of salespeople on any revisions made in the compensation plan. The revised compensation plan can be automated after evaluating the feedback of employees. However, using analytics to drive sales performance is important for promoting efficiency and visibility across sales operations. It is necessary to integrate CRM and compensation management applications for greater insight into sales opportunities.

Performance Control Management

A control feedback system is one of the core components of international marketing management and it serves to assess performance. Monitoring is one of the tools to measure the degree of the success of international marketing and needs to be incorporated in the plan itself. The sales plan of a firm needs to specify the periodicity of monitoring and evaluation of sales performance and the performance appraisals should be aligned with the corporate objectives of the firm. The monitoring calendar for international marketing firms may be designed keeping the following checks in mind:

- Budgetary control
- Plan implementation
- Performance of marketing functions (11Ps) which include product, price, place, promotion, packaging, pace, people, performance, psychodynamics, posture and proliferation
- Periodical appraisals of marketing information
- Social, cultural and political changes.

The overall objective of these checks and controls is to determine the achievement of targeted results on time. These issues need to be administered from the corporate office of the business firm in a centralized manner in order to enable effective planning and execution process. The standardization of marketing-mix is usually centralized to ensure the quality of all the components of the mix across markets in the operational region. Besides, it is important to provide a common business language across markets which would help in understanding local markets more analytically. The checks need to be exercised at different levels of the marketing plan execution and to build-up a strong communication and information system. A consolidated document of the target group index (TGI) may be an appropriate tool for information processing and analysis. The variables which need to be covered in the TGI include consumer goods, industrial goods, services, spatial and temporal trend of demand and price, distribution patterns, marketing budgets, response to advertising, communication services and the like. International marketing research needs to be conducted on specific issues of interest and inferences may be tagged along with the Monitoring and Evaluation (M&E) process. Nevertheless, M&E should be conducted periodically as a tool of control.

There is a close relation between the types of activities undertaken in different countries and their institutional structures. A distinguishing characteristic of financing high technology firms is their evolving pattern of control by different investor groups. While stock markets are an important component of the development of the most successful firms, they are not the most common. Regulation has a significant influence on institutional structure. The degree of risk taking by financial institutions and the diversity of their investments are affected by trade-offs between competition and stability and the emphasis placed on minority investor protection [14].

Task Administration and Control

Managers may define clear sales tasks to ensure significant results from the salespeople and evaluate timely performance feedback. A firm may focus on administering sales activities in four major processes including territory-target based deployment of sales force, efficiently managing client accounts, working with improved information systems, and minimizing field level task conflicts. Hence, limiting the number of accounts per salesperson may work out as a good option to improve the efficiency of sales activities in the market. Field sales managers may be encouraged to follow 'management by objectives' approach for improving their performance. In addition, managers should also conduct performance reviews of salespeople periodically.

Personal Development and Motivational Factors

Effective management of sales force typically requires both effective planning and action to motivate salespeople, establish performance measures and reward systems to reinforce specific goals of sales force. Firms engaged in selling products and services in a competitive marketplace should be able to develop mind-sets of salespeople by creating a sense of shared responsibility in performing tasks. Firms must consider the relationship between the changing roles of internal salespeople and external selling partners, the size of the sales force, its degree of specialization, and the way salespeople share their efforts among different customers, products, and activities. These variables are critical because they determine how quickly sales force of a firm responds to market opportunities, influences performance of salespeople, and affects revenues, costs, and profitability of the firm [15]. The training of salesmen should be directed at making them aware of the advantages of adopting relational behaviours and boosting their capabilities and skills preparatory to an effective implementation of such behaviours. The systems for controlling and appraising the sales force may be designed for the long term and provide incentives for relational behaviours if behaviour-based control systems are used, and utilize relational performance indicators [16].

SALES PLANNING

Sales plan guides the implementation and control process in reference to predetermined sales objectives and directs strategic and tactical approaches to enhance sales performance. In growing competitive marketplace new concepts of strategic thinking are reshaping sales strategies in high-performing companies. Implementing high-level corporate strategy to the ground-level sales effort is a challenging task for most sales organizations and also risk averse at the same time. Most companies drive such strategies through high-performing sales teams that carry strong commitment to the company goals and priorities. The risk in implementing sales strategies is visualized by many companies in terms of process gap that may exist between the way strategy is implemented in most parts of the organization and its role in sales affecting a great deal of potential value. Recognizing that, some companies are infusing their sales efforts with a more vigorous sense of strategy and shifting their functional paradigm towards low cost-high volume sales [17]. Sales firms should also address to the issues on uncertainty during the implementing strategic sales planning addressing the problem of aligning demand and supply in configure-to-order systems. Using stochastic programming methods, a study demonstrates the value of accounting for the uncertainty associated with purchase orders is configured. It is observed that there can be significant improvements in revenue and serviceability by appropriately accounting for the uncertainty associated with order configurations [18]. Many firms look at immediate managerial concerns in the developing the strategic sales process to gain positive results in the marketplace. Managerial concerns close to the strategizing process include involvement in business and marketing strategy decisions, market intelligenceas a value added input, integration of cross-functional contributions to customer value, internal marketing of customer priorities to internal departments and employees, and upgrading the infrastructure of existing sales organization that will support new strategic role. Long-term organizational effects on implementing the strategic sales plan consist of inspiration or leadership at several levels, influence over important issues of strategic direction, integrity in ethical standards, client services initiatives, and, an international perspective on managing sales and customers [19]. The core philosophy of strategic sales planning is exhibited in Figure 8.2.

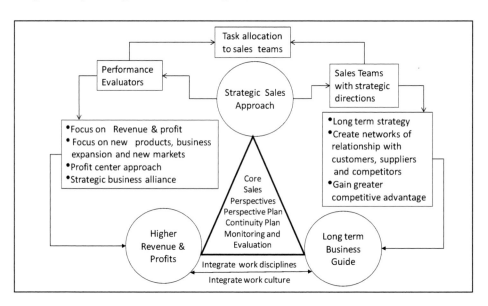

Figure 8.2. Strategic Sales Planning: Core Philosophy.

The sales strategyconsists of perspective plan and action plan as illustrated in Figure 8.2. The perspective plan is developed with long term objectives while the action plan is formed by phasing out the activities delineated in the long term perspective plan in order to monitor the annual progress of activities and achieve the annual targets. The planning cycle is continuous. Plans are developed, implemented, evaluated, and revised to keep the sales strategy on target. Since a strategy typically extends beyond one year, the action plan is used to guide short-term sales activities. The planning process is a series of action plans guided by the sales strategy. Determining an annual planning period is necessary since several of the activities require action within stipulated period and budget need to be planned accordingly. In large companies the product managers are responsible for coordinating the preparation of plans. A planning workshop is conducted midyear for the kick off of the next year's plans. Top management as well as product, research, sales, and finance managers attend the workshop. Accounts manager of the firm's advertising agency also participates in the workshop. The current year's plans are reviewed and each product manager presents the proposed sales plan for the next year. The workshop members analyze each plan and suggest changes. Because the requested budget may exceed available funds, priorities are placed on major budget components. Each product manager must provide strong support for requested funds. The same group meets again in 90 days to review the revised plans. In this meeting, the plans are finalized and approved for implementation. Each product manager is responsible for coordinating and implementing the plan. Progress is reviewed throughout the plan year, and the plan is revised when necessary. Successful sales continuity plans allow an organization to perform its business processes effectively. The effectiveness of plan implementation relies on personality attributes of salespeople, skills, and commitment in managing client accounts and retail customers. In growing competition sales strategies tend to be highly technical given both the nature of the business and the attributes of customers. Considering shifts in the competitor strategy and preferences of customers, salespeople often requires counterintuitive thinking, turning over conventional thought processes and discarding otherwise irrational assumptions. Such complex issues and previously unconventional thought processes are accounted for in developing strategic sales planning, which is known as a sales impact analysis [20].

Firms need to develop an organizational culture that supports sales continuity plan as an outgrowth of longer term perspective sales plan. However, organizational culture nurtured by the salespeople at the bottom of the pyramid does not imply that the entire organizational culture needs to be responsive for driving sales continuity planning process effective. However, sales continuity plans need to be developed and implemented within the sphere of influence of various levels of managers. Organizational culture determines the degree to which resistance to change affects the rollout of any plan and also projects in success [21]. A short-term sales plan forms the core of an overseas subsidiary's planning effort and covers sales operations usually for about a year. The complexity of planning varies among companies. In some cases, it may amount to simple preparation of sales budgets. In multinational firms, planning involves multiple considerations to consolidate mutual interdependence between different overseas affiliates and the parent corporation. A good implementation process spells out the activities to be implemented, who is responsible for implementation, the time and location of implementation, and how implementation will be achieved. Sales representatives should target all accounts currently using a competitive product. A desirable implementation skill is exhibited include:

- The ability to understand how others feel
- Good bargaining skills
- The strength to be tough and fair in putting people and resources where they will be most effective

- Effectiveness in focusing on the critical aspects of performance in managing sales activities
- The ability to create a necessarily informal organization or network, and
- Problem solving ability when they are confronted.

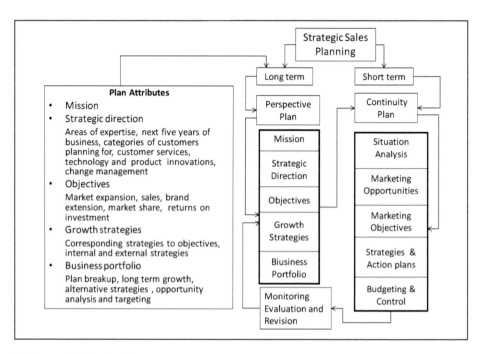

Figure 8.3.Strategic Planning Process.

A plan should be developed to convert 5 percent of these accounts to the company brand during the year. Account listings should be prepared and distributed by product management. The sales plan can be used to identify the organizational units and managers that are responsible for implementing the various activities in the plan. Deadlines indicate the time available for implementation. The sales manager is responsible for implementation of the plan through the sales force. Planners and implementers often have different strengths and weaknesses. An effective planner may not be good at implementing plans. The effective process of planning and implementation of strategic sales plans is exhibited in Figure 8.3.

One way to deal with difficulty in the implementation of the sales plan is to administer the balanced scorecard methods. This process is a formalized management control system that implements a given business-unit strategy by means of activities across four areas: financial, customer, internal business process, and learning and growth. The balanced scorecard provides the framework to minimize such an occurrence by encouraging implementation of a common strategy that is communicated and coordinated across all major areas of the organization. The 'balanced' component of the balanced scorecard reflects the need to consider how all areas of the organization function together to achieve a common goal of strategy implementation. The major benefit of the balanced scorecard is that a frequently aggregate, broadly defined strategy is translated into very specific actions. Through execution and monitoring these actions, management can assess the success of the strategy and, if necessary, modify and adjust it. Another major benefit of the balanced scorecard methodology is that it is feasible for any strategy at the business-unit level and provides a means to link performance evaluation to strategy implementation. The balanced scorecard is a strategic management system that categorically presents the strategy implementation process and translates it into action. It is widely used by organizations as a tool to assess and manage the organizational and functional performance like marketing, brand performance, sales, and services of a firm. However, firms need to align their scorecards with their strategy in order to gain maximum benefits. In this process, the risk is that managers' expectations may be raised too high, leading to potential disappointment. The balanced scorecard tool can be and is used in different ways involving many different functional areas and indicators. Different ways of implementing and using this strategic tool may have different effects on performance of the firm [22].

MPSI Systems Inc. provides proprietary retail planning software and builds market research databases, primarily to assist large petroleum companies in selecting retail sites. Customers use its market-specific studies to predict sales volume, supply and demand, and competition while researching the potential performance of new properties, as well as existing locations. MPSI also offers such professional services as database maintenance, consulting, support, training, key competition identification, and price tracking. The company offers services for marketing and sales planning in terms of providing and analyzing marketplace data, modeling tools and analytical solutions to customers in over 30 different retail sectors around the world. Improvements in technology, increased opportunities in retail marketing and the changing needs of customers have played a major role in the ongoing advancements in the competitive marketplace. The strengths of the company include client services on site selection, network planning, network optimization, sales projections, trade area analysis, market and intersection ranking, and customer profiling. The planning for a network, will aim at generating volume projections for new sites and changes to existing sites, through computer simulation in exploring the right solution. MPSI Systems develops optimal network plan by determining the optimal retail network for achieving market objectives of client firms. The company conducts the performance mapping and volume projections through its acquisition analysis tools, which provides customized results to meet specific needs of clients using retail consultants to guide the functional projects like sales, pricing, cost and needed investment [23].

The corporate headquarter shares with its subsidiary the perspectives of its mission and objective. This input helps the subsidiary, find its overall goals and specific sales objectives. Corporate headquarter in order to develop homogeneity among the sales plans of different subsidiaries prescribes a standard procedure conducting the planning process. It is necessary for a firm to consider external factors like demands in emerging markets, regulations of the host country, international trade policies, and foreign exchange regulations etc. On the basis of the above information, planning starts with a review of past sales and projections into the future. The projected forecasts are duly revised in planning inputs. The sales plan should clearly address various constituents of sales-mix (11 Ps) including product, price, place, promotion, packaging, pace (competitor analysis), people (sales front liners), performance, psychodynamics, and posture of the firm and proliferations of products, services and markets. Sales budget should be included in the sales plan for each of these constituents. The sales budget must be reviewed by the subsidiary management to add and accommodate the company-wide outlook.Strategic evaluation requires analyzing information to gauge performance and take the actions necessary to keep results on track. Managers need to monitor performance continuously and, when necessary, revise their strategies according to changing conditions. The finance function should reflect on impact of the fluctuations in local currency value, and a capital-expenditure and capital plan need to be appended.

Strategic evaluation, the last stage in the sales strategyprocess, is really the starting point. Strategic sales planning requires information from ongoing monitoring and evaluation. Evaluation consumes a high proportion of sales executives' time and energy. Evaluation may be done to find new opportunities or avoid threats, keep performance in line with management's expectations, and to solve specific problems that exist. Areas of evaluation include environmental scanning, product-market analysis, brand equity analysis, sales program evaluation, and gauging the effectiveness of the selling process. Planning is the approach to making decisions concerning systematic allocation of resources. It is worth emphasizing that planning is a process, not an event. The sales planning process should have the flexibility in fixing the physical and financial sales targets. Flexibility can be defined as the extent to which new and alternative decisions are generated and considered in strategic planning, allowing for positive organizational change. It is observed that flexible planning bridges the relationship between strategic planning and financial performance [24].

Strategic planning is systematic means of making the firm successful through the discipline of strategic thinking and vision used as a framework for all other decisions in the firm. Strategic planning requires an honest evaluation of the company's current situation. Market planning demands the commitment of the top management for its success. It requires commitment of resources, both financial and personnel, for its development. It demands complete follow up. Market plan which is not carried out due to lack of leadership or the required tools needed for completion is a total failure and a waste of time and money.

SALES COMPENSATION PLANNING

Compensation management is becoming increasingly more difficult for organizations to control because sales employees want more and more. Managers need to find out what sales employees want and give it to them in a way that is fair and specific. Developing competitive compensation and incentive plans is becoming the new method for managers to follow, at the same time promoting a team atmosphere among sales employees. The study reveals that compensation management of commissioned sales employees does not point to one best method, but managers are encouraging sales employees to work as a group and not against each other. This would create confidence among employees, thus enhancing the work environment and increasing quality and quantity of sales [25]. The team selling would also be advantageous when an account requires special treatment or large number of people are involved in the process of buying decision. In addition, team selling is more likely to be employed when the potential sale is large for the representative firm and when the product is new to the product line of salespeople. Analytical sales force compensation research offers only limited answers to sales managers who try to devise effective compensation plans, because it often rests on restrictive assumptions, and it considers only simple compensation plan structures. In practice, sales managers need to predict how alternative and relatively complex compensation schemes would affect sales revenues and profits, as well as their likely impacts on sales force morale and turnover. This is why they typically obtain key salespeople's prior reactions to a new scheme, or pretest the new plan on a limited scale. These procedures, however, may not provide accurate long-run predictions, and they can be applied to only one or two schemes at a time [26].

Financial compensation has long been held as the primary motivator of salespeople. Motivation however may be achieved differently in various countries, as the large disparities in pay schemes across countries seem to indicate [27]. Management of a company determines the proportion of incentive pay to total compensation which may be described as the rewarding dimension of the sales forces in a control process. It operates independently from managers' monitoring, directing, and evaluating activities, and needs to be assessed separately in management practice. The incentive compensation as an essential sales management responsibility may be reorganized accordingly. It has been argued that the state of variable pay compensation plans for salespeople in business-to-business markets and, in particular, reviews proposed methods for rewarding key account managers [28]. Recent studies on salespeople compensation reveals that there is a growing use of bonus payments in sales force compensation plans to improve sales productivity and towards achieving variety of organizational objectives. It is observed that bonuses may be effective in inducing higher sales productivity because of their flexibility in tying rewards to performance. Further, planning for monetary incentives to the salespeople may also be effective in directing salespeople's efforts toward specific organizational objectives such as promoting new product sales or sales to new customer groups, and enhancing customer satisfaction and retention [29].

Managers consider management control and pay as related, yet distinct forms of control. Determining the level of fixed salary and incentive payment policies is an important sales management decision. Extensive organizational studies on reward systems support the proposal that compensation is an important, strategic management decision variable that operates independent of other forms of management control [30]. Firms should be able to apply the time-based philosophy of revenue management to their sales forces. A different type of proposed measure, revenue per available salesperson hour, is proposed to better integrate the value of the salesperson's time as a factor in sales potential and revenue calculation. The sales unit effectiveness is an overall evaluation of outcomes (e.g., sales, profit contribution) of the sales unit for which the field sales manager is responsible. Evaluating effectiveness is important since it provides an indication of the performance of the manager's sales unit [31].

Small business owners and marketing managers encounter the threat of higher employee turnover as most of them do not realize the firms' limitations and are not familiar with the fundamentals of the practice or the theory of sales. In this situation, they tend to copy sales force compensation plans from each other. A persistent theme in business has been the search for the one solution that unlocks the mystery of sales force effectiveness. Sales organizations focus much attention on compensation as the means to drive performance. In most cases, companies are caught in a seemingly endless cycle in which disappointing sales performance prompts a redesign of the commission plan. This article examines companies that have broken the cycle. These companies go beyond the content of plan design to concentrate on the process underlying the redesign effort [32]. A new profit-center approach for sales force compensation allows firms to replace quotas with an innovative commission structure that ties compensation to corporate expenses and desired profitability. Sales representatives contribute their share of corporate expenses and profit, after which they can keep most of the moneythey bring into the company. In addition, sales associates choose their compensation plan and decide how to contribute to the company's expenses. With this method, firms can more effectively motivate sales representatives, strengthen recruiting, increase market share and ensure that the business meets profitability goals [33].

Modes of Compensation

Straight salaries are attractive to the salespeople but most companies avoid enrolling salespeople in this compensation plan. It is argued that a straight salary compensation plan is an effective measure for motivating salespeople to achieve higher goals than short-run sales volume and on situations when the performance of an individual salesperson in reference to sales volume is difficult to measure in a given time. As more and more companies prefer to pay less compensation on straight salary and look to other financial options, managers of such firms have challenging task to find the best approaches to take. In this process there are many a slips between the cup and lip. Indeed, much of the conventional wisdom and public discussion about pay today is misleading, incorrect, or both. The result is that business people are adopting wrong notions about how to pay people and why. In particular, there are six dangerous myths [34] about pay:

- Labor rates are the same as labor costs,
- Cutting labor rates will lower labor costs,
- Labor costs represent a large portion of a company's total costs,
- Keeping labor costs low creates a potent and sustainable competitive edge.
- Managers also often misunderstand that individual incentive pay improves performance, and
- People work primarily for the money.

The primary advantage of a straight salary is that management can require salespeople to spend their time on activities that may not result in immediate sales. Therefore, a salary plan offering a large proportion of fixed salary is appropriate when the salesperson is expected to perform many servicing account or other non-selling activities such as market research, customer problem analysis, inventory management, or various types of sales promotions. Straight salary plans provide salespeople with a steady, guaranteed income. Thus, salary based compensation plans are often used when the salesperson's ability to generate immediate sales is uncertain or when a firm is introducing a new products in new territories. Previous studies have found that compensation plans include only salary, only commission and a combination of both modes of compensations. It is observed that salary plans have fixed pay rates and are appropriate when measurements of performance are difficult to ascertain while, commission plans pay salespeople in direct proportion to their sales and are suitable in maximizing incentives and for predicting sales costs in direct relationship to sales volume. However, a combination plan of salary and commission includes all variations of salary plans plus other monetary incentive plans. This plan is more complex to administer, though it allows for greater incentive to the salespeople and operational flexibility [35]. The major limitation of straight salary compensation scheme is that financial rewards are not tagged directly to any specific aspect of job performance. Consequently, salespeople are likely to get low motivation in accepting challenging tasks and possess low instrumentality in carrying the tasks. Salaried salespeople develop higher value for money perceptions and in the long run cultivate the practice of weighing the challenge of the assigned tasks in terms of monetary benefits.

Compensation management is becoming difficult for organizations due to complex account process and rising demand of salespeople. Managers need to analyze carefully the demand of salespeople to deliver commissions to them in a way that is fair and specific. Companies should be able to manage commission based sales compensation schemes in a way that contributes to maximize the profit and enhance the productivity of salespeople. Commission plan may be designed and customized for employees based on their hiring contract. Prior to developing a compensation plan managers of the firm need to find out the financial determinants that motivate salespeople to increase their performance. The most common way is to offer salespeople variable commissions, where relatively high commissions are paid for sales of the most profitable products or sales to the most profitable accounts. Variable commission rates can also be used to direct the sales force's efforts toward other straight sales objectives like introduction of new products, expansion of sales activities in new territories, and managing key accounts.

The major key drivers for augmenting sales performance are motivationand perceived value of salespeople towards the organizational compensation policies. The intrinsic motivation suggests that an individual will be driven to put forth effort on a given task based on the notion that an acceptable performance level will lead to the reward. Thus, the intrinsically motivated salesperson values things such as personal achievement and success as well as putting additional thrust in selling to improve the level of performance. Since intrinsic motivation results in a series of moods and abilities that are consistent with relationship building, evidence suggests that intrinsic motivation is consistent with successful performance supported with deserving reward [36]. There is a direct link between sales performance and commission earned by a salesperson. Consequently, salespeople are strongly motivated to improve their sales productivity to increase their level of commission until they reach such high pay that further increases become less attractive to draw commissions. Commission plans need to be developed as an integrated part of human resource management. The commission plans needs automation to ensure that good performers are automatically rewarded, whereas poor performers are discouraged from continuing their low productivity. However, straight commission compensation plans have some important limitations that have caused many firms to abandon them. Perhaps the most critical weakness is that management has little control over the sales force.

In addition to the above mentioned modes of compensation, combination plans are drawn on an integrated base salary with commissions, bonuses, or both. When salary with a predefined commission is used, the commissions are tied to sales volume or profitability, just as with a straight commission plan. The only difference is that the commissions are smaller in a combination plan than when the salesperson is compensated solely by commission. The combination plans also offer bonus implemented at the discretion of management for achieving the level of performance set by the management from time to time. However, commissions are typically paid for each sale that is made; a bonus is typically not paid until the salesperson reaches the predetermined level of total sales. When the salesperson reaches the minimum level of performance required to earn a bonus, it is determined by the volume of quota fixed for an individual or for a team of salespeople in a given region. Generally, such targets are set for strategic business units or profit centers identified by the company. Hence payment of bonus is considered as an additional incentive to motivate salespeople to reach high levels of performance, rather than part of the basic compensation plan.

Developing an equitable and functional compensation plan that combines balance, consistent motivation and flexibility is extremely challenging in international operations. This is especially true when a company operates in a number of countries, when it has individuals who work in a number of countries, or when the sales force is composed of expatriate and local personnel. Fringe benefits play a major role in many countries [37]. The expatriates working in high-tax countries prefer liberal expense accounts and fringe benefits that are non-taxable instead of direct income subject to high taxes. A successful sales commission plan should possess three characteristics as given below:

- The plan should offer an immediate positive reward to the salesperson for his achievements.
- The compensation plan linked to the career path should be clear and simple to understand.
- It should be relatively free of influence from factors outside the salesperson's control.

A sales commission plan based on these three principles encourage the profitable selling by the sales force. Salespeople operate under the gun most of the time. They're pressured to produce by the company, by their families, and by their own egos. Paying salespeople a straight salary is uncommon. Such plans are well suited for paying sales support personnel and sales trainees. Sales support personnel, including missionaries and prospectors, are involved in situations in which it is difficult to determine who really makes the sale. Because missionaries and prospectors are concerned primarily with dissemination of information rather than direct solicitation of orders, a salary can equitably compensate for effort. Compensation based on sales results might not be fair.

Table 8.1. Compensation Plans for Salespersons

Components	Needs
Salary	Motivate effort on non-selling activities
	Adjust for differences in territory potential
	Reward experience and competence
Commissions	Motivate a high level of selling effort
	Encourage sales success
Incentive payments	Direct effort toward strategic objectives
	Provide additional rewards for top (bonus) performers
	Encourage sales success
Sales contests	Stimulate additional effort targeted at specific short-term objectives
Personal benefits	Satisfy salespeople's security needs
	Match competitive offers

Salaries can provide control over salespeople's activities, and reassigning salespeople and changing sales territories is lesser a problem with salary plans than with other financial compensation plans. Salary plans also make it easier to encourage teamwork and customer service. Other methods used by the multinational companies include straight commission and a combination of salary and commission plan. Table 8.1 exhibits the reasons for adopting the specific compensation plans.

The straight commission offers strong financial incentives to the salespersons to maximize performance. However, it also limits the control of the sales force. Some industries like real estate, insurance, automobiles, and securities traditionally pay the salespeople straight commission. In these industries, the primary responsibility of salespeople is simply to close sales; non-selling activities are less important to the employer than in some other industries. The huge direct-sales industry, including such companies as Mary Kay Cosmetics, Tupperware, and Avon, also pays by straight commission. Sales managers have recognized that money is the primary means for motivating salespeople. Firms should focus on compensation plans, in reference to objectives, type of plan, size of reward, and frequency of paying compensation to salespeople. The attributes of a successful compensation plan is exhibited in Table 8.2.

Table 8.2. Frequency of Incentives Payment to Salespeople

Frequent Payment Advantages (Monthly/Quarterly)	Infrequent Payment Advantages (Semiannually/Annually)
Sustainable policy of incentives payment need to be designed Salespeople receive frequent feedback and rewards when selling cycle is short. Rewards are close in time proximity to the successes that provided the reward. Frequency of financial rewards develop strong link between successful behavior and reward system Payment of incentives increases individual motivation and develops organizational loyalty	Payments at bonus time are larger and have greater impact. Occasional payments of incentives cause uncertainty and lower the occupational spirit Performance is more stable because short-term sales variations are smoothed over the longer time horizon. Incentives are not paid till end of year that ensures smoother cash flow.

Research from non-managerial disciplines suggests that the use of different schedules of compensation may positively affect sales-force motivation. It is observed that compensation plans and schedules have implications on sales-force productivity. Emerging new companies face challenges in developing their organization as they need the best managers on one hand and lack financial stability to match the compensation and benefits the managers on the other. These firms should evaluate compensation and benefits options by introspecting their firm in reference to compensation and cash flow effects, tax implications, impact on accounting and strategies of competitors in acquiring talents using various compensation packages [38]. Firms need to consider the following perspectives while developing compensation plan for salespeople:

- Type of compensation plan
- Categorical benefits – monetary and non-monetary
- Strategic Fix
  - Revenue/Volume of sales/cost based compensation
  - Product category/Territory Attributes
  - Alliance selling/Routes to market
  - Client focused
  - Customer segment/customer value based
  - Discrete compensation plan
  - Competition linked compensation-Market share based
- Rewards and bonus
- Group compensation.

In many companies, sales management pursues several objectives through the sales force compensation scheme some of which are qualitative and long-run. On the other hand, salesmen within the same sales force are likely to display heterogeneous preference patterns for compensation plan developed by the company. Salespeople need additional money apart from the regular compensation to keep their motivation on for continuous achievement. A study revealed that bonuses may be effective in inducing higher sales productivity because of their flexibility in tying rewards to performance. Further, they may be effective in directing salespeople's efforts toward specific organizational objectives such as promoting new product sales or sales to new customer groups, and enhancing customer satisfaction and retention [39]. Firms need to develop suitable reward systems to offer monetary incentives to the salespeople besides the regular compensation. In developing a reward system managers are required to consider the following attributes:

- Acceptable ratio of costs to sales force output in volume, profit, or other objectives
- Encourage activities consistent with firm's overall, marketing, and sales force objectives and strategies
- Attract and retain competent salespeople, thereby enhancing long-term customer relationships
- Be clear and flexible enough to allow adjustments that facilitate administration.

It has been observed that when companies set new goals, or implement change in initiatives, a domino effect often affects the expectation of employees. New priorities for the company result in new priorities for divisions, departments, and individuals. It is expected that by instituting rewards salespeople assume new tasks or take on additional responsibilities and sales teams are reconfigured. Indeed, companies implementing even the most sweeping of change programs often neglect to address the way people are paid. The result can be that while employees are being directed toward one end, they are being inadvertently motivated to engage in counterproductive behaviors. Manager may pay attention towards learning why any corporation implementing a new initiative should use the occasion to take a close look at its reward structure and the behavior it motivates [40].

Managers need to find compensation methods that benefit both the organization and employees. However methods for creating compensation should be cultivated within the organization considering the organizational culture and general disposition of salespeople. Compensation plans may vary widely and often reflect industry-specific standards. However, some ground rules cut across industries and they can provide a framework for establishing or reviewing the company's sales incentive package. Some of the more important rules include:

- Start with the outcomes and behaviors needed to be fostered
- Prioritize behaviors
- Keep the incentives flexible so that they can evolve along with the company's goals
- Make the compensation plan easy for everyone to understand
- Benchmark competition to stay competitive, and
- Review regularly to measure the relevance and excellence of method of incentives employed.

The compensation plan needs to be specified in reference to commissions earned, in case the salespeople are employed on a commission basis or a combination of salary and commission. Some options include payment at the time of invoicing or product shipment, when the customer payment is received or a combination thereof. Paying full commission when a sale is invoiced or when a product is shipped has the inherent risk of paying out the salesperson on what may be a slow or nonpaying customer. To alleviate this risk, some companies pay commissions only after customer payment is received. Another strategy is to pay half of the commission at the time of invoicing as long as certain conditions are met, with the remainder payable at cash collection. Another condition might be pricing in accordance with the internal discount guidelines. If salespeople get any additional discount approved beyond the guidelines, a proportionately penalized commission percentage should be paid or payment should be made only when the cash is received, or both. However, paying commission against orders causes salespeople to move quickly on to their next opportunity and reduces their interest in delivery and implementation. Deferring payment until the customer pays disconnects sales success from the reward, devaluing the commission scheme. With some schemes, salespeople may not receive the commission associated with a sale for six months or more.

It has been observed in previous research studies that performance of the firms increases with right managerial approaches to balance the task-compensation relationship. The study revealed that compensation for the salespeople should be a mix of salary and commission policy. Managers may emphasize more on commission oriented tasks as they are linked with performance [41].

Global firms rely on designing compensation plans for the salespeople with high base salaries and small commission combination that encourage long-term focus and good corporate behavior. It has been observed that this method of compensation settlement attracts good employee relationships and corporate reputation that help in maintaining a sustainable competitive lead in the marketplace. High base salary and small compensation scheme as an incentive plan drives strategic sales opportunities and encourages salespeople to take more loads of sales and often overachieve their targets. The bonus payments based on company performance and individual salary also provide team oriented motivation. It is observed that salespeople largely demonstrate relationship selling behavior in reference to interaction intensity, preferential treatment of customers, and developing trust. Different relationship selling behaviors indirectly increase the social relationship value and functional relationship value in light of different types of trust, such as the interaction intensity primarily influences social relationship value [42].

HYBRID SALES DESIGNS

In the *international sales division structure*, the firm's activities are separated into two units comprising domestic and international operations. The main function of such an international division is to draw a distinction between its domestic and international business. A worldwide *geographic organization* can overcome the problems associated with the international division structure. In this structure, foreign and domestic operations are not isolated, but are integrated as if foreign boundaries did not exist Worldwide markets are segregated into geographic areas. Operational responsibility goes to area line managers, whereas corporate headquarters maintains responsibility for wide planning and control. Major attributes of the geographic organizational design of multinational companies are as below:

- Product lines are less diverse
- Products are sold to end users
- Sales is a critical variable
- Similar channel is used for sales of all products
- Products are based on the local consumer needs.

This organizational design has various advantages markedly delegation of line of authority and explicit responsibility. Specifically, the merits of this system include:

- Responsibility and delegation of line of authority
- Manufacturing and product sales coordination
- Large number of executives
- Conflicts of roles and responsibilities
- Lack of specialists in the product sales line.

An important disadvantage in geographic organizational design may be the large number of top level executives involved in operational tasks which leads to conflict in power play and command execution in the organization. Besides the agglomeration of top management personnel, the individual products may suffer, as responsibilities cannot be fixed easily on the operational executives. A product organization design is different than geographic design wherein a worldwide responsibility to product group executives at the line management level is assigned and emphasis is placed on the product line rather than on geographic differences. The coordination of activities in a geographic area is handled through specialists at the corporate staff level whereas in the product organizations focus is laid on the performance of product-mix in a given area. Multinational companies operating within this structure have variety of end users, handle diversified product lines with high technological capability and logistics cost are diverted to the local manufacturers. This type of organizational design has several benefits including:

- Decentralization of authority
- High motivation of the divisional heads
- Adding or dropping new products have marginal impact on operations
- Control of product through the product life cycle.

In this organizational structure, a firm is segregated along product lines considering each division as a separate profit center with the division head directly accountable for profitability. Decentralization of operations is critical in this structure and more decisions are likely to be left to the local manager, who is then usually more highly motivated. This structure allows the product managers to add new products and product lines and withdraw old ones with only marginal effect on overall operations. Another advantage of this structure is that the control of a product through the product life cycle can be managed more readily and securely. However, firms following this organization structure often face the problem of coordination among product and territory managers. In addition, it is felt that executives quickly get biased towards the regional and corporate staff in managing any product process.

In recent years, a synergy of all the above organizational structures has emerged among the multinational companies which are defined as matrix structure. The matrix structure offers greater flexibility than the single lineof-command structures already discussed and reconciles this flexibility with coordination and economies of scale to keep the strength of large organizations. The attributes and advantages of matrix organization include:

- Multiple command line
- Product and geographic coordination
- Product lines in a national setting
- Organization design reacts quickly to the local environment demand.

For the multinational firm, the matrix organization is a solution to the problem responding to both economic and political environments. General Electric Company in Asia operates with matrix structure and has been successful. A matrix organization can encompass geographic and product-management components. However, some of the disadvantages in following this organizational design are power struggles among the supervisory personnel and parallel decision making.

Global Local Marketing

The global growth products of multinational companies are mostly centralized in the country of origin and the products that emerge tend to have features, advantages and benefits specified by central sales system of the company. Hence, key technologies and major product introductions cater primarily to customers in that geographical region. Sales and customers in other regions are relegated to acceptance of custom modifications; or they have the choice to buy from other local suppliers. The product targeting goes beyond the perceived use values of the customers, local preferences and local language. Expectations regarding size, shape, customized items, price and availability vary widely. Hence regional markets tend to be dominated by local companies. Often the companies offer locally engineered or customized products at a differential price to win market share. For growth and success in the new global economy, the guiding principle must be- *Go global and think Local*. Automation suppliers must become truly global by allowing local development of products for local markets. The best approach is to develop technology (hardware & software) through global alliances – preferably with relatively small, fast-moving local companies. In a global market, there are following key factors that constitute the winning difference:

- Sales abilities that assess correctly the local needs in a global arena.
- Proprietary technology and products targeted specifically for local markets.
- High-value-added services offered through effective local service providers.

In the global village of the new economy, automation companies have little choice – they must find more ways and means to expand globally. To do this they need to minimize domination of the central corporate culture, and maximize responsiveness to local customer needs.

Ever since Fujifilm began actively exporting its products throughout the world in the early 1960s, it has been one of Japan's leading companies regarding overseas operations. Besides establishing a world-spanning network of local sales bases, it was among the first Japanese companies to initiate overseas manufacturing, starting with the 1974 construction of a color photographic paper processing plant in Brazil. Since the 1980s, Fujifilm has arranged for Japanese and local technical staff to cooperatively design and construct overseas manufacturing facilities that make appropriate use of the Company's unique technologies in harmony with local conditions. These efforts have enabled the steady expansion and strengthening of Fujifilm's global production, sales, and service networks. Having consistently placed strong emphasis on understanding regional characteristics and on respecting and adapting to local cultures, Fujifilm has been highly praised for its localization efforts and contributions to the communities in which it operates. The Company views globalization and localization as two equally crucial elements of its overall business strategy. The localization has been done by the companies to serve the markets in Americas, Europe and Asia-Pacific.

As continued growth in the digitization of diagnostic imaging has also supported rising demand for dry medical imaging film in North America, Fuji Photo Film, Inc., based in Greenwood, South Carolina in the United States, has been proceeding with the expansion of its medical-use film factory. The company completed that facility and began operating an integrated manufacturing system that performs a full range of manufacturing processes, from coating to processing. Because the range of products manufactured at the facility has been expanded to include dry medical imaging film, Fujifilm has significantly upgraded its systems for efficiently supplying customers throughout North America with its high-quality medical imaging products and related services. Fujifilm's headquarters in Europe, Fuji Photo Film (Europe) GmbH, recently established sales companies in Poland, the Czech Republic, Slovakia, and Italy. In view of the huge changes under way in Europe, Fuji Photo Film (Europe) has been seeking to expand and strengthen its sales systems so that it can accurately respond to local needs and expeditiously supply products and services. Fujifilm began sales the DocuCentre Color series of digital color multifunction machines in the Asia-Oceania region. The Company also proactively proceeded with measures to expand its manufacturing and sales systems in the region, particularly in China, which has markets that are expected to grow greatly [43].

The world's most recognized companies and brands like Coca-Cola continues to prosper by innovating and adapting to the local needs of its customers and consumers throughout the world. Despite ferocious competition, significant currency devaluations in key markets and major acquisition-related write-downs, the company reported first quarter 2002 sales of US$4.08 billion and has predicted long term growth approaching six percent. Global Sales helps set global sales strategies and product positioning, which is then implemented locally and adapted to local sales needs.

PLAYING IN THE COMPETITION

Organizations that seek to win the sales by modeling their strategy in a suitable way should build their energy on two platforms. First, they need to endeavor to model the competitive game in view of the various entities involved such as organizational players, sales arenas, information builders and scorers. Second, the company should get acquainted to the selling rules as how the customers, channels, factors and institutions are attracted, won and retained in the business. The competitors strive to attract, obtain, and retain the support, commitment, and involvement of end customers, channel members, factor suppliers, and institutions within the context or conditions of strategy games. Such elements include the structure of the game combatants, the arenas, the nature of the stakes they hold, and the entire composition of the business domain. The marketplace rules are governed by the factors such as how to introduce best packaging and distribution of products, create brand image and corporate reputation, deliver service, and build sustainable relationships. Many firms may intend to redefine or rescale the customer satisfaction from narrow product functionality to include all aspects of interactions with customers, for example, the Japanese automobile manufacturers have redefined *quality* for many consumers to distinguish their quality, services and customer values against the competing overseas brands. In the competitive marketplace some firms may look for introducing frequently the product upgrades to cope up with pace of sales competition. Besides, some firms may initiate major efforts to develop new capabilities and competency and monitor their progress in doing so against specific competitors as one element in their scorecard to determine who will reserve their plans to win in the future marketplace.

The companies try to alter the number of players by creating situation for deposing them from the market or change their own position relative to other players. The strategies such as alliances, mergers and acquisitions are the direct means of reducing the competition or outperforming the existing rivals from the market. The Hindustan Lever Ltd., the giant in the FMCG segment in India, has largely reduced the number of players in the consumer goods segment by continuous acquisitions and mergers of potential firms. The development of networks, linking suppliers, manufacturers and consumers is another popular strategy to discourage competition in the particular segment of goods and services. The quick implementation of research and development, new products development and brand extensions indirectly break the existing competition in the sales and allow the new company to employ its selling strategies. The pace of rivalry is such that no firm can now afford to take its resources for granted. Some firms may perceive to be in the low-tech business segment such as textiles, shoes etc. Furniture, paint, and books, are feverishly pursuing new knowledge that might radically reshape established products or traditional ways of manufacturing and distributing them. In high-tech businesses such as electronics, firms including IBM, Apple, Motorola, Intel, and Microsoft have formed multiple alliances with entities all across the activity/value chain and share innovative knowledge, skills and capabilities. Several modes of competition can be employed within the end-customer and channel arenas to get and keep customers. Although rivals compete in many distinct ways, firms can make eight key choices to distinguish and differentiate themselves in the eyes of customers and channels, including product-mix, product features, functionality, service, distribution or availability, image and reputation, selling and relationships, and price [44].

Companies also maneuver the arena of customers, channels, institutions and the geographical coverage in order to reconfigure their competitive strategy. The software companies like Intel, Microsoft and 3M always keep extending the product line implementing the R&D results and never let the competition stagnate in the end-customer arena. The healthy companies feel that the greater the competition the higher will be challenge to establish the brand among the buyers. The Suzuki collaboration with Maruti Udyog Limited in India has changed the dominance of the popular brand holder Premier Automobiles Ltd and created new competitive context of small city cars. The channels of supply for any company are always vulnerable to competition. The common practice followed by the competitors is breaking the supply chain by offering more perks and margins than the leading brand. The collaboration between the Proctor and Gamble and Wal-Mart involving strong integration of the product ordering, inventory control and logistic may be a classic example in this context [45]. The factor advantage in the competition may be defined as the relationship of the manufacturing, marketing or sales oriented company with the service providers who develop loyalty towards them. The service providers may be the suppliers of raw materials, packaging services, hiring of machines and the like. Many companies use the legal support, government patronage etc to shape the competitive conditions to their advantage while building the institutional arena in the business.

Satisfying buyer needs may be a prerequisite for industry profitability. One of the underlying issues in developing the competitive strategy is to address the profitability in reference to the capability of the firm whether it can capture the value in the process for retaining the buyers, or whether this value is competed away to others. The buying power of customers determines the extent to which they retain most of the value created for themselves. The threat of substitutes determines the extent to which some other product can meet the same buyer needs, and thus places a ceiling on the amount a buyer is willing to pay for an industry's product. The power of suppliers determines the extent to which value created for buyers will be appropriated by suppliers rather than by firms in an industry. The intensity of rivalry acts similar to the threat of entry. It determines the extent to which firms already in an industry will compete away the value they create for buyers among themselves, passing it on to buyers in lower prices or dissipating it in higher costs of competing. Conceptual framework of competitive forces in the marketplace has been provided by Porter as a five-force model for industry analysis is shown in Exhibit 8.4.

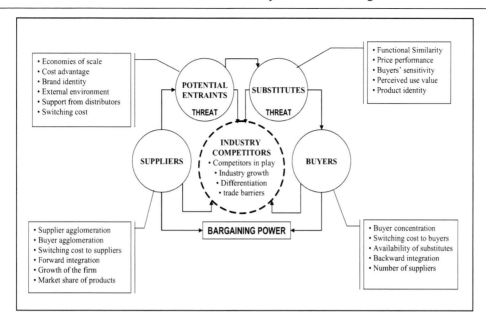

Figure 8.4. Competitive Strategy: Network of Sales Forces.

The strongest forces become the dominant factors in determining industry profitability and the focal points of strategy formulation. The model identifies the key structural features that determine the strength of the competitive forces within an industry in reference to profitability. It may be explained through the model that the degree of rivalry among different firms is a function of the number of competitors, industry growth, asset intensity, product differentiation, and exit barriers. Among these the most influential variables may be identified as the number of competitors and industry growth. The industries with high fixed costs tend to be more competitive because competing firms are forced to cut price to enable them to operate at the economies of scale. However, with the differentiation strategy rivalry is reduced among the products and services offered by the competitors, in both real and perceived senses.

Another significant concept which may be explained through this model is the associated difficulty of exit from an industry, which may result into struggle for survival. Further, there remains the threat of entry of new firms which may enhance competition in the industry. Several barriers, however, make it difficult to enter an industry. Two cost-related entry barriers are economies of scale and absolute cost advantage. In this process of competition the new entrants face an up-hill challenge of scaling at a high level of production or to accept a cost advantage. However, absolute cost advantage remains with the long standing firms in the market which possess technological and brand advantages for their products and services. The substitute products in the market affect the industry potential adversely as well as pose threat to the customer preferences. Bargaining power of buyers refers to the direct or indirect pressure tactics to force the industry to reduce prices or increase product features, in view to optimize the customer value. Buyers gain power when they have choices-when their needs can be met by a substitute product or by the same product offered by another supplier. In addition, high buyer concentration, the threat of backward integration, and low switching costs add to buyer power. Similarly the bargaining power of suppliers refers to their ability to force the industry to accept higher margins or reduced services, in the interest of augmenting or securing their profits. The factors influencing supplier power are the same as that of buyer power. In this case, however, industry members act as buyers.

SELECT READINGS

Boyatzis, R (1982), *The Competent Manager: A Mode for Effective Performance*, John Wiley & Sons: New York.
Fahey, L. (1999), *Competitors: Outwitting, Outmaneuvering and Outperforming*, John Wiley and Sons, Canada.
McGonagle, J. J. (1996), *A New Archetype for Competitive Intelligence*, Quorum Books, Westport, CT.
Porter, M. E. (1980), *Competitive Strategy: Techniques for Analyzing Industry and Competitors*, The Free press, New York.
Singe, P. (1990), *The Fifth Discipline*, Doubleday/Currency, New York, 1990.
Warren, K. (2002), *Competitive Strategy Dynamics*, Wiley, New York.

REFERENCES

[1]   Zoltners, A. A., Sinha, P., and Lorimer, S. E. (2006), Match Your Sales Force Structure to Your Business Life Cycle, *Harvard Business Review*, 84 (7-8), 80-89.
[2]   Likierman, A. (2009), The five traps of performance measurements, *Harvard Business Review*, 87 (10). 96-101.
[3]   Darr, A. (2003), Control and autonomy among knowledge workers in sales: an employee perspective, *Employee Relations*, 25 (1), 31-41.
[4]   Carter, T. (2006), Sales management coaching: A model for improved insurance company performance, *Journal of Hospital Marketing & Public Relations*, 16(1), 113-125.
[5]   Hultink, E. J. and Atuahene,G. K. (2000), The effect of sales force adoption on new product selling performance, *Journal of Product Innovation Management*, 17 (6), November, 435-450.
[6]   Farrell, M. (2005), The effect of a market-oriented organizational culture on sales-force behavior and attitudes, *Journal of Strategic Marketing*, 13 (4), 261-273.
[7]   Lloyd, C. and Newell, H. (2001), Capture and transfer: improving the performance of the pharmaceutical sales representatives, *The International Journal of Human Resource Management*, 12(3), 464-483.
[8]   Panagopoulos, N. G., and Avlonitis, G.J. (2010), Performance implications of sales strategy: The moderating effects of leadership and environment, *International Journal of Research in Marketing*, 27 (1), 46-57.
[9]   Kandybin, A. (2009), Which innovation efforts will play, *Sloan Management Review,* 51 (1), 53-60.
[10] LaForge, R. W., Ingram, T. N. and Cravens, D. W. (2009), Strategic alignment for sales organization transformation, *Journal of Strategic Marketing*, 17(3), 199-219.

[11] Rajagopal (2006), Measuring Customer Value and Market Dynamics for New Products of a Firm: An Analytical Construct for Gaining Competitive Advantage, *Global Business and Economics Review*, 8 (3-4), 187-204.

[12] Magrath, A. J. (1997), From the practitioner's desk: A comment on personal selling and sales management in the new millennium, *Journal of Personal Selling and Sales Management*, 17 (1), 45-47.

[13] Piercy, N. and Rich, N. (2009), The implications of lean operations for sales strategy: from sales-force to marketing-force. *Journal of Strategic Marketing*, 17(3), 237-255.

[14] Mayer C (2002), Financing the New Economy: financial institutions and corporate governance, *Information Economics and Policy*, 14 (2), 311-326.

[15] Zoltners, A. A., Sinha, P., and Lorimer, S. E. (2006), Match your sales force structure to your business life cycle, *Harvard Business Review*, 84 (7), 80-89.

[16] Shultz, R. J. and Good, D. J. (2000), Impact of consideration of future sales consequences and customer oriented selling on long term buyer-seller relationship, *Journal of Business and Industrial Marketing*, 15 (4), 200-215.

[17] Kinni, T. (2004), *How strategic is your sales strategy*, Harvard Business Publishing Newsletter, Harvard Business School, February 01.

[18] Chen-Ritzo, C. H., Ervolina, T., Harrison, T. P., and Gupta, B. (2010), Sales and operations planning in systems with order configuration uncertainty, *European Journal of Operational Research*, 205 (3), 604-614.

[19] Lane, N. and Piercy, N. (2009), Strategizing the sales organization, *Journal of Strategic Marketing*, 17(3), 307-322.

[20] Goldberg, E. M. (2008), Sustainable Utility Business Continuity Planning: A Primer, an Overview and a Proven Culture-Based Approach, *The Electricity Journal*, 21 (10), 67-74.

[21] Schein, E. (1992), *Organizational culture and leadership*, San Francisco, Jossey-Bass

[22] Braam, G. J. M., and Nijssen, E. J. (2004), Performance effects of using the Balanced Scorecard: a note on the Dutch experience, *Long Range Planning*, 37 (4), 335-349.

[23] For details see corporate web site of MPSI Systems Inc. http://www.mpsisolutions.com

[24] Rudd, J. M., Greenley, G. E., Beatson, A. T., and Lings, I. N. (2008), Strategic planning and performance: Extending the debate, *Journal of Business Research*, 61 (2), 99-108.

[25] [Shipley, C. J. and Kleiner, B. H. (2005), Compensation Management of Commissioned Sales Employees, *Management Research News*, 28 (2), 2-10.

[26] Darmon R Y (1997), Predicting the long-run profit impact of a contemplated sales force compensation plan, *Journal of the Operational Research Society*, 48 (1), 1215-1225.

[27] Rouzies, D., Segalla, M., and Besson, M. (1999), *The Impact of Cultural Dimensions on Sales Force Compensation*, Working Paper, No. 670, HEC Business School, France, 1-45.

[28] Ryals, L. J. and Rogers, B. (2005), Sales compensation plans-One size does not fit all, *Journal of Targeting, Measurement and Analysis for Marketing*, 13 (4), 354-362.

[29] Joseph, K., and Kalwani, M. U.(1998), The role of bonus pay in sales force compensation plans, *Industrial Marketing Management*, 27 (2), 147-159.

[30] Lawler E E (1990), *Strategic pay: Aligning organizational strategies and pay systems*, San Francisco, CA, Jossey-Bass.

[31] Siguaw, J. A., Kimes, S. E., and Gassenheimer, J. B. (2003), B to B Sales Force Productivity: Applications of Revenue Management Strategies to Sales Management, *Industrial Marketing Management*, 32 (7), 539-551.

[32] Gundy, P. (2002), Sales compensation programs: Built to last, *Compensation & Benefits Review*, 34 (5), 21-28.

[33] Cocks, D.J. and Gould, D. (2001), Sales Compensation: A New Technology-Enabled Strategy, *Compensation & Benefits Review*, 33 (1), 27-31.

[34] Pfeffer, J. (1998), Six dangerous myths about pay, *Harvard Business Review*, 76 (3), 108-109.

[35] Steinbrink, J. P. (1978), How to pay your sales force, *Harvard Business Review*, 56 (4), 111-122.

[36] Mallin, M. L.,and Pullins, E. B. (2009), The moderating effect of control systems on the relationship between commission and salesperson intrinsic motivation in a customer oriented environment, *Industrial Marketing Management*, 38 (7), 769-777.

[37] Peterson, R. B., Napier, N.K., and Shim, W. S. (2000), Expatriate Management: A Comparison of MNCs across Four Parent Countries, *Thunderbird International Business Review*, 42(2), 145-166.

[38] Dubinsky, A. J., and Berkowitz, E. N. (1979), The frequency of monetary compensation for salesmen, *Industrial Marketing Management*, 8 (1), 12-23.

[39] Joseph, K., and Kalwani, M. U. (1998), The role of bonus pay in sales force compensation plans, *Industrial Marketing Management*, 27 (2), 147-159.

[40] Sisk, M. (2005), *Reward system that really Works*, Harvard Business Publishing Newsletter, Harvard Business School, September.

[41] Rajagopal, Pitt, M and Price, S (2010), Measuring sales performance of home decor products, Journal of Retail and Leisure Property, 9 (2).

[42] Chen, T., Chen, C., and Tsung, C. (2007), Promoting relationship selling behaviors to establish relationship value- The case of international airlines, *Journal of Relationship Marketing*, 5(4), 43-62.

[43] Fujifilm Corporate Hoame page: http://home.fujifilm.com/info/profile/operation.html 27 May 2004.

[44] For detailed discussion on the corporate key choices to withstand competition, see Peter Senge: *The Fifth Discipline*, New York, Doubleday/Currency, 1990.

[45] Fahey, L. (1999), *Competitors: Outwitting, Outmaneuvering and Outperforming*, John Wiley and Sons, Canada.

Chapter 9

SALES PROMOTION

There are varieties of sales promotions targeted at consumers, which include coupons, sweepstakes, and free offers to attract consumers and enhance the volume of sales. Such promotions are becoming a large and growing part of marketing budgets among large and multinational companies. Sales promotions have many distinct aspects comprising an economic aspect that provides incentives to purchase a brand, and motivates purchase decisions of consumers, and an affective aspect that influences how consumers feel about their shopping transactions. It is necessary to design a customer value based promotional offer and communicate it widely through various modes of publicity and social networks to maximize the perceived economic value of promotion and its profitability[1]. The common methods followed by sales oriented companies to promote the sales of its products and services include cross selling, price discounts, volume packs etc. However there are some innovative ideas, which are used by the multinational companies as described below:

- Coupons- Assuring benefit
- Bonus pack- more product for regular price
- In-pack, On-pack, Near-pack
- Specialty Container- Value added benefits
- Continuity program- rewards system for multiple buying
- Refund- Price draw-back
- Sweepstakes- Random chance of winning
- Contests
- Mail premium- Post-purchase benefits
- Sampling
- Price-off-discounts
- Trade deal- Incentives to retailers
- Cause marketing- social benefit oriented (like soliciting contribution to UNICEF)

Besides these services, retail stores provide many pre- and post-purchase services to satisfy the customers. The *pre-purchase services* include accepting orders via telephone and by mail, in-store and outside store advertising, interior and window display, consumer convenience, fitting rooms, shopping hours, and organizing consumer entertainment such as the integrated cable TV, fashion shows, recipe contest, baby shows etc. The retailers provide selected *post-purchase* services to consumer such as home delivery of goods, gift packaging, returns and exchanges, tailoring, installation, demonstrations, and accepting credit transactions. In addition to the services, listed large and chain retail stores provide leisure and general amenities to the consumers like rest rooms for consumers, baby sitters, restaurants, etc.

Managers go through a long process of inducting a new product or service in a market by investing in sales promotions like sweepstakes, coupons, time-limited price discounts, free gifts or samples, special events, displays, membership rewards, and consumer-directed promotions that may develop a unique selling proposition to new product. In a competitive business environment firms intend to stay ahead in promotional strategies as strong promotions also help in building consumer loyalty. The sales promotions absorb a significant portion of a company's investment in marketing budgets. However, it is observed that the stronger the promotion, the higher the competitive advantage. It is found that when all these strategic factors are aligned, the result is a successful promotion, as evident with successful promotions conducted by General Motors, Home Depot, and Procter & Gamble. However, such promotional strategies neither require inventiveness, nor originality, nor swift action. Managers in such organizations must not only successfully tailor a promotion to its intended market, but also skillfully drive it through internal barriers[2].

Firms engaged in marketingof industrial and consumer goods tend to improve the management of their complex distribution channels by using various promotional techniques known as the "promotional mix". Marketing organizations can use sales promotion to move merchandise through the distribution pipeline by carefully organizing the channels of distribution, selling through and not to distributors and dealers, and determining product objectives. This concept has a focused scope that is highly cost-effective and applicable to a wide range of products and situations[3]. Promotional strategies have been considered as an effective tool for improving the sales performance in a marketplace. A firm's promotion plan is basically comprised of the four Ps including product, people, perceived value, and psychodynamics, combined with the Attention, Interest, Desire and Action (AIDA) factors, A decade of company-based research suggests, however, that it is time to rebuild the marketing machine by focusing on the key strategic issues that companies, and marketers, face in today's rapidly evolving, digitized marketplace. If marketing is to become a way of doing business rather than merely one of several organizational functions, marketers must recognize that the promotional-mix toolkit is important for success[4]. The promotion planning has to be done considering the following parameters:

- Who are the customers to reach?
- What are the determinants of consumer behavior?
- What is the goal of sales program?
  - Loyal users
  - Short-run profit
  - Reinforcing the market
  - Enhance sales territories
  - Cross selling

The benefit of sales promotions is that they induce choice. However, this benefit may drive slow moving brands that are not commonly preferred by the consumers. Despite the fact that sales promotions have long been employed in marketing practice and researched academically, a clear understanding of the impact of sales promotion on post-promotion brand preference continues to support the sales of products. Some studies suggest that, on average, sales promotions drive brand preference among consumers. However, depending upon characteristic of thesales promotion and the promoted product, promotions can lay impact on preference for a brand[5].

PROMOTING RETAIL SALES

Large self-service retail stores or supermarkets are potential outlets where customers experience innovative promotions on variety of products and services which drive the buying decisions. Among various promotional offers price discounts, free samples, bonus packs, and in-store display are associated with product trial. Trial determines repurchase behavior and also mediates in the relationship between sales promotions and repeat buying behavior[6]. Repeat buying behavior of customers is largely determined by the values acquired on the product. The Attributes, Awareness, Trial, Availability and Repeat (AATAR) factors influence the customers towards making re-buying decisions in reference to the marketing strategies of the firm. The decision of customers on repeat buying is also affected by the level of satisfaction derived on the products and number of customers attracted towards buying the same product, as a behavioral determinant[7]. Among growing competition in retailing consumer products, innovative point of sales promotions offered by super markets are aimed at boosting sales and augmenting the store brand value. Purchase acceleration and product trial are found to be the two most influential variables of retail point of sales promotions. It has been found that there exists significant association between the four consumer promotional approaches including coupons, price discounts, samples and *buy-one-get-one-free,* and compulsive buying behavior[8]. The occurrence and the choice of appropriate retail sales promotion techniques are important decisions for retailers. It is crucial for the retailing firms to apprehend the mechanisms involved at the consumer level regarding these sales promotions. Variables such as variety seeking, perceived financial benefit, brand loyalty and store loyalty towards point of sales promotions have specific influences on the buying behavior and volume of retail sales[9]. Retail sales promotions should not only be targeted to the potential buyers and designed to drive the following type of consumers in particular:

- Competitively Loyal consumers
  - Intense loyalists
  - Value buyers
  - Habit-bound buyers
- Switchers
  - Availability
  - Value
  - Occasion usage
  - Variety

  - Price buyers
- Non-users
  - Price
  - Value
  - Lack of need

Leisure shopping is influenced by time and attractiveness of sales offers which include variables*viz.* hours of work, public holidays, and paid leave entitlements, point of sales promotions and effectiveness of customer relations. These factors vary widely in reference to consumer segments and markets attractiveness, and induce compulsive buying behavior among customers. The shopping behavior of consumers is largely governed by the satisfaction in spending and perceiving pleasure of buying occasionally exercising choice and passing time in knowing new products, services, technologies, and understanding fellow customers[10]. Retailers using a "store as the brand" strategy invest in creating a specific, unique shopping experience for their target customer and encourage leisure and group buying behavior where delivery of customer satisfaction seems to be an effective source of differentiation. Change-of-season sales are most frequently introduced with attractive sales promotions in reference to price discounts or two for one price basis and linked with objectives of moving a volume of stock. Retail promotional sales also include general sales, and these are linked with other promotional objectives and activities such as increasing profit and inventory management[11].

Promotions Driving Compulsive Buying

Compulsive buying is closely associated with the obsessive behavior of customers who orient to their mind to acquire certain products or services. There exists a close relationship between compulsive buyers and specific types of external stimuli such as sales promotions and bargains offered in the large self-service retail stores. Customers who have a higher tendency to buy compulsively are more prone to promotions and are more likely to yield to innovative sales promotions in retail stores. Such customers have a greater likelihood to use promotion tools such as Electronic Cash Cards (ECC), Shopping Advantage Cards (SAC) and Bulk Purchase Price offers (BPP) offered by retail stores, and subsequently have a greater incidence of compulsive shopping [12]. Clinically, compulsive buying is closely related to major depression, obsessive–compulsive disorder, and in particular, compulsive hoarding. Like compulsive hoarding, compulsive buying is thought to be influenced by a range of cognitive domains including deficits in decision-making, emotional attachments to objects and erroneous beliefs about possessions, and other maladaptive beliefs[13]. Sales promotions should be designed to effectively converge with consumer choice and brand preference. The process of influence of sales promotion on consumer behavior and loyalty is exhibited in Figure 9.1.

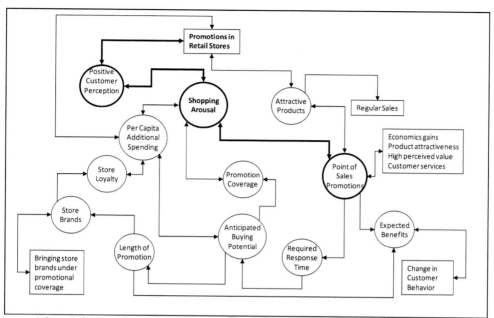

Source: Adapted from Rajagopal (2008), Point of sales promotions and buying stimulation in retail
stores, *Journal of Database Marketing and Customer Strategy Management, 15 (4), 249-266*

Figure 9.1.Covergence of Sales Promotions, Consumer Behavior and Loyalty Attitude.

Five essential qualities of aesthetic judgment, which include *interest, subjectivity,
exclusivity, thoughtfulness, and internality,* need to be nurtured among consumers to develop
conviction in buying and generating shopping arousal among consumers as exhibited in
Figure 9.1. The quality of aesthetic judgment driven by in-store aura and arousal on new
products, exercised by the customers in association with the sales promoters, determines the
extent to which new products and brands promoted enhance quality of life[14]. Convergence
of sales promotion, customer's perceptions, value for money and product features drive
arousal among customers. The nature of customer-retailer relationship functions as the key in
selling and buying process in reference to in-store promotions. However, in this process the
perceptional problems with customers can greatly devalue the customer-promoter relationship
and brand as a whole. Consumer appreciation of premium-based promotional offers is more
positive when the premium is offered through an easy process and in combination of
relatively lower quantity of products to purchase. It has also been found that when value of
the premium is mentioned and brand perception is positive, compulsive buying tendencies are
higher among customer[15].

The in-store environment variables driving impulsive buying behavior include display of
point of sales posters, exhibiting promoting discounts and cheaper prices, while the
atmosphere engagement referring to enjoyment, elegance, and attractiveness is conveyed by
in-store point of sales posters. Such behavioral drivers may also be referred as in-store
promotional effect and atmospheric effect[16]. It has been observed in some studies that
consumers who intend to do shopping in short notice, generally lean towards impulsive or
compulsive buying behavior driven by arousal effect in the retail stores. Gender, age, leaning
towards unplanned purchases, and tendency to buy products not on shopping lists, serve to
predict compulsive tendencies. However, there are some common strategies adopted by

retailers to overcome the problems of fickle consumers, price-slashing competitors, and mood swings in the economy. Such wishful thinking holds that sales promoters can thrive only if they communicate better with consumers during pre-purchase situations and assist in product demonstrations involving consumers to help their purchase decisions[17].

Hispanic Consumers Shopping Behavior

Human personality traits are determined by multi-dimensional factors like the individual's behavior, appearance, attitude and beliefs, and demographic characteristics. Based on the trait theory, researchers have concluded that there are five stable personality dimensions, also called the 'Big Five' human personality dimensions[18]. The Big five factors include extraversion, agreeableness, conscientiousness, neuroticism and openness to experience. Relationship between the point of sales promotions and retail buying decisions is largely governed by the psychographic variables that can be measured broadly by the closeness and farness of the personalities of brand and customer. The type of relationship that customers possess with the point of sales promotions offered by retail stores is largely based on the loyalty levels[19]. The new generation marketing approaches include customer focused, market-driven, outside-in, one-to-one marketing, data-driven marketing, relationship marketing, integrated marketing, and integrated marketing communications that emphasize two-way communication through better listening to customers and the idea that communication before, during and after transactions can build or destroy important brand relationships. It has been observed that Hispanic consumers are sensitive to the price while making buying decisions have higher tendency to buy compulsively, are more prone to promotions and are more likely to use online sales promotions[20].

Hispanic consumers are found to be attached firmly to the ethnic culture and tend to shop at the same store, especially those stores owned by members of the subculture and stores with Spanish-speaking salespeople. Marketers reinforce the relationship between consumers and their stores by introducing periodical sales promotions. In general, Hispanic consumers show the tendency of buying products offered in sales promotions. In Latin America consumers' diversity is apparent and so is their attitude towards promotions. Consumers consider relative advantages in perceived price and product promotions, and prefer big-bargains offered by the retail stores to buy. Retailers accrued higher benefit from such buying behavior of consumers while defining their promotional strategies, especially emphasizing on Everyday Low Prices (EDLP) strategy like Wal-Mart[21]. Practically consumers react favorably to leisure sales campaigns of goods and services. However, customer value plays a decisive role in the shopping process. Sales promoters instill emotions among customers in terms of merchandise choice, visual merchandising, store environment, attitude of salespeople, pricing policies and promotional activities during the pre-purchase stage. These factors are the very foundations of consumer satisfaction and decision drivers towards buying products[22].

Point of sales promotion strategy towards prospecting new customers and generating shopping arousal among the existing customers through in-store ECC, SAC and BPP involves impulsive and compulsive buying process. The effects of location convenience, one-stop shopping convenience, reputation of retail stores, in-store ambience, and direct mailings generate shopping arousal among the leisure shoppers. Satisfaction and trust developed by the retail stores during pre-purchase phase help persuading the sales promotions among customers[23]. Consumers respond encouragingly to point of sales campaigns run by the self-service retail stores in Mexico, however, customer services associated with point of sales promotions and perceived promotional advantage play a decisive role in the buying process. Information on the current point of sale promotion and previous experience of customer with the promotional offers of the retail store stimulate consumer feelings and prompt their decision towards experimenting new products on promotional offers[24]. The strategy of point of sales promotion in acquiring new customers, retaining existing customer and increase the volume of sales by shopping arousal has been derived from the previous research studies reviewed in the pretext and exhibited in Figure 9.2.

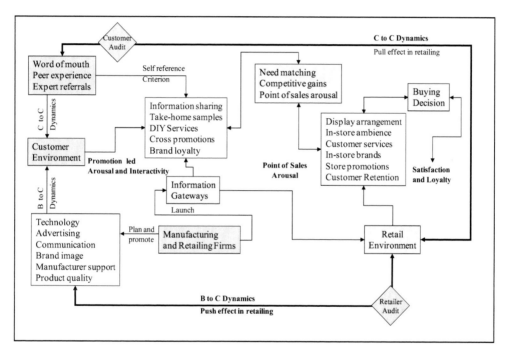

Source: Adapted from Rajagopal (2008), Point of sales promotions and buying stimulation in retail stores, *Journal of Database Marketing and Customer Strategy Management, 15 (4), 249-266*

Figure 9.2.Interactive Sales Promotion Strategies in Customer Prospecting and Developing Shopping Arousal.

The above figure illustrates that shopping arousal and compulsive buying behavior are developed by the retailing firms through point of sales promotions engaging innovative promotional offers. Such promotional attractions to oversell products drive business-to-consumer (B-to-C) marketing approaches. Retailing firms develop the point of sales satisfaction and post-sales satisfaction which prompt consumer-to-consumer (C-to-C) interactions and help in building store loyalty. Many companies are successfully following the art of customer relationship management through the point of sales promotions. These firms design the sales promotion programs according to preferences of consumers and buying behavior. Sales promotions are implemented in finer segments of buyers and offer promotions on slow moving products and services to accelerate the sales. But all too often those efforts are too narrow as they concentrate only on the promotion schemes where the customer comes into contact with the company. Few businesses have looked at the customer scenario in the broad context in which customers select, buy, and use products and services with customized sales promotions. For example Schneider Electric Company a manufacturer of telecommunication transponders, looked beyond the purchasing agents that buy in bulk to find ways to make it easier for engineers to design components into their specifications for mobile telephones. By developing a customer scenario that describes how people actually shop for groceries, Tesco Direct, an online and catalogue based sales company, learned the importance of Web shopping and how to reduce search costs and increase satisfaction of online customers[25]. Sales promotion strategies should be subject to the following rationale for optimizing cost effectiveness and consumer response:

- Sales potential
- Expected sales growth rate
- Ability to exert sales leadership
- Scope of developing new brand segments
- Cost of market entry
- Cost to serve
- Channel abilities and strategic punch
- Channel configuration cost
- Competitive environment
- Strategic fit

As new and exciting products are introduced, firms prospect the consumers through inter-personal negotiations managed by the sales promoters and inculcate high arousal among customers towards buying these products. The *visual merchandising* and computer aided simulations act as stimuli to consumers who intend to elicit apositiveresponse. This creates shopping arousal among customers in reference to merchandise choice, store ambience, attributes of promotional products, perceived use value, pricing policies and promotional activities. These factors may be considered as foundations of consumer behaviortowards point of sales promotions offered in retail stores[26]. Visual effects and economic advantage associated with promotional products in the retail stores often stimulate the compulsive buying behavior. Point of sales brochures, catalogues and posters build assumption on perceived use value and motivational relevance of buying decisions of product. Emotional visuals exhibited on contextual factors such as proximity or stimulus size, drive perception and subjective reactions on utility and expected satisfaction of the products[27]. In addition, a pleasant store ambience where attractive displays, music, hands-on experience facilities and recreation are integrated helps in maximizing the consumer arousal towards buying. It has been observed that consumers perceive positive effect during interaction with sales promoters if arousal is high[28]. The impact of initial interactions among fellow customers about the point of sales promotions can be measured in reference to the degree of stimulation gained by customers. Interactive tools on product learning provided by the retailers significantly affect the level of arousal and pleasure which contribute towards experience, and thereby influence the buying behavior. As higher stimulation or interactive learning provided by the sales promoters focuses on gaining initial experience on the product use, consumers tend to engage in higher arousing activities and adhere to the sale promotions offered by the retail stores[29].

CLEARANCE SALES

Theoretically, it issuggested that consumers willreact favorably to aclearance sale however customer values go negative for the high-technology and high use value products in proportion tothe consumer's individual behavior to the discount sales option.It may be argued that individual consumer behavior to a buyingoption is a functionof preference for theoption, whether it isa considered option or any choice constraintis personally directed. The consumerresponse is also affected bysearch associated with making adecision on a clearance sale. Consequently, when aclearance sale leads to anincrease in decision difficultyconsumers will respond morenegatively or indifferently, while if theclearance sale leads to adecrease in difficulty and offers a wider product line, theresponse to a clearance salecan be positive. Theresults across a seriesof micro-studies discussed in this paper evidence that consumer responseto clearance sales is positivelyrelated to the importanceof price advantage while is inversely relatedto the change indecision difficulty or intimidation of search costs. The studiesshow that consumers respondto clearance sales by changingtheir evaluations of satisfactionwith the decision processand by changing theirstore-loyalty or brand-loyalty behavior, but notby changing their satisfactionwith the consumption ofthe product ultimately selected.Firms should consider the clearance strategy at the as an ultimate tool to drive consumers attractions and dispose of the products to overcome the high inventory cost. Thus, firms should develop go-to-market strategies considering the following aspects:

- Understanding customers' needs, expectation and behavior
  - Products, Channels, Value Proposition and Markets
- Aggressive use of low cost channels
  - Short run market penetration, profit maximization
- Fit between selling approach and selling contents
  - Product-channel fit, process fit, value fit
- Trade-off between market coverage and control
  - Reaching the range of potential customers
  - Exercising control on sales relationship and deliveries
  - Market expansion and new channel partners

The implication of consumerresponse to clearance sales issubstantial; particularly given theprevalence of clearance sales with predetermined cyclicality inself-service stores (*e.g.* Clearance sale levels of25 to 50 percent in retailsettings have found tobe the predetermined expectation of the consumers, ratherthan an exceptional offer in the retail self-service stores. More recently,clearance sales have become amajor problem for onlinemerchants, due to bothtraditional forecasting problems andpoor links between theirinventory systems and theirweb sites. The retail self-service stores, which largely operate in chain are based on the rationale of *touch, feel and pick* which provides consumers a wide range of options to make buying decisions. The in-stores promotions and *do it yourself* (DIY) opportunities constitute the major motivation for the buyers and also support their decision making process. Motivational forces are commonly accepted to have a key influencing role in the explanation of shopping behavior. Personal shopping motives, values and perceived shopping alternatives are often considered independent inputs into a choice model, we argue that shopping motives influence the perception of retail store attributes as well as the attitude towards retail stores[30]. Hence for planning effective selling strategies in multi-channel business environment should develop in reference to the following attributes:

- Customer value enhancement
  - Fit between the product and customer value
  - Workplace for sales
  - Third-party services and value adjustments (Application Service Providers-ASP)
- Channel co-operation
- Develop, control and improve your distributor network
  - Select the right channel partner mix to create a portfolio that consistently meets or exceeds
  - Provide channel partners with the tools and support they need to achieve their revenue targets
  - Avoid channel conflict and provide rapid problem resolution
  - Increase sales, improve profitability and build customer loyalty by better managing the relationships, processes and practices in the sales/distribution channel.

In retail self-service store where consumer exercises in-store brand options, both service and merchandise quality exert significant influence on store performance, measured by sales growth and customer growth, and their impact is mediated by customer satisfaction. The liberal environment of the self-service stores for merchandise decisions, service quality and learning about competitive brands are the major attributes of retail self-service stores[31]. The retail self-service stores offer an environment of three distinct dimensions of emotions e.g. pleasantness, arousal and dominance. The retail self-service stores operate on a market size effect and a price cutting effect[32]. In the growing market competition, most firms depend in channel partners in increasing the sales volume. In motivating channel partners to augment sales, firm should develop strategies on price and non-price factors as illustrated in Figure 9.3.

The long-term effects of promotions on sales can be increased by developing customer loyalty through channel partners. Firms can also lean on drawing the effects of network selling in building loyalty. However, increase in sales can be more effective through channel by implementing the price and non-price related promotion strategies as exhibited in Figure 9.3. Promotions through various selling channels are increasingly linked to the supposed shift of consumer perceptions and buying power of consumers. However, high volume of sales through channel partners also leads to increase their bargaining power. The formal knowledge about how they influence channel decisions under different promotional arrangements and the distribution of channel profits remains very sparse. Results of some studies indicate that sales firms always invest in retailer promotions, while manufacturers may find it optimal to invest in consumer promotions directly to build their brand[33].

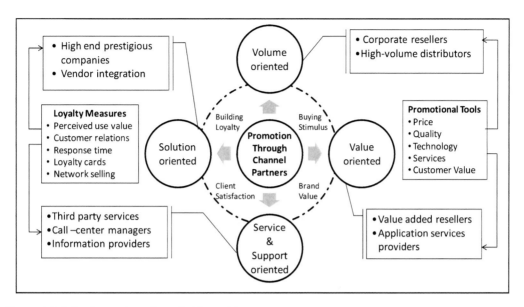

Figure 9.3.Sales Promotions through Channel Partners.

As the retail self-service stores display wide range of multi-brand products the consumers enjoy higher chance of finding preferred products (a market size effect). On the other hand, concentration of stores leads to fierce price competition (a price cutting effect). A number of researchershave dealt indirectly withconsumer response to clearance sales. As a marketing tool it has always had an important and legitimate role in creating consumer excitement and in clearing surplus or obsolete stock. However, in recent years the sale has become ubiquitous and pervasive and, apparently, used to excess, threatening profit margins and even the survival of certain retailers that rely too much on them. There are some critical issues associated to the price sensitive consumer behavior, whether customers are equally price-sensitive while purchasing products for functional (*e.g.* purchasing frozen vegetables, toiletries or paper towels) versus hedonic (*e.g.* purchasing a high technology computer or a video camera) consumption situations and whether perceived value derived during consuming the product influences price sensitivity. It may also be stated that higher price volatility makes consumers more sensitive to gains and less sensitive to losses, while intense price promotion by competing brands makes consumers more sensitive to losses but does not influence consumers' sensitivity to gains[34].

The prospect theory laid by Tversky and Khanman proposes that the intensity of gains plays strategic role in value enhancement as $G_{xt} = g_{pt}\left(\partial_x / \partial_p\right)$. In this equation, G_{xt} offers competitive advantage in a given time t and the promotional strategies are implemented to enhance the customer values in reference to product specific gains as expressed by (g_{pt}). In the above expression x represents the volume of goods while p denote the price of goods[35]. Research on consumer reaction to price has been largely confined to examining consumers' price information search, evaluation of price alternatives, and individual purchase behaviors without regard to situational influences. At the same time, consumption has often been dichotomized in terms of its functional-hedonic nature and closely associated with the level of satisfaction leading to determine the customer value influence[36]. As the new products are introduced, a firm may routinely pass these costs on to consumers resulting into high prices. However a less obvious strategy in a competitive situation may be to maintain price, in order to drive the new product in the market with more emphasis on quality, brand name, post-sales services and customer relations management as non-price factors. In many ways, such strategies of a firm with the new products may drive the consumer behaviortowards being sensitive to the price increase when it comes to making a buying decision. Some of the marketplace and experimental studies show that consumers are more sensitive to changes in price than to innovation and new products introduced by the firm[37].

SALES PROMOTION AND LOYALTY DEVELOPMENT

It is argued that the attribute rating covariance between two alternatives low-price brand switching and high price branded products without discounts leads to differences in the difficulty to choose. Should manufacturers of products such as automobiles and household appliances offer cash rebates to all consumers at the time of purchase, or offer trade deals to retailers? Some studies reveal that choosing between the two types of price promotion critically depends on the consumer sensitivity to both regular and promotional prices. More specifically, when consumers are more sensitive to promotions than to regular prices, manufacturers and sales channels are better off offering trade deals to increase brand equity among consumers[38]. It is observed that each of the new generation marketing approaches include customer focused, market-driven, outside-in, one-to-one marketing, data-driven marketing, relationship marketing, integrated marketing, and integrated marketing communications that emphasize two-way communication through better listening to customers and the idea that communication before, during and after transactions can build or destroy important brand relationships[39]. Behavioralresponse to clearance sales wasmeasured by tracking therate of store loyaltyacross conditions. The relationship between the clearance sales offer and customer decision is largely governed by the psychographic variables that can be measured broadly by the closeness and farness of the personalities of brand, product attractiveness, store loyalty and customer personality. The relationship attributes between the brand and customer relationship is exhibited in Figure 9.4.

The psychographic bond between the brand and customer would be stronger and attributed with emotions when the store loyalty and customer personality match closely as explained in the quadrant B in Figure 9.4. On the contrary when the personality attributes of the store loyalty and customers do not match with each other the relationship gets detached (quadrant C) and needs to be rebuilt for reviving the same. The companies attempt persuasive measures to bridge the gap between store loyalty and customer personalities when it is observed that the brands are thriving to create image but are unable to live to the customers' expectation (quadrant A). The relationship becomes discrete when the attributes of customer personality attempt to get acquaintance with non-responding brands as discussed in quadrant D of the above referred figure.

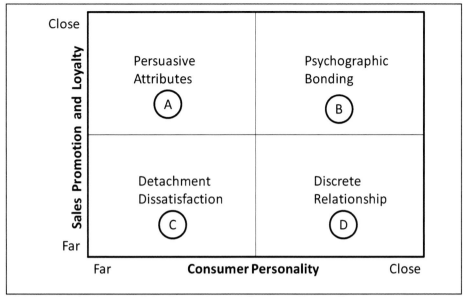

Figure 9.4.Relationship between Sales Promotion and Consumer Personality.

The magnitude of consumer response to clearance sales is weighed in two ways - evaluative and behavioral. Firstly, consumer satisfaction with the decision process leading to the expected level of satisfaction is measured, which may be expressed as one of a number of cognitive and affective responses that may result from a clearance sale. In a study of consumer durable purchases Westbrook, Newman, and Taylor introduced the basic concept of satisfaction with consumers' experience in arriving at purchase decision. They argued that while substantial research had been done on consumer satisfaction with the use or consumption of a good, little research had addressed consumers' experiences of learning about brands and product categories or deciding which option to purchase[40]. It has been observed that greater success in sales of fashion goods sales is significantly associated with the clearance sales of older inventory to allow the inflow of new fashion goods effectively. The retail sales performance and the customer value approach are conceptually and methodically analogous. The satisfaction is the customer's perception of the value received in a transaction or relationship and it helps in making re-patronage decisions on the basis of their predictions concerning the value of a future product. Hence the customer value paradigm is contemporary, which includes many elements of the customer satisfaction paradigm and is being more widely adopted and deployed by the firms[41].

The inventory control strategies, return policies, and clearance salespolicies are also considered as indirect sales promotion strategies which lower the cost of inventory of the firm and encourage clearance sales policy instead of motivating consumers to return the products upon dissatisfaction. In a clearance sales policy for a distribution channel consisting of a supplier, a retailer, and a Discount Sales Outlet (DSO) are integrated. It is observed in some studies that the products are initially sold in a retail outlet and after the selling season, the leftovers are moved to a DSO for a permanent clearance sale. When the retailer and the DSO coordinate, they share information on the demand forecast and jointly decide the stocking, markdown sales, and return policies to maximize mutual profit[42]. Clearance sales are largely driven by the price factor. Consumers prefer to buy products in clearance sales as they get heavy discount on the regular price of the product. In such sales managers miss out on significant profits because they shy away from pricing decisions for fear that they will alienate their customers. But if management is not controlling its pricing policies, the customers probably would demand for low prices. Two basic principles, the pocket price waterfall and the pocket price band, show managers how to control the pricing puzzle. The pocket price waterfall reveals how price erodes between a company's invoice figure and the actual amount paid by the customer--the transaction price. It tracks volume purchase discounts, early payment bonuses, and frequent customer incentives that squeeze a company's profits. The pocket price band plots the range of pocket prices over which any given unit volume of a single product sells. Wide price bands are common and transaction prices also have long ranges for many manufacturers. Using the pocket price bank enables a manager to control the price range to greater profits[43].

PROMOTION-MIX

Promotion strategies are concerned with the planning, implementation, and control of persuasive communication with customers. These strategies may be designed around advertising, personal selling, sales promotion, or any combination of these functions. One of the major strategic issues associated with the development of effective promotion strategy is the availability of financial resources for a specific product and market. The distribution of the notional budget among advertising, personal selling, and sales promotion is another strategic matter. The formulation of strategies dealing with these determines the role that each type of promotion plays in a particular situation. Promotion strategy consists of planning, implementing and controlling communications from an organization to its customers and other target audiences. The function of promotion in the sales program is to achieve various communications objectives in the market segment. An important sales responsibility is to plan and coordinate an integrated promotion strategy and to select the specific strategies for the promotion components. It is important to recognize that word-of-mouth communications among buyers and the communications of other organizations may also influence the target audience of the company. The promotion-mix has the following components:

- Advertising
- Personal Selling
- Sales Promotion
- Direct Marketing, and
- Publicity

Advertising may be defined as the strategy of communicating a sales message to potential customers. Advertising is one segment of a well-organized, continuous sales plan. Effective advertising is a cumulative process that maintains current customers, attracts new customers and establishes a favorable position for the business with competitors. Advertising will not cure slow business growthor low profits, nor will it create a better business person or a well-organized business. Advertising offers specified benefits to a specific or target audience. As part of a sound sales plan, advertising becomes an investment in the future of the business, instead of one more expense. An effective advertisement is based on a careful analysis of the situation before money is spent. "Advertising and promotions" is bringing a service to the attention of potential and current customers. Many products or services have failed in the market, not because of their quality, packaging or pricing, but because the potential customers didn't know they were there, and if they did, they didn't know what those were or how to use them. In order to sell a product or service the firm must promote it. One effective method of promotion is advertising. The goals of the plan should very much depend on the overall goals and strategies of the organization, and the results of the sales analysis, including the positioning statement. The plan usually includes what target markets the firm wants to reach, what features and benefits it wants to convey to them, how the firm will convey it to them (this is often called the firms advertising campaign), who is responsible to carry the various activities in the plan and how much money is budgeted for this effort. Successful advertising depends on knowing the preferred methods and styles of communications of the target markets that the firm wants to reach with its advertisements. A media plan and calendar can be very useful, which specify what advertising methods are to be used and when.

The Fashion Advertisements (FAds) and strategies building for optimum sales realization are prominent among them. As discussed earlier, the FAds have a greater impact on the elite clientele group as compared to other measures used for raising the sales. The product branding and packaging technology is the core input for FAds. Attractive packaging and popular branding have a significant role in the market expansion and product promotion. In a competitive market economy, the brands are hired by the manufactures for product marketing. In this system, new product managers have to face an uphill task. In sales of new products, it is essential to take potential as well as existing customers into confidence through an effective communication management. In absence of building up such awareness, the new product manager gets fringe benefits while the brand owner gets a higher share in the consumer's money. As such, these companies may not be in a position to establish their own brand due to many weaknesses pertaining to capital, technical know-how and market guidance. The future threat in this regard can be visualized in the light of selling their product. In the long run their identity will be only as a manufacturing unit, but not as a product seller. Packaging in the competitive product market is an important determinant as far as the buyer's behavior is concerned. The more attractive and durable the packaging of any product, the greater would be the product resistance and market demand. 'The New packs' may be the hard core of some fashion advertisements which could be more appealing to the target customers. It requires enough capital to invest in the packaging technology.

SALES PROMOTION MIX

Personal selling consists of verbal communication between a salesperson or selling team and one or more prospective purchasers with the objective of making or influencing a sale. Many companies feel that the personal selling is a better strategy to manage the interface of buyer and seller and so annual expenditures on personal selling are larger than advertising. However, advertising and personal selling strategies share some common features, including creating awareness of the product, transmitting information, and persuading people to buy. The personal selling is an expensive way of persuading the buyers as compared to various ways of advertising. Salespeople can interact with buyers to answer questions and overcome objections, they can target buyers, and they have the capacity to accumulate market knowledge and provide feedback. Sales promotion consists of various promotional activities, including trade shows, contests, samples, point-of-purchase displays, trade incentives, and coupons. Sales promotion expenditures are substantially greater than the amount spent on advertising. Direct salesinclude the various communication channels that enable companies to make direct contact with individual buyers. The common direct sales techniques are catalogues, direct mail, telemarketing, television commercials, radio, magazine, newspaper, electronic shopping and kiosk shopping etc. The distinguishing feature of direct sales is the opportunity for the marketer to gain direct access to the buyer. Direct sales expenditures account for a large portion of promotion expenditures. Electronic shopping is one of the newer forms of direct marketing. Publicity or public relations for a company's product, service, or idea involves communications placed commercially in the media. The objective of public relations is to encourage the media to include company-released information in media communication.

Development of an optimum promotion mix is by no means easy. Many companies often undermine the roles of advertising, personal selling, and sales promotion in a given product or market situation. Decisions about the promotional mix are often diffused among the decision makers, impeding the formation of a unified promotion strategy. The personal selling plans are sometimes divorced from the planning of advertising a promotion. A variety of factors should be considered to determine the correct promotion mix in a particular product/market situation. These factors may be classified as product factors, market factors, customer factors, budget factors, and sales-mix factors as outlined in the Table 9.1.

Table 9.1. Criteria for Determining Sales Promotion Mix

Product Factors	Market Factors	Customer Factors	Budget Factors	Sales-Mix Factors
Nature of product Perceived risk Durable versus non-durable Typical purchase amount	Position in its life cycle Market share Industry concentration Intensity of competition Demand perspectives	Household versus business customers Number of customers Concentration of customers	Financial resources of the organization Traditional promotional perspectives	Relative price/relative quality Distribution strategy Brand life cycle Geographic scope of the market

The following strategic planning rules may be helpful in finalizing promotion mix decisions [44].

- A company that has a higher share of sales must generally spend more on advertising to maintain its share.
- If a company has a high percentage of its sales resulting from new products, it must spend more on advertising compared to companies that have well-established products.

- Companies competing in fast-growing markets should spend comparatively more on promotion.
- The lower the unit price of a company's products, the more it should spend on promotion because of the greater likelihood of brand switching.
- Products that constitute a lower proportion of customers' purchases generally require higher promotion expenditures.
- Both very high-priced (and premium) products and very low priced (or discount) products require higher ad expenditures because. in both the cases, price is an important factor in the buying decision and the buyer must be convinced (through advertising) that the product is a good value.
- Higher-quality products require a greater promotion effort because of the need to convince the consumer that the product is unique.
- Companies with a broad line of products must spend more on promotion compared to companies with specialized product lines.
- Standardized products produced in large quantities should be backed by higher promotion outlays because they are likely to have more competition in the market.

The expansion of product market largely depends on the way the product sales is organized. There are many techniques adopted in organizing effective sales. They are departmental stores, supermarkets, mobile sale units, emporia, exhibitions and fun sales. Among these systems, the departmental stores and supermarkets have received considerable attention in towns, urban and semi-urban areas, while the mobile shops and fun sales have induced the buyers in the rural areas. The fun sales is organized in less developed areas through the entertainment program. Such a system is being observed in different Asian countries. In future more flexible methods of sales may appear to cover a larger consumer segments under product market.

Promotion strategies are concerned with the planning, implementation, and control of persuasive communication with customers. Promotion includes advertising, personal selling, sales promotion, and publicity of goods and services. Sales promotions also refers to the corporate-sponsored messages transmitted through the mass media channels including electronic and print media and static communication sources like billboards, wall paintings etc. Personal selling involves strategies of establishing person-to-person business relations with the customers. The sales promotion encompasses different techniques (for example, samples, trading stamps, point-of-purchase motion, coupons, contests, gifts, allowances, and displays) that support and complement promotion and personal selling. Publicity includes seeking favorable comments on product or service and/or the firm itself through a write-up or presentation in mass for which the sponsor is not charged. These strategies may be designed around personal selling, sales promotion, or any combination of functions of these. Global advertising is largely uniform across many countries though outreach and frequency of communication may vary across market segments. In many cases complete uniformity is unobtainable because of linguistic and regulatory differences between nations or differences in media availability, but with regard to products and services, the localized promotion can still be considered as uniform in a given region. In contrast, multi-domestic promotion is a type of customized promotions adapted to particular markets and audiences. The development of an effective sales promotion-mix is a strategic process which needs to be derived from analyzing the internal and external fit. The external fit of the promotional programs should be viewed from the perspective of competitor's sales strategies and long terms competitive lead upon investing resources in the sales promotion programs. The layout of sales promotional programs is exhibited in Figure 9.5.

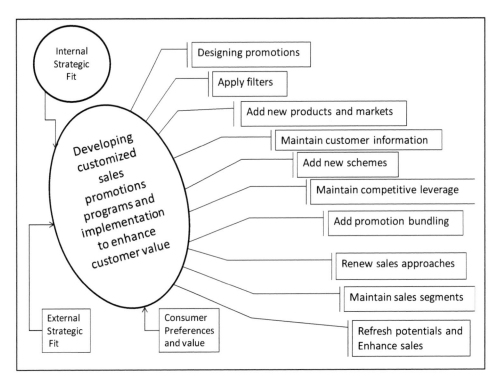

Figure 9.5.Layout of Sales Promotion Programs.

One of the major strategic issues associated with the development of effective promotion strategy is the availability of financial resources for a specific product/market. The distribution of the marginal budget among advertising, personal selling, and sales promotion is another strategic matter. The formulation of strategies dealing with these determines the role that each type of promotion plays in a particular situation. Promotion strategy consists of planning, implementing and controlling communications from an organization to its customers and other target audiences. The function of promotion in the sales program is to achieve various communications objectives in the market segment. An important sales responsibility is to plan and coordinate an integrated promotion strategy and to select the specific strategies for the promotion components. It is important to recognize that word-of-mouth communications among buyers and the communications of other organizations may also influence the target audience of the company. The promotion-mix has the following components:

- Advertising
- Personal Selling
- Sales Promotion

- Direct Marketing, and
- Publicity.

Sales promotions may be defined as the strategy of communicating a sales message to potential customers. Advertising plays a crucial role in international business worldwide and it is the critical factor in achieving sales goals under tough competitive environment. It has been observed that in the globalization era, the national and multinational companies are increasingly considering successful advertising as a prerequisite to profitable global operations. Sales promotion is one segment of a well-organized, continuous sales plan. Effective sales promotion is a cumulative process that maintains current customers, attracts new customers and establishes a favorable position for the business with competitors. Sales promotion offers specified benefits to a specific or target audience. The common sales promotion methods include those listed below:

- Price discounts
- Tie-up promotions
- Cross promotions, and
- Loyalty promotion.

Goals of the sales promotions plan should very much depend on the overall goals and strategies of the organization, and the results of the sales analysis, including the positioning statement. The plan usually includes what target markets you want to reach, what features and benefits you want to convey to them, how you will convey it to them (this is often called your advertising campaign), who is responsible to carry out the various activities in the plan and how much money is budgeted for this effort. Successful advertising depends on knowing the preferred methods and styles of communications of the target markets that the firm wants to reach with the effective advertisements. A media plan and calendar can be very useful, which specify what advertising methods are to be used and when. Sales promotions should be implemented in the market or to the clients directly using the following methods:

- Cold canvassing
- Endless chain of communications
- Turning around orphaned customers
- Establishing sales lead clubs
- Checking all consumers in the prospect lists
- Public exhibitions and demonstrations
- Identifying buying center and factors influencing consumer decisions
- Contacting customers through direct mail
- Extensive usage of telephone and telemarketing
- On site observation about the consumer behavior, and
- Social networking and generating strong word of mouth influence.

Store promotions are competitive for retailers, as more than the brands, the retail store compete in the marketplace in Latin American countries. Hence, retailers are engaged in extensive promotional activity by advertising through all media. Radio advertisements are largely targeted to the urban commuters. Besides media, retailers also outsource sales promoters to deliver gifts and price lists to people at strategic traffic points. Such promotion campaigns allow the store to increase its turnover by achieving a higher volume of sale in the market area, an increase in the frequency of visits, and stimulate spending by consumers in the store. Store-level promotions through radio advertisements help urban commuters to acquire information and take decision on buying or induce family and friends to help in visiting stores to witness promotions and buy. The radio advertisements reinforce a low-price positioning, a key to attract customers of price sensitive segment using an 'everyday low price' or 'everyday new promotion' strategy. However, such a strategy leads to an increase in sales at the expense of a substantial loss in profit in long run[45].

Radio commercials hold a marginal share among the main media categories, for example newspaper and television. However, it is still regarded as an important and useful medium in marketing and advertising in large cities and metropolitan areas. The broadcast of commercials on radio needs real feel orientation and voice is the single major determinant that draws the attention of listeners. Programs on sales and market news are the principal preferences of urban commuters for large metro radio stations[46]. The majority of short and informative advertisements for consumer products in an urban setting use radio advertising with communication appeal related to the efficacy of products and psychosocial enhancement of consumers at retail outlets. Promotional efforts by manufacturing and retailing companies appear to focus on positive emotional appeal to influence consumers through radio advertisements[47]. The design of effective communications depends upon an adequate model of the communication process. In traditional model speech conveys semantic information and gesture conveys information about emotion and interpersonal attitudes. Radio advertisements which have more than one voice induce higher emotions as message is split between voice and language which drives significant differences in communicative effectiveness[48]. The emotional content of a retail promotion advertisement is a stronger predictor of buying behavior reflected in message recall rate among audience. In addition, advertisements with high emotional intensity achieve increase in message recall ability among audience. Emotionally arousing advertisements require fewer broadcasts to achieve the same level of recall, and hence are likely to be less expensive to a campaign[49].

DEVELOPING INTERNATIONAL ADVERTISING AND SALES PROMOTION STRATEGY

Advertising and sales promotions are key instruments in promoting the products and services in the international markets. However, the rationale of advertising and media communication strategy has regional variations and among industries. Advertising carries the corporate products and services communication among the consumer segments. However, sometime companies provide deceptive information in the advertisements to gain quick consumer response and faster growth in market share. Hence, many firms have code of ethics for advertising and sales promotions. Advertisinghas significant contribution in international business as it involves a commitment of funds. Though the cost of effective and ineffective advertising varies, both incur high expenses. An effective advertising and sales promotion campaign represents a tangible resource and is transferable internationally from one market to another. Further advertising is regarded as the sole representative by many international companies in establishing and maintaining a desired position of products in the international market. Once a desired position for a product or service has been achieved through advertising, any local market interventions such as price or promotion related effects, make low impact. Thus, global advertising and sales promotion strategies need certain degree of centralization in terms of controlling the expenditure and carrying the sustainable impact of communication in the markets world over.

Demand for advertising and marketing services comes largely from businesses that sell consumer products, entertainment, financial services, technology, and telecommunications. The profitability of individual companies depends on creative skills and maintaining client relationships. Large companies benefit from being able to serve the varied needs of major customers. Small companies can compete by focusing on niche markets or by offering lower pricing. New competition for advertising and marketing business has come from database marketing companies, systems integrators, and telemarketers. Inventa Sales & Promotions helps clients create branding experiences. The company also provides field marketing services to the client organizations. The company promotes products through public interaction such as games and contests targeting consumers from desirable markets to build positive impressions of the company and its products. Other services include sampling, event marketing, merchandising, and mystery shopping. Based in Canada, Inventa has worked in the past with international food and beverage companies, clothing companies, and Internet banking firms, including Campbell Soup, Mattel, Old Navy, and Procter & Gamble[50].

The multinational companies may derive important strategic decisions for international advertising and sales promotion activities to make their brand image stronger. The concept of an advertising and sales promotions campaign developed at home can be transferred to other countries by translating them into local languages or there is a need to customize the advertising as per the socio-cultural and market requirements in different countries. Many firms strongly believe that a successful advertising concept has no geographic boundaries and may do well anywhere. Critics of standardization in advertising argue however, that cultural differences require a campaign to be tailored to each country. However, customization of advertising for a particular country may be justified on the plea of cultural differences and public communication regulations among countries. A standardized advertising and sales promotion approach seems particularly unsuited in developing countries where exist differences in lifestyle, per capita income, market structure, and various aspects of the environment as compared with developed countries.

Nike is a champion brand builder. Its advertising slogans *Just Do It, There Is no Finish Line* have moved beyond advertising into popular expression. Nike's athletic footwear and clothing has become a commonly talked product of America and its brand name is as well known around the world as Honda and Federal Express. Consumers know Nike for attention-grabbing commercials that feature athletes like Bo Jackson and Michael Jordan. They anxiously await the debut of new Nike advertising, and are equally enthusiastic in their response. "Nike's advertising is laden with the solitary, self-involved spirit some associate with jogging". Bo knows advertising. Wieden & Kennedy's Nike ads, multi-talented baseball and football star Bo Jackson, stretched the boundaries of jock worship and neatly crossed the bridge from sports marketing into pop culture. The eight spots helped build a Bo Jackson myth, then deconstructed that legend and found humor in chastising those who got caught up in it, all the while peddling high-tech, cross-training shoes. The first Jackson ads were part of the 1988 debut of Nike's famous tagline, "Just do it." A trio of light hearted spots showed Jackson cycling, running and playing basketball. An advertisement shows a radiant woman running past two clogged lanes of suffocating traffic, the drivers gray and faceless. Under the title, "Man vs. Machine," the text notes that machines have "put our bodies out of a job". A series of posters shows a solitary runner in a majestic natural scene over the headline, "There Is No Finish Line." Nike expands in the text: "Sooner or later the serious runner goes through a special, very personal experience that is unknown to most people. Some call it euphoria. Others say it's a new kind of mystical experience..." and so on. TV advertising is a large part of Nike's marketing success. Originally, Nike became a billion dollar company without television. For years Nike relied on the athletes to wear their shoes, and ran a limited number of print ads in specialized magazines like Runner's World. They did not complete the advertising spectrum until 1987 when they used television for the first time [51].

It is a complex issue for the multinational companies to opt for the sales promotion strategy between the standardization and localization. Among many factors that help the firms to decide on the type of sales promotion strategies to be adopted, the analysis choice criteria, sales promotion transferability and organizational support may be useful. The choice criteria comprise of the factors associated with environmental factors, sales promotion objectives of the company and target market where the advertisement needs to be delivered. A range of environmental variables affect sales promotion transferability across national boundaries. Of these some are listed as below:

- Rate of economic growth of country
- Per capita income and distribution of income
- Average size of household
- Level of education
- Level of nationalistic feeling among the consumers
- Attitudes towards work and leisure
- Attitudes toward wealth and returns on investment
- Assessment of satisfaction towards spending
- Ethical and moral standards
- Availability of time on commercial broadcast media
- Adequacy of coverage of market by broadcast media
- Availability of satisfactory electronic, print and outdoor media
- Extent of Independence of media
- Government control on public and commercial communications
- Import/export perspectives
- Legal restraints on sales promotion within the country
- Import duties and quotas in the country.

It has been observed that *sales promotion objectives* vary in each country. Though, sales promotion of products and services do not lead directly to sales, transferring the customer from one phase to the next. The sales promotion objectives for the products in the developing countries are largely oriented towards perceived benefits, value for money and better quality. For example, a DVD player manufacturer's sales promotion objectives for India would have to be different from those for the UK and USA, as in these countries the DVD player market has attained its maturity in the product life cycle, while in India it is in its growth stage. Hence, the major focus of sales promotion in India may be to comprehend the customer about the values associated with the product whereas in the developed countries like UK and USA the advertisements may focus on moving the consumers from conviction to action. Thus, the sales promotion concepts for marketing DVD players would not be effective in India. The sales promotion strategy also depends on the *target market* in the host country. If the proposed sales promotion campaign for another country is aimed towards a segment that is more or less similar in nature to the segment served in the home country, standardized sales promotion would appear satisfactory. Further, the attributes of a product involve also determines whether such sales promotion is appropriate. The product attributes include buying and usage patterns, psychological attributes and cultural factors. The *media availability* and the *cost benefit relationship* are other factors that determine the nature of sales promotion to be implemented in the host country. If the cost of adoption of sales promotion to the local conditions exceeds the benefits that an adaptation might provide, it would be wise for a firm to choose standardization of sales promotion contents.

The global economic downturn intensified the competition with other forms of entertainment. Not surprisingly, sports that have been traditionally popular in the United States or Europe which are the most lucrative television-advertising markets enjoy an inherent advantage and are grabbing the biggest slices of a competitive pie. Soccer and US football together take more than a third of the revenues in the global TV sports market (exhibit). But commercial success isn't just about popularity. The sports industry's core sources of revenue can be divided into three main categories: broadcast rights, sponsorships (including licensing and merchandising), and ticketing and hospitality (such as entertainment and catering in sports venues). Each category accounts for roughly one-third of the industry's revenues, and each has enjoyed strong growth. Total revenues from television rights are particularly important because of the exposure television generates for a sport and its sponsors rose by 14 percent a year from 1997 to 2001. The advertising revenues of the US networks couldn't match the amount they had paid to acquire broadcast rights at the business cycle's peak, so they lost an estimated $4 billion from sports programming that year. When sports rights holders renegotiate their television contracts, prices are unlikely to go on rising sharply. Golf and tennis are major sports but not huge to be regarded as global sports. Their audiences consist largely of affluent people who often play tennis or golf themselves. Such characteristics make the audience not only attractive targets for advertisers but also potential purchasers of goods licensed by the governing bodies of the two sports or by event organizers. Yet golf has been more successful than tennis has in exploiting this high-income niche. The decline of consumer interest is another sobering, and more structural, development. In Europe, the number of sports-TV viewers fell by 15 percent from 1996 to 2001 as other forms of entertainment programming, such as reality TV, came into fashion. In the United States, even the traditionally popular Monday Night Football game of the National Football League (NFL) has lost 17 percent of its TV viewers since 1999. And fewer amateur athletes are playing certain major sports—the number of participants in US baseball games dropped by 10 percent from 1991 to 2001, for example. This decline potentially reduces the level of interest in watching these sports on TV and in spending money on them over the longer term. Consequently, in some areas the sports business is becoming a buyer's market in which broadcasters and sponsors not only resist any increase in price but also insist that games be tailored to their needs [52].

The *transferability* of a commercial advertisement has various barriers comprising the cultural, communication styles, legislative, competitive and implementation factors. The cultural barriers also carry the social and cultural limitations laid by the religion. The local phonetics for the standardized advertisement that may provide an awful sense sometime hinder the free transferability of commercials. The competition for a product varies from one country to another country and some time requires changing the viewpoint for its proper positioning. The legislative barriers may be those imposed under the public laws in a country like child and gender abuse. However, some companies develop the global prototypes which are developed considering all possible barriers in other country and can be easily transferred. Such global prototypes comprise the facilities of voice-over and the provisions of visual change to avoid language and cultural problems. These advertisements may also be re-shot with local spokespeople but using the same visualization. Drakkar Noir, a man's fragrance, in an Arab print advertisement shows a woman's hand caressing a man's hand holding the product; in the United States the same hand grasps the man's wrist. Colgate-Palmolive, Proctor & Gamble, Nestlé and Coca-Cola often use prototypes of actual commercial and advertisement samples that demonstrate what headquarters want in the advertisements with specific written guidelines for acceptable deviations from the prototypes in terms of story and message (usually limited flexibility) and creative aspects (layout, color, symbols-usually more flexibility), with suggestions for appropriate media.

DIRECT RESPONSE ADVERTISING

Media advertising largely attempts to generate an indirect response towards the merchandise or services advertised. *Direct response advertising* includes mail-media advertising, catalogues, departmental store's advertising yellow pages, handouts, and window displays. Media marketing is an effective buyer-seller interactive system in which the merchandise advertised is brought close to the buyer using one or more advertising styles, and the response is measured with reference to the location and volume of transaction. Mail-media advertising involves promoting the merchandise market by establishing contact with potential and existing consumer through mail orders, publicity materials and telephone service. In this process no personal selling is performed. Direct mail advertising has many advantages. It attempts to build goodwill between sellers and buyers. Hence mail advertising is often identified as productive advertising technique.

The numerous advantages of this system are listed below:

- It is highly selective
- This form of advertising is elastic as retailer can add or delete the name of consumer at his discretion
- A wide variety of merchandise or services can be advertised to the same consumer
- Privacy on consumer preference / order can be maintained
- Market competition can be avoided instantly
- Direct mail advertising is personal specific
- Home delivery of goods and service can be assured
- Performance of merchandise / service sales can be monitored and evaluated.

Despite the many advantages of the direct mail advertising, there exist some demerits also. The most commonly observed problem in mailing business orders is the high expenditure involved in the process. The periodical updating of the mailing list is a major task in direct-mail advertising and consumes large business time. Besides these advantages, it sometimes becomes disinteresting and irritable to the persons addressed and they feel offended as the 'copy' the mail-order may not match with the profile of person to whom it is sent. The different types of mail-order advertising comprise of a comprehensive text and a visual copy which attempts to make the interesting reading. The various types of mail-order advertising are:

- Business reply mail with pre-paid postage fee
- Information enclosures, circulars etc
- Postal cards without an order form
- Self-mailing folders and
- Booklets and catalogues.

Mail-order advertising is a quantitative exercise and requires the systematic processing of data. Hence computerization is the basic requirement to handle the data with reference to classifying consumers, sorting types of orders, making a record of compliance to order and other functions.

DISPLAY SALES PROMOTION

Window display is the most prominent style of direct response advertising as it establishes ready information, product impression, and impulse of buying that helps in decision making. Display advertising can be indoor and outdoor. Indoor display advertising consists of showcase advertising and indoor displays in departmental stores. There are some common kinds of *indoor display* of merchandise. They are:

- Display of merchandise in showcase of departmental stores
- Display of merchandise in a decorative style in the showcase/window
- Display of prestige copy under a simulated environment
- Display of merchandise on a dummy
- Theme display and
- Demonstration of the use of the merchandise.

The *outdoor display* of merchandise or services may be done in the form of sign boards, commercial hoardings, posters, neon signs, vehicle sign boards, train posters, electronic sign boards, balloons, fiber optical billboard and other special effects. Merchandise advertising on vehicles is called transit advertising which carries the message from place to place. It is a good way to reach specific markets and can be tailored according to the geographical market segmentation.

Advocacy advertising on the other hand, attempts to highlight contemporary arguments directed at specific general clients like political activists, consumer groups, and media and government agencies. Advocacy advertising consists of:

- Ideological advertising which is principle oriented and attempts to highlight the ethics of an institution,
- Defense advertising which argues to protect the image of institution against contemporary controversies,
- Reply bound advertising seeking quick and ready response to issues highlighted in the advertisement,
- Position taking advertising emphasizing the point of view of an institute and thus of an emerging issue with a strong arguments to seek public acceptance or a referendum, and
- Ally recruitment advertising asking interested persons to present their views in support of the ethics of an institution in order to strengthen its position before their joining a position in the institute.

Advocacy advertising is also a part of sales promotion programs and has the advantage of exhibiting the message in a controlled situation, which then helps in dealing with complex issues. An institution can plan a series of advertisements for a campaign, supporting its views and an image building simultaneously among the clientele group. Institutional advertisements are generally released on multi-media and cover a substantially larger segment of the target audience.

SALES PROMOTION OBJECTIVES

It is necessary to pinpoint the objectives to build a good sales promotion program of the ad campaign. It would be wrong to assume that all sales promotions lead directly to sales. A sale is a multiphase phenomenon, and sales promotion can be used to transfer the customer from one phase to the next: from unawareness of a product or service, to awareness, to comprehension, to conviction, to action. The objectives of sales promotion may be defined by any one of the following approaches: inventory approach, hierarchy approach, or attitudinal approach.

A number of scholars have articulated functions performed by sales promotion. The objectives of an ad campaign may be defined from an inventory based on a firm's overall marketing perspective. The *inventory approach* to set-up objectives of the sales promotion may be set considering many aspects as presented in Table 9.2.

The inventory approach is helpful in emphasizing different objectives in sales promotion. These sales promotion objectives may be selected with reference to the overall marketing plan. This approach helps the advertiser choose a better functional platform for projecting the advertisement.

Table 9.2.Sales Promotion Attributes

Areas of Influence	Attributes
Increase sales	Encouraging potential purchasers to visit the company or its dealers Obtaining leads Inducing professional people (e.g., doctors, architects) to recommend the product Securing new distributors Prompting immediate purchases through announcements of special sales and contests
Creating awareness	Informing potential customers about product features Announcing new models Highlighting the unique features of the product Informing customers about product availability Announcing price changes Demonstrating the product in use

In *hierarchy approach*,the objectives of sales promotion should be stated in an action oriented psychological form. Accordingly, the objectives of sales promotion may be defined as (i) gaining customers' initial attention, perception, continued favorable attention, and interest; or (ii) affecting customers' comprehension, feeling, emotion, motivation, belief, intentions, decision, imagery, association, recall, and recognition. The concept of this approach is that customers move from one psychological state to another before actually buying a product. The purpose of sales promotion should be to change the customers from the state of indifference or negative to the positive approach and ultimately towards purchasing the product.

The *attitudinal approach* of sales promotion is instrumental in producing changes in attitudes; therefore, sales promotion goals should be defined to influence attitudinal structures. This strategy is helpful in developing a positive consumer attitude towards the company, brand, product class and attributes, competitive advantages and post sales services. The attitudinal approach is an improvement over the hierarchical approach because it attempts to relate sales promotion objectives to product/market objectives. Managers completely familiar with all product/market perspectives should define sales promotion objectives. A good definition of objectives aids in the writing of appropriate ad copy and in selecting the right media. It should be recognized that different ad campaigns for the same product can have varied objectives.

SELECT READINGS

Belch, G. and Belch, M. (2008), Advertising and promotion: An integrated marketing communications perspective, McGraw Hill, New York, NY.

Mullin, R. and Cummins, J. (2008),*Sales promotion: How to create, implement and integrate campaigns that really work*, Kogan Page, London.

Neslin, S.A. (2002), *Sales promotion*, Marketing Science Institute, Cambridge, MA.

Rigby, D. (2009), *Winning in turbulence*, Harvard Business School Press, Harvard Business School, Boston, MA.

Silk, A. J. (2006), *What is marketing*, Harvard Business School Press, Harvard Business School, Boston, MA.

Sugars, B. and Sugars, B. (2005), *Instant promotions: Instant success stories*, McGraw Hill, New York, NY.

REFERENCES

[1]   Raghubir, P., Inman, J. J.,and Grande, H. (2004), Three faces of consumer promotions, *California Management Review*, 46 (4), 23-42.

[2]   Gelb, B., Andrews, D., and Lam, S. (2007), Strategic perspectives on sales promotions, *Sloan Management Review*, 48 (4), 1-7.

[3]   Shapiro, B. P. (1977), Improve distribution with your promotional mix, *Harvard Business Review*, 55 (2), 115-123.

[4]   Crittenden, V. L. (2005), Rebuilt marketing machine, *Business Horizons*, 48 (5), 409-420.

[5]   DelVecchio, D., Henard, D. H.,and Traci H. F. (2006), The effect of sales promotion on post-promotion brand preference: A meta-analysis, *Journal of Retailing*, 82 (3), 203-213.

[6]   Ndubisi, N. O. and Moi, C. T. (2005), Customers behavioral responses to sales promotion: the role of fear of losing face,*Asia Pacific Journal of Marketing and Logistics*, 17 (1), 32-49.

[7]   Rajagopal (2005), Buying Decisions towards Organic Products: Analysis of Customer Value and Brand Drivers, *International Journal of Emerging Markets*, 2 (3), 236-251.

[8]   Gilbert, D. C. and Jackaria, N. (2002), The efficacy of sales promotions in UK supermarkets: a consumer view, *International Journal of Retail & Distribution Management*, 30 (6), 315-322.

[9]   Laroche, M., Pons, F., Zgolli, N., Cervellon, M. C. and Kim, C. (2003), A model of consumer response to two retail sales promotion techniques,*Journal of Business Research*, 56 (7), 513-522.

[10] Watkins, M. and Bond, C. (2007), Ways of experiencing leisure, *Leisure Sciences*, 29 (3), 287-307.

[11] Carpenter, J. M., Moore, M. and Fairhurst, A. E. (2005), Consumer shopping value for retail brands, *Journal of Fashion Marketing and Management*, 9 (1), 45-53.

[12] Vicdan, H., Chapa, S., and de Los, S. G. (2007), Understanding compulsive buyers' online shopping incidence: a closer look at the effects of sales promotions and bargains on Hispanic Americans,*Journal of Customer Behaviour*, 6 (1), 57-74.

[13] Kyrios, M., Frost, R. O., and Steketee, G. (2004), Cognitions in Compulsive Buying and Acquisition,*Cognitive Therapy and Research*, 28 (2), 241-258.

[14] Dobson, J. (2007), Aesthetics as a Foundation for Business Activity,*Journal of Business Ethics*, 72 (1), 41-46.

[15] Platz, L. A. and Temponi, C. (2007), Defining the most desirable outsourcing contract between customer and vendor, *Management Decision*, 45 (10), 1656-1666.

[16] Zhou, L. and Wong, A. (2004), Consumer Impulse Buying and In-Store Stimuli in Chinese Supermarkets, *Journal of International Consumer Marketing*, 16 (2), 37-53.

[17]   Berry, L. L. (2001), The old pillars of new retailing, *Harvard Business Review*, 79 (4), 131-137.

[18]   Batra, R., Lehmann, D. and Singh, D. (1993), The Brand Personality Component of Brand Goodwill: Some Antecedents and Consequences, in Aaker & Bie: *Brand Equity and Advertising: Advertising's Role in Building Strong Brands*, Hillsdale: Lawrence Erlbaum Associate Publishers, 83-95.

[19]   Rajagopal (2007), Influence of Brand Name in Variety Seeking Behavior of Consumers: An Empirical Study, *International Journal of Management Practice*, 2 (4), 306-323.

[20]   Vicdan, H., Chapa, S., and de Los, S. G. (2007), Understanding compulsive buyers' online shopping incidence: a closer look at the effects of sales promotions and bargains on Hispanic Americans,*Journal of Customer Behaviour*, 6 (1), 57-74.

[21]   d'Andrea, G., Schleicher, M. and Lunardini, F. (2006), The role of promotions and other factors affecting overall store price image in Latin America,*International Journal of Retail & Distribution Management*, 34 (9), 688-700.

[22]   Otieno, R., Harrow, C. and Lea, G. G. (2005), The unhappy shopper, a retail experience: exploring fashion, fit and affordability, *International Journal of Retail & Distribution Management*, 33 (4), 298-309.

[23]   Liu, T. C. and Wu, L. W. (2007), Customer retention and cross-buying in the banking industry: An integration of service attributes, satisfaction and trust, *Journal of Financial Services Marketing*, 12 (2), 132-145.

[24]   Backstrom, K. (2006), Understanding Recreational Shopping, *The International Review of Retail, Distribution and Consumer Research*, 16 (2), 143-158.

[25]   Seybold, P. B.(2001), Get Inside the Lives of Your Customers, *Harvard Business Review*, 79 (5), 80-89.

[26]   Otieno, R., Harrow, C. and Lea, G. G. (2005), The unhappy shopper, a retail experience: exploring fashion, fit and affordability, *International Journal of Retail & Distribution Management*, 33 (4), 298-309.

[27]   Codispoti, M. and De Cesarei, A. (2007), Arousal and attention: Picture size and emotional reactions, *Psychophysiology*, 44 (5), 680-686.

[28]   Wirtz, J., Mattila, A. S. and Tan, R. L. P. (2007), The role of arousal congruency in influencing consumers' satisfaction evaluations and in-store behaviors, *International Journal of Service Industry Management*, 18 (1), 6-24.

[29]   Rajagopal (2008), Outsourcing Salespeople in Building Arousal towards Retail Buying, *Journal of Database Marketing and Customer Strategy Management*, 15 (2), 106-118.

[30]   Morschett, D., Bernhard, S., and Thomas, F. (2005), Perception of Store Attributes and Overall Attitude towards Grocery Retailers: The Role of Shopping Motives, *The International Review of Retail, Distribution and Consumer Research*, 15 (4), 423-447.

[31]   Babakus, E., Bienstock, C. C., and van Scotter, J. R. (2004), Linking Perceived Quality and Customer Satisfaction to Store Traffic and Revenue Growth, *Decision Sciences*, 35 (4), 713-737.

[32]   Lazear, E. P. (1986), Retail Pricing and Clearance Sales, *The American Economic Review*, 76 (1), 14-32.

[33]   Sigue, S. P. (2008), Consumer and Retailer Promotions: Who is Better Off?,*Journal of Retailing*, 84 (4), 449-460.

[34] Han, S., Gupta, S., and Lehmann, D. R. (2001), Consumer Price Sensitivity and Price Thresholds, *Journal of Retailing*, 77 (4), Winter, 435-456.

[35] Tversky, A., and Kahnman, D. (1981), The Framing Decisions and Psychology of Choice, *Science*, No.211, 453-458.

[36] Wakefield, K. L., and Inman, J. J. (2003), Situational Price Sensitivity: The Role of Consumption Occasion, Social Context and Income, *Journal of Retailing*, 79 (4), 199-212.

[37] Gourville, J. T. and Koehler, J. J. (2004), *Downsizing Price Increases: A Greater Sensitivity to Price than Quantity in Consumer Markets*, HBS Marketing Research Paper No. 04-01, January.

[38] Herran, G. M., Sigue, S. P.,and Zaccour, G. (2010), The Dilemma of Pull and Push-Price Promotions, *Journal of Retailing*, 86 (1), 51-68.

[39] Duncan, T., and Moriarty, S. E. (1998), A Communication Based Marketing Model for Managing Relationships, *Journal of Marketing*, 62 (4), 1-13.

[40] Westbrook, R. A., Newman, J. W., and Taylor, J. R. (1978), Satisfaction /Dissatisfaction in the Purchase Decision Process, *Journal of Marketing*, 42 (4), 54-60.

[41] Oakley, P. (1996), High Tech NPD Success through Faster Overseas Launch, *European Journal of Marketing*, 30 (8), 75-91.

[42] Lee, C. H. (2001), Coordinated stocking, clearance sales, and return policies for a supply chain, *European Journal of Operational Research*, 131 (3), 491-513.

[43] Marn, M. V., and Rosiello, R. L. (1992), Managing price, gaining profit, *Harvard Business Review*, 70 (5), 84-94.

[44] Refer Workbook for Estimating your Advertising Budget: Cahners Publishing Co., Boston, 1984.

[45] Hoch, S. J., Drèze, X., and Purk, M. E. (1994), EDLP, hi–lo and margin arithmetic, *Journal of Marketing*, 58 (4), 16–27.

[46] Parker, S. J. (1993), A prescription for AM radio programming, *Journal of Radio & Audio Media*, 2 (1), 29-33.

[47] Yusuff, K. B. and Yusuf, A. (2009), Advertising of OTC products in a Nigerian urban setting: Content analysis for indications, targets, and advertising appeal, *Journal of the American Pharmacists Association*, 49(3), 432-435.

[48] Beattie, G., and Shovelton, H. (2005), Why the spontaneous images created by the hands during talk can help make TV advertisements more effective, *British Journal of Psychology*, 96(1), 21-37.

[49] Biener, L., Wakefield, M., Shiner, C. M., and Siegel, M. (2008), How broadcast volume and emotional content affect youth recall of anti-tobacco advertising, *American Journal of Preventive Medicine*, 35(1), 14-19.

[50] For details see Inventa Sales and Promotion Company, British Columbia, Canada http://www.inventaworld.com/

[51] for detailed discussion on Nike please see Robert J Dolan: Nike Inc, in 1990's: The New Directions, Harvard Business School Case, Harvard Business School, April 1995; also see official web site of Nike Inc. www.nike.com

[52] Tekla V. B., Blatter, P., and Bughin, J. R. (2004), Playing to Win in the Business of Sports, *The McKinsey Quarterly*, Internet Edition, July.

Chapter 10

SALES AUTOMATION

The new information technology is becoming an important factor in the future development of sales and retailing industry. Forward-looking companies had begun sales automation process since late eighties in the twentieth century. These firms had activated sales automation process by introducing marketing and sales productivity (MSP) systems, which boosted about one third growth in sales and sales force productivity. MSP systems helped automating routine tasks and gather and interpret data that was either scattered or uncollected before. They not only upgraded sales and marketing efficiency but also improved the timeliness and quality of executives' decision making. The sales automation is viewed as a corporate strategic investment, companies can exploit the synergies possible from linkages with other parts of the organization [1]. The growth in sales of consumer products and services has significantly influenced customers interacting with salespeople to a greater extent through remote technological channels. Though there is much variation in online sales and adoption levels, little is understood about actual customer motivators and perceived barriers towards buying online sales and services. The impact of the Internet on bank - customer relationships has emerged as a key determinant amidst global competition which had driven customers, motivators and inhibitors towards adapting to online sales and services [2]. The developments information and communication technology have significantly contributed to the exponential growth and profits of the manufacturing, sales and retailing services worldwide. This evolution had transformed the way sales firms deliver products and services, using technologies such as automated vending machines, phones, the Internet, credit cards, and electronic cash. However, banks face a number of important questions on strategies for deriving full advantage of new technology opportunities and tracking electronic development changes affecting interactions with the customers. Designing and implementation of sales automation process should be considered because it support following growth options:

- Sales process/activity management
- Sales and territory management
- Contact management
- Lead management
- Configuration support
- Knowledge management

Sales force automation (SFA) technologies are increasingly used to support customer relationship management strategies in order to improve the effectiveness of sales process on one hand and enhance the customer value on the other. However, there is a need to examine the impact of SFA usage on both customer relationship quality and sales performance. Additionally, the mediating roles of learning and adaptive selling behaviors on the outcomes of SFA usage need to be meticulously established by the firms. It is observed that the role of salesperson learning and adaptive selling behaviors in the SFA increases sales performance and customer relationship. In general terms, increasing convenience is a way of raising consumers' surplus provided new technology is adopted by the selling firms in order to offer convenience to the customers, may be through an electronic transaction as a substitute for a trip to the retail outlets for buying. The technology based services imply different combinations of accessibility attributes (time, distance, and search costs), ease of use and price. Another factor in determining the magnitude of the surplus that the bank can seize is the relative importance of cross-selling. The bundle of services provided electronically is usually not the same as the one available at a store. For this reason new technology based sales services with high customer value may offer better service conditions to harmonize the flow of information and services across the spatial and temporal dimensions [3].

Roche-Syntex Mexico was engaged in selling the diagnostic reagents and equipments to the government and private clinics. The company also provided the diagnostic equipments to these health institutions and hospitals on lease. The geographic expansion of Roche Synetx Mexico was concentrated in central Mexico followed by north and pacific regions of country. The call center concept was introduced in the Roche Diagnostics Mexico in May 2002 with a broad objective of integrating the customer services with sales force performance in the market. The company had developed the call center approach for increasing the performance of the customer services. Conventionally during the early 90s, logging and monitoring in the call center was relatively simple: incoming or outgoing calls were recorded to verify transactions or to gather samples for quality monitoring. The company had new management dimensions to achieve its goal of building a competitive advantage through superior customer service. The project enabled Roche to streamline its service workflow, improve business information, and integrate its field and call center organizations, resulting in reduced cost of service and improved customer satisfaction. The new call center approach has been designed to provide the basis for Roche's new technology infrastructure. Each call center of the company was committed to supporting its customers through tailored service offerings based on local customer needs. The company was also engaged in doing continuous monitoring of product quality and service performance, which allows rapid response to customer concerns and changing market trends. The software "Clarify" had been designated by the company for use in the call center to make effective customer relationship. The software was used by the call center at Roche Diagnostics Mexico. It had been expressed in general that the call center was imposed on the organization and is practically under hibernation due to lack of adequate diffusion of the concept and application of the center objectives and tools. The center had not been perceived by the employees as an efficient tool in day-today work and providing customer attention. Most of the employees agree that the role of call center is instrumental in customer relationship management program and so the company has invested more money to strengthen call center administration [4].

Salespeople are in a unique position of being in very close proximity to customers and thus have the ability to create and maintain long-term relationships with them. Sales force automation systems have the ability to help or hinder this relationship building process. Many recent studies have revealed that SFA suffers from inefficiency issues within the firm unless particular attention is being paid to technology implementation and reasons for high SFA failure rates. However, there exist strong opportunities for improving the effectiveness of SFA rather than the efficiency side of SFA. Although efficiency is still of great importance to SFA researchers and practitioners, the customer or relationship focus of SFA effectiveness is also needed [5]. The growth of SFA in the hospitality industry in recent years has led to virtually every hotel salesperson having a computer at desk in order to perform sales responsibilities. However, simply providing a computer and software is not going to miraculously increase sales volume or productivity. The adoption of SFA has been separated in two stages. The first stage is at the organizational level and the second stage is at the individual salesperson level. There are significant differences in the level of both organizational and individual adoption of SFA in the hotel sales profession. SFA can be analyzed by the firms to determine its effectiveness in achieving the goal of improved productivity [6].

ING Comercial America was a Mexican company and a leader in the insurance, pension benefits and guaranties industry, providing financial services to people who were engaged in developing new projects in Mexico. The company had implemented sales automation through the *Consult-Ing* project as a tool accelerate the customer prospecting activity and fence the switching options with the customers. However, sales force did not respond enthusiastically to the automation process and laid back on the traditional approaches of prospecting customers and selling of services. It was observed that the sales promoters were more comfortable following the traditional tracking approaches than surfing through the web to get instant access to their status [7]. The major problems identified for the slow receptivity to the sales automation process were more cognitive than organizational. The automation support to the agents of ING helps the agents to drive the prospects to their quick and sustainable decision on buying the services of the company. Every company knows that it costs far less to hold on to a customer than to acquire a new one and this notion is gradually getting implanted in the services industries. Yet defector customers are far less of a problem than the customers who change their buying patterns and incline towards switching. Developing accounts is inherently more important than customer acquisition. That means a customer focused B-to-B site will need a sales force that is differently trained, structured, and compensated. The sales force must orient itself around the goal of finding products for its customers by using efficiently the *Consult-Ing* web site [8].

In view of growing competition firms are creating a digitized selling capability by developing Web sites designed to provide information and conduct transactions with customers, replacing many routine sales force activities. Firms use the motivation ability framework to shape a conceptual model that enhances the effects of the digitization of selling activity and performance of salesperson. Some studies have observed that digitization has the paradoxical effect of improving salesperson effectiveness and heightening job insecurity concerns. However, managers can improve the technology-enabled multichannel capabilities of the firm by giving priority attention to human capital improvement, sales force control systems, and communication of the digitization strategy [9]. The improvements in various activities of sales process due to implementation of sales force automation process is exhibited in Figure 10.1.

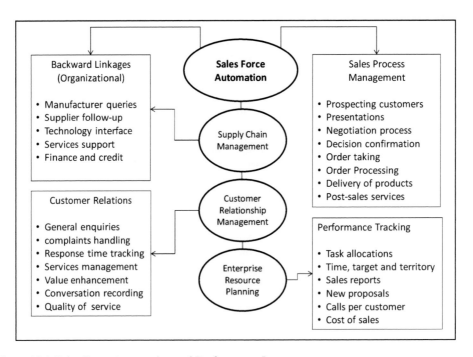

Figure 10.1.Sales Force Automation and Performance Improvement.

SFA applications enable emerging manufacturing and sales oriented firms to automate sales activities and monitor administrative responsibilities for the sales professional, enhancing the productivity of salespeople. It has been observed in a study that the global spending on SFA tools has grown at an annual rate of 27 percent to reach $3.2 billion in 2007 and is forecast to reach almost $9 billion in 2012 reflecting the increasingly important role of SFA adoption [10]. The importance of successful SFA adoption is highlighted through its effect on both SFA implementation and overall firm performance. Adoption of SFA systems by the sales force is an important determinant of a successful SFA system implementation. Implementation of SFA is a function of individual perceptions and commitment of the firm's management. Thus, leadership commitment alignment may be the key to the development of an environment that provides a welcome reception for sales force technology adoption. Additionally, failure on the salespersons' part to adopt the technology might be construed unfavorably by both immediate and top management since both levels are committed to SFA technology adoption. This should enhance salesperson adoption of recommended technologies since it results in satisfaction at multiple levels of the organization's management team. High levels of congruence in commitment at both levels of management should convince salespeople that the whole organization is committed to and serious about prioritizing the adoption of the technology [11]. Most firms continue to struggle with the implementation of sales force technology tools and to determine the role they play in improving the performance of salespeople. The concept of sales force automation and customer relationship management (CRM) need to be integrated by converging technology services to optimize performance by the salespeople. A study conducted to measure the impact of SFA on sales performance indicate that SFA usage has a direct impact on effort, thereby reducing number of hours worked, and CRM usage has a direct positive impact on adaptive selling behaviors. Moreover, experience of firms in using sales automation processes reveals that SFA moderates the relationship between CRM usage and adaptive selling techniques [12].

AUTOMATION VS. CONVENTIONAL WISDOM

The maxims of technology spread in the operations of financial institutions may have relational effect on the size and volume of operations of the organization. It is observed that conventionally sales firms were using heterogeneous technologies that resulted in slow returns to scale in different strategic financial business units. In contrast, when firms engaged in selling products and services followed conventional technologies in developing countries, the low cost technology was followed in production, operations and marketing management [13]. Whenever the innovation is initially introduced, large sales firms have an advantage to adopt it first and enjoy further growth of size. Over time, as the innovation diffuses into smaller sales firms, the aggregate market size distribution increases stochastically towards a wider area with more number of customers. Applying the theory to a panel study of virtual shopping diffusion across developed countries, it has been observed that technological, economic and institutional factors largely govern the transaction process supported with technology. The empirical findings disentangle the interrelationship between internet driven purchasing behavior and growth of firms, and explain the variation of diffusion rates across geographic regions [14].

As competition intensifies in many markets, firms intend to expand geographically. Such decision becomes more challenging for salespeople as it increases their workload. Unfortunately, despite the apparent need for more salespeople to be used in many marketing activities that increase costs of deploying salespeople. It is therefore not surprising that greater attention is being paid recently on how to make salespeople's activities more efficient. Toward this end, many companies are introducing diverse sales force automation technologies such as SFA-a dedicated software and information systems. Even though in theory such SFA technologies should bring about only positive results, the reality is that it leads to some negative results as well. For example, if salespeople do not use SFA technologies in a positive manner or have psychological resistance to them after their adoption, then the introduction of these technologies could end up failing and wasting money for companies [15].

The most important electronic commerce design goal was developed to address electronic commerce as a whole, securing the essential steps of each e-commerce transaction including the offer, the order and the payment. Globe-ID(R) is a system based on an intermediation server which acts as a trusted third party for merchants and consumers. It contributes directly to the security and notarization of the transactions, manages the e-commerce player accounts, and acts as a gateway to the private networks of traditional financial instruments [16]. In order to get along with increasing competition, retailing firms need to encourage usage of advanced technology. However, adoption of new technology may require substantive cognitive efforts from both retailers and consumers. Therefore, firms should solicit cooperation of customers towards automation of sales and services, and infuse concept of high-tech marketing successfully. This process has so far been used only marginally as most prominent high-technologies are not fully dominating the marketplace [17]. The use of e-commerce is basically unaffected by the size of the city where the household lives. Geographically remote consumers are discouraged from purchasing goods by the fact that they cannot inspect them beforehand. Leisure activities and cultural items (i.e., books, CDs, and tickets for museums and theaters) are the only goods and services for which e-commerce is used more in isolated areas [18].

In increasing global competition, the retailers whether running a physical store, a catalog business, an e-commerce site, or a combination of the three, need to offer customers superior solutions to their needs, treat them with respect, and connect with them on an emotional level. Retailers also have to set prices fairly and make it easy for people to find what they need, pay for it quickly, and then move on. Hence, e-commerce is shifting from making purchases online to going shopping online, a social experience in which people interact in a 3-D Web space. Moving ahead with the growing technology in the days to come in all sorts of markets, customers will use choice-boards, an interactive buying platform and manufacturing companies and retailers would be able to use on-line systems that let people develop design of their own products by choosing from a menu of attributes, prices, and delivery options. It is found that self-service technologies, which can be customized by the users, build higher sense of belongingness and safety with the banking industry. The association with the self-service technologies in a financial operations leads to three forms of positive attachment based on three different foundations, which include the credibility of the organization, compatibility between the values of the organization and those of the consumer, and interpersonal or relational considerations [19].

In the global financial markets, various initiatives of using the mobile phone to provide financial services to the customers to gain value added services benefits over the traditional sales firms which introduced the m-banking or m-payments systems. This innovation in banking system has revealed three cross-cutting customer centric benefits including amplification of services, simultaneous payments, and a multi-dimensional trust [20]. Trust has been identified as the key to e-commerce because it is crucial wherever uncertainty and interdependence exist. It is observed that that trust and perceived risk are direct antecedents of intention, and trust is a multi-dimensional construct with three antecedents including perceived trustworthiness, perceived security, and perceived privacy [21].When a major technology innovation arrives, a wave of new firms enters the market implementing the innovation for profits. However, if the innovation complements existing technology, some new entrants will later be forced out as more and more incumbent firms succeed in adopting the innovation. Such situation has revealed that the diffusion of Internet technology among traditional brick-and-mortar firms is indeed the driving force behind the rise and fall of dotcoms as well as the sustained growth of e-commerce [22]. However in reference to banking reforms in India, technology has been found to be the major input in driving competition which has been evidenced in a study revealing a positive relationship between the level of competition and marketing efficiency.

One of the common tools of sales automation, which has been followed by most sales oriented firms, is e-commerce with a virtual consumer interface. With e-commerce practices maturing across sales firms and countries, firms are being able to sell higher volume of products or services online. However some studies reveal that virtual shopping no longer gives competitive advantage to consumers. Any company is now able to quickly build an e-commerce infrastructure and launch services with readily available technology solutions and the use of development experts at low cost. One of the efforts that many companies have made in order to survive in this competitive environment is to develop various customer aid functions that can provide a more satisfying shopping experience to customers when compared to rival Internet stores [23].There are many such customer aid functions that range from simply assorting products in various ways to providing intelligent personalized services. For example, many on-line stores such as Amazon.com recommend products in several different ways, according to individual customers' activities or transaction history. Many stores also have personalized pages that provide a list of recommended products, information, or advertisements that are prepared according to the estimated preference of each customer, based on their activities or transaction records. There are also simpler and non-personalized recommendations provided in many stores that show the lists of best selling products, top-rated products, most-clicked products, etc. Simple customer aid functions are relatively easy to build, but sophisticated methods such as personalization often require advanced techniques based on statistics or artificial intelligence [24]. Sales process automation helps in developing efficiency of following activities:

- Offer calendars to assist in planning of key customer events
- Automates both individual and organizational to-do list
- Team collaboration tools to synchronize activities and coordinate with various market clients
- Provides valuable post facto analysis of a sales cycle; which allows the team to examine the duration and procedures involved in critical tasks, and
- Sales process and activity management are only as good as their ability to be tailored to intervals sales methods.

The recent developments in the mobile technology (M-Commerce), has increased scope of penetration of consumer products as an ultimate marketing vehicle, which enabled business entities to establish a pervasive electronic presence alongside their customers anytime, anywhere. As soon as global companies realized the business potential in this channel, mobile services have infiltrated virtually in the consumer space to influence virtual shopping [25]. Many marketing and sales institutions have built websites to inform and attract customers. Information technology affects banking in two main ways. First, it may reduce costs by replacing paper-based, labor intensive methods with automated processes. Second, it may modify the ways in which consumers have access to banks' services and products and, hence, may enhance the contestability of markets, especially in retail banking. Due to deregulation and technological advances, new opportunities become available, but the skill needed to exploit them effectively may be unknown. Early entry of financial institutions into the technology expanding activities may have learning benefits that are manifested in discovery of the skill needed to operate effectively [26]. Sales products and services are getting more and more advanced and increasing in variety by providing information at the early stage to providing transactional activities. The average e-marketing penetration for developing countries by the end of 1999 was close to 5 percent. The access of computers at homes in different countries during 2000-2008 is exhibited in Table 10.1.

Table 10.1. Households with access to a computer at home

Country	2000	2001	2002	2003	2004	2005	2006	2007	2008
Iceland	..	..	..	..	85.7	89.3	84.6	89.1	91.9
Netherlands	..	..	69.0	70.8	..	77.9	80.0	86.3	87.7
Sweden	59.9	69.2	..	..	..	79.7	82.5	82.9	87.1
Japan	50.5	58.0	71.7	78.2	77.5	80.5	80.8	85.0	85.9
Norway	..	..	..	71.2	71.5	74.2	75.4	82.4	85.8
Denmark	65.0	69.6	72.2	78.5	79.3	83.8	85.0	83.0	85.5
Luxembourg	..	..	52.6	58.0	67.3	74.5	77.3	80.0	82.8
Germany	47.3	53.0	61.0	65.2	68.7	69.9	76.9	78.6	81.8
Korea	71.0	76.9	78.6	77.9	77.8	78.9	79.6	80.5	80.9
Canada (2007)	55.2	59.8	64.1	66.6	68.7	72.0	75.4	78.4	..
United Kingdom	38.0	49.0	57.9	63.2	65.3	70.0	71.5	75.4	78.0
Switzerland (2006)	57.7	62.2	65.4	68.9	70.6	76.5	77.4	..	..
Austria	34.0	..	49.2	50.8	58.6	63.1	67.1	70.7	75.9
Finland	47.0	52.9	54.5	57.4	57.0	64.0	71.1	74.0	75.8
Australia (2007)	53.0	58.0	61.0	66.0	67.0	70.0	73.0	75.0	..
New Zealand (2006)	..	46.6	..	62.0	..	..	71.6	..	..
Ireland	32.4	..	..	42.2	46.3	54.9	58.6	65.5	70.3
Belgium	..	..	..	..	..	..	57.5	67.2	70.0
France	27.0	32.4	36.6	45.7	49.8	..	..	61.6	68.4
Spain	30.4	..	..	47.1	52.1	54.6	57.2	60.4	63.6
Slovak Republic	..	..	..	..	38.5	46.7	50.1	55.4	63.2
United States (2003)	51.0	56.2	..	61.8	..	..	..	..	..
Poland	..	..	..	..	36.1	40.1	45.4	53.7	58.9
Hungary	..	..	..	..	31.9	42.3	49.6	53.5	58.8
Italy	29.4	..	39.9	47.7	47.4	45.7	51.6	53.4	56.0
Czech Republic	..	..	..	23.8	29.5	30.0	39.0	43.4	52.4
Portugal	27.0	39.0	26.8	38.3	41.3	42.5	45.6	48.3	49.8
Greece	..	..	25.3	28.7	29.0	32.6	36.7	40.2	44.0
Mexico	..	11.6	15.4	..	18.4	18.3	21.2	22.0	26.1

Source: OECD, ICT database and Eurostat, Community Survey on ICT usage in enterprises, May 2009.
Generally, data from the EU Community Survey on household use of ICT, which covers EU countries plus Iceland, Norway and Turkey, relate to the first quarter of the reference year. For the Czech Republic, data relate to the fourth quarter of the reference year.
For Korea: Previously, data for Korea were based on the Computer and Internet Use Survey conducted by the Korean National Statistical Office. Certain items of that survey are no longer collected and data are now sourced from the Survey on Computer and Internet Usage conducted by the National Internet Development Agency (NIDA) of Korea. The NIDA series shows larger shares than the previous survey. Up until 2005, data included only desktop and notebook PCs. From 2006 onwards, data includes portable and handheld PCs.
For New Zealand: The information is based on households in private occupied dwellings. Visitor-only dwellings, such as hotels, are excluded.

There are many socio-cultural factors associated with adoption of technology in services industry which should have the compatibility to the customer culture. However, as a concept, self-service technology (SST) has been generally limited to technical or functional factors and with regard to value compatibility among the SST providing organization and customers. Cultural disparities among the customers also affect the performance of self- service technology offered by the retail sales industry [27]. Building an effective trust is an important way to prevent fraudulent behavior in M-commerce. Firms need to develop ways to strengthen trust in mobile transaction process on the basis of the credibility of mobile marketing alliance companies. The trusted transactions make the M-commerce safe by preventing frauds beforehand [28]. Retail sales firms are increasingly relying on self-service technology (SST) applications to augment efficiency in rendering financial services to their customers and in such information and communication technology environment, customers serve themselves at the convenience of these service options more than ever before. Firms engaged in retailing operations try to create and sustain a competitive advantage by offering customer value based on the utilization of SST options, while customers try to receive the best value for their participation in service production and delivery [29]. The Point of Sales Terminals (POST) and terminals of the Electronic Fund Transfer at Point of Sale (EFTPOS) provide refinements on a dual evolutionary process underpinning the development of a system of innovation in reference to the growing technology and the emergence of new forms of coordination across the players in retailing and banking industry [30].

> ABC Parcel Services is a leader in package delivery via ground and air shipping. It has provided services internationally as well as in all the 50 United States for almost 100 years. At the beginning of 2001, the CEO and the management of the company were faced with an unusually fierce competitive market and rapidly changing customer requirements. The automation implemented so far did not justify premium prices from the customers anymore. Customers were able to receive the same services that ABC was able to provide using this automation, at competitive prices from other providers in the market. It was obvious to the management that a value added service had to be provided to keep the edge over the competition. Looking at the changing business landscape, one area that was identified as critical was a customer centric approach using automation. The results from the implementation of such a system were expected to be tremendous for the organization to counter the competitive pressure of rival organizations and maintain market leadership [31].

The e-marketing services include e-remittances, e-payments, e-trades, and e-credit. However, many e-businesses have been forced out of market due to the low customer trust. Internet-based transactions require their own security measures for which private solutions may not be sufficient. For example, government procurements are needed to set up a framework for digital signatures and to designate agencies or processes to authenticate public keys associated with transactions. Consequently, internet-only sales firms have been substantially less profitable. They generate lower business volumes and any savings generated by lower physical overheads appear to be offset by other types of non-interest expenditures, notably marketing to attract new customers. However, e-banking develops automated credit authorization system by developing appropriate credit scoring system and cash-flow scoring system to reduce operating costs, improve asset quality, and increase client profitability. One of the major benefits of credit scoring system is that lenders can make credit decisions without necessarily obtaining financial statements, credit reports, or other time-consuming and hard-to-get information [32].

A large number of vending machines have occupied major place in sales in China since 1999, when this self-service technology channel was in its very first stages of development. By 2004, the popularity of vending food and beverages through automatic machines had grown immensely in China. The vending machines were strategically located at airports, and in subways, stores, parks and many other major public places in big cities. However, most of the population at the bottom of the pyramid segment is not yet accustomed to buying goods from vending machines, and, sometimes, consumers purchase a product just to satisfy their curiosity. In addition, the usage of coins as a method of payment for daily shopping is relatively low in China, where it is more common to pay with banknotes. In Brazil, Coca-Cola was a pioneer of vending machines, and the first "Coke Machines", initially operated with tokens, were set up in São Paulo in May 1992, thus initiating modern operations in the vending channel. "Coke Machines" rapidly became popular in key locations in the country's main cities, and the number of machines increased manifold since 2004. However, vending is a relatively new distribution channel in Russia. Even though it existed during the Soviet era, it disappeared at the beginning of the market reform period in Russia, as frequent fraud and vandalism caused significant losses for vending operators and suppliers. However, following the launch of new coin denominations in 1998, which increased the value of coins 1,000 times, new vending machines could work with coins again, which opened new opportunities for operators. In China, vending machines selling goods that can be paid for by mobile phones recently appeared in GuangZhou. Industry sources reported that this kind of machine is very popular among young people. Similarly, in Brazil, mobile phone operators have already tested payment via mobile phone in the vending channel, although its development is still hampered by the extra cost of the machines. Within hot drinks, Nestlé's Nescafé brand became widely available in vending machines in Russia. In Mexico, the company started to install vending machines in stores in Guadalajara and Monterrey in 2003. In India, hot drinks manufacturers have expanded their presence rapidly, with Nestlé and Café Coffee Day setting up numerous token-operated machines with an attendant[33].

Business negotiations often involve multiple issues and multiple parties and synchronization of deal is important to arrive at a common business conclusion. The required amount of information processing and communication in those cases can easily exceed a single human being's capability, especially when there are many negotiation issues and partners. Therefore, systems supporting and automating business negotiations have a great potential value. Considering the degree of automation achieved in other business process domains, the importance of managing negotiations to achieve more efficient business operations has become even higher. Since most business negotiations are in the context of corporate procurement or sales process, negotiation systems must be approached from the business process management perspective. A business negotiation process is often ad hoc in nature because the relationships between negotiation partners are dynamic and the most appropriate negotiation process is dependent on the specific case. Five issues in managing temporary workflows in business negotiations have been identified that include ease of generating new processes, flexibility, understandability, information sharing, and processes for decision support and collaboration [34]. Advances in technologies have allowed service providers to incorporate many different technologies into the delivery of their services. Previous studies on consumer attitudes towards adoption of Self-Service Technologies (SSTs) across various sales and services firms reveal that trust, quality and time are the major attributes that influence buying behavior of consumers [35].

CUSTOMER VALUE MANAGEMENT

Customer centric approaches are practiced efficiently by the call centers to connect the customer issues with appropriate and interactive solutions. Call centers not only offer personalized attention to their problems but also help in building customer loyalty. Customers rely on call centers which have high value work force towards services scheduling methods such as queuing system models to achieve optimal performance. Most of these models assume a homogeneous population of servers, or at least a static service capacity per service agent. The customer values for banking services are shaped more by habits, reinforcement effects, and situational influences than strongly-held attitudes. However, the aggregate returns on the customer value towards the new product from the perspective financial institution may be manifested in enhancing the market share, services coverage and augmenting profit in a competitive environment. Academics, consultants and business people had speculated that marketing in the new century would be very different from the time when much of the pioneering work on customer loyalty was undertaken. Yet there exists the scope for improving the applied concepts as there have been many changes over conventional ideologies [36]. The value concept in the above relationship governs the customer portfolio decision in terms of formulation of recursive utility over time. It shows that the optimal portfolio demand for products under competition varies strongly with the values associated with the brand, industry attractiveness, knowledge management and ethical issues of the organization. The extent of business values determines the relative risk aversion in terms of functional and logistical efficiency between the organization and supplier while the switching attitude may influence the customers if the organizational values are not strong and sustainable in the given competitive environment [37]. Customer value can be created by responding to the customers in a short period with all needed information to support the customer-seller relationship. The coordination between consumer demand and information flow can be effectively managed through the sales automation tools. Salespeople can establish successful contacts with the prospecting and existing customers using appropriate automation tools. Sales automation enable firms to:

- Organize and manage data across and within a company's client and prospect organization
- Maintain local client databases, display updated organization charts, and allow salespeople to maintain notes on specific clients or prospects
- Track consumers on real time and offer customized sales and services to enhance their customer value
- Drive qualified leads through marketing campaigns or lead referrals and track other prospect attributes
- Allow a salesperson to input client information and prospect and match it with the corporate policies on-line, in order to determine the most suitable sales offers to the prospecting buyers
- Perform CRM activities well
- Develop effective sales presentations and graphical analysis of performance
- Perform price quoting, order taking, order processing and order follow-up activities.
- Deliver intra-organization electronic communication.

Some studies reveal that customers perceive benefits on four main dimensions of their interaction with salespersons that include salespeople's professionalism, customer interaction frequency, salesperson responsiveness, and salesperson-customer relationship quality [38]. A perspective from resource-advantage theory is used to formulate expectations on the degree to which the use of information on customer value, competition, and costs contribute to the success of a price decision. It is argued that the success of these practices is contingent on the relative customer value the firm has created and the degree to which this position of relative value is sustainable in the competitive market place. These expectations are empirically tested on pricing decisions with respect to the introduction of new industrial capital goods [39]. Enterprise resource planning (ERP) systems have some inherent disadvantages as businesses become more complex and competition more intense. The mismatch between business processes that an organization desires and the business process provided by ERP systems can be unproductive and dysfunctional due to the inherent lack of flexibility offered by ERP systems. Business components-based design provides the flexibility to enhance better matches between 'real' business processes and 'systems' embedded business processes. Although business component-based design affords more flexibility, there exist significant implementation challenges including the lack of availability of business components in the market. Traditional business components-based designs use object oriented design techniques that use a bottom-up approach. The sales automation process helps firms in streamlining the customer relationship management (CRM) aspects and allows effective coordination between CRM and measuring the performance of salespeople using ERP system [40].

The role of customer value has been largely recognized over time by the financial institutions as an instrument towards stimulating market share and profit optimization. The customer values for a new product of financial institution in competitive markets are shaped more by habits, reinforcement effects, and situational influences than strongly-held attitudes. A strong and sustainable customer value associated with a new product launched by a financial institution may also lead to build the customer loyalty in the long run. An analysis of the new product-market structuring based on customer value may be developed well within the microeconomic framework of financial institutions. The value of a customer may be defined in reference to a firm as the expected performance measures are based on key assumptions concerning retention rate and profit margin and the customer value also tracks market value of these firms over time. The value of all customers is determined by the acquisition rate and cost of acquiring new customers [41].

Salesforce.com knows the power of good customer relations. The company offers hosted applications that manage customer information for sales, marketing, and customer support, providing clients with a rapidly deployable alternative to buying and maintaining enterprise software. The applications of this company are used by more than 65,000 clients for generating sales leads, maintaining customer information, and tracking customer interactions. The company's applications can be accessed from PCs and mobile devices. Customers of Salesforce.com company come from a variety of industries, including financial services, telecommunications, manufacturing, and entertainment. Large businesses could afford to make investments in enterprise resource planning, or ERP, and CRM applications, to gain an enterprise-wide view of all business information and automate and improve basic processes. In addition, large businesses attempted to customize and connect various incompatible packaged applications through time-consuming, extensive and costly integration efforts. The company has flagship Salesforce CRM applications, which were initially introduced in

February 2000, help companies better record, track, manage, analyze and share information regarding sales, customer service and support, and marketing operations. Salesforce.com cloud computing platform, which was introduced in 2007, allows customers and partners to more extensively customize and integrate Salesforce CRM applications or build entirely new SAAS applications beyond CRM without having to invest in new software, hardware and related infrastructure. These newly developed applications, which are hosted and run on our infrastructure, can then be used for internal operations or sold to third parties. The company also offers the AppExchange (application exchange), an online directory for SAAS applications, where customers can browse, test-drive and install applications from salesforce.com and their business partners. Since the introduction of AppExchange directory in 2006, more than 800 applications developed mostly by the business partners have been posted to the directory [42].

In the process of enhancing the customer value for the new products, a financial institution may simultaneously use intensive customer value for technology based banking services and intensive customer relationship management (CRM) strategies to the competitive sales and marketing strategies. The integrated impact of CRM, sales and marketing strategies at different stages of service attractiveness would contribute to the customer value and influence the aggregate returns on the customer value derived at various stages of service attractiveness of the financial nstitutions. However, a financial institution may need to compute the trend of customer value for all the services in its product line, and measure the variability in the customer values perceived for its new services. The customer values are broadly reflected in the competitive gains, perceived values, and extent of association with the financial services and level of quintessence with the customer relationship management services of the organization.

Systematically explored concepts in the field of customer value, and market driven approach towards new products would be beneficial for a company to derive long term profit optimization strategy over the period. Hence, a comprehensive framework for estimating both the value of a customer and profit optimization need to be developed. The self-service technology is distinguished by consumer participation in service production and delivery which is independent of service personnel and marginal interpersonal interaction between consumers and service organization. As reports of consumer dissatisfaction with SSTs become increasingly common, it is important, therefore, to investigate how organizations with SST-based offerings can encourage consumers to voice their dissatisfaction directly to the organization. Although the antecedents of consumer voice are well documented in the interpersonal services context, in the context of SSTs they have been subject to very little conceptual or empirical examination. However, on a tactical level, managers need to consider the optimum spread of customers on a matrix of product attractiveness and market coverage. This needs careful attention and the application of managerial judgment and experience to measure the value driven performance of the product of the firm. It is necessary for the managers to understand that customer value is context dependent and there exists a whole value network to measure, not just a value chain. This value network will contain important entities far beyond the ones commonly taken into consideration in financial projections and business analyses.

Technology intervention in retailing has put forth the concept and application of digital marketing that has reflected in consumer buying behavior with fundamental changes. Consumers are increasingly engaged in usage of Web to explore products and services instead of books, yellow pages, libraries, car dealers, department stores, or real-estate agents, and to search for information. In this process consumers often become aware of new products, companies and prices which are used as important variables in making purchase decision. Trust emerging from integrity, benevolence, and ability is the central dimension of e-commerce systems adoption. Based on customer self-service systems, business relationship is developed through human-computer interaction between the company and consumers. The inculcation of e-trust among consumers is a long term process [43].A large number of consumer goods and durables manufacturing and retailing companies are using these tools in marketing areas which they deem most important. Collaborative tools such as blogs, wikis, and social networks are being used in advertising, product development, and customer service [44].

Buying decision of consumers depend on various independent variables which range for perceived benefits to the quality of marketing services provided by the manufacturing and retailing firms. Thus companies reorient their marketing and sales strategies frequently in reference to the consumer perceptions on buying indicators. The manufacturing and retailing firms focus on brand portfolios, rethink spending approaches, generate more fine-grained customer insights, overhaul pricing and segment management, and restructure sales, service, and channel strategies. Implementation of each strategy is a challenge in its own, and some companies keep tackling the effectiveness of more than one strategy. General Electric, for example, has been trying simultaneously to improve the way it approaches innovation, brand management, and customer care. This level of consumer patronage represents a commercial transformation which explains a transformation of the company's broad-based marketing and sales elements [45].

Tata Motors ranks as the world's fifth-largest manufacturer of medium and heavy trucks—it has a 61 percent domestic market share in this segment—and has taken the number-two position for sales of passenger vehicles in the Indian market. It has also built a significant global presence, both through sales efforts in overseas markets (such as the former Soviet republics, the Middle East, South Africa, South Asia, and Turkey) and through acquisitions such as the takeover of Daewoo's commercia-vehicle business in South Korea and the purchase of a 21 percent stake in the Spanish bus manufacturer Hispano Carrocera. Tata Motors plunged into the low demand and low sales cycles in 2001 and the company had to revive its brand icon in the market and develop new strategies to acquire customers. Accordingly, the company attempted to concentrate its strategies on cost reduction, but while this was going on to think about taking action in areas that would have an impact during the other phases. Cost reduction strategy was laid on improving the product quality and upgrading product features so as to make the products more competitive. Tata Motors also started working on new products that would be required by the market after three to five years and strengthened the firm's position in the marketplace by setting up a new sales-planning process, tightening credit norms, improving the liquidity and profitability of the dealers, reorienting toward customer satisfaction, and extending the reach of its distribution network. Besides, to empower customer, the representatives of the company went on a "know the customer" campaign by listening to customers, talk about the problems they were facing and the suggestions they had for product improvement. Tata Motors also exposed its employees to the products of competitors in order to analyze good and bad about the competitor's products and compare them with their own company, so that salespeople of Tata Motors may understand clearly why customers buy someone else's products rather than its own brands [46].

The growth of team retailing is changing conventional ideas about sales and customer service. Team selling, also known as cross-functional selling, involves the efforts of different people from different areas within an organization working together to sell to and service a customer. This strategy of selling began in early '70s when sales teams were formed in high-technology organizations such as IBM to service the complex needs of their larger customers. It was necessary to have technical support people and application specialists involved in inventory control, manufacturing and production planning, along with a salesperson. Besides IBM, General Motors also adopted the team retailing strategy and developed it as part of the corporate business culture. Executives of the company felt that organizing a sales team is a major task to sell the high technology-high value products. The sales team should necessarily have a technical and finance person. Interestingly team retailing has filtered into less-technology-oriented businesses such as advertising, consumer packaged goods, and financial services. Several issues may emerge before they are formally addressed as the sales teams generally start to form naturally and informally. Communication among team members is difficult to establish, but is one of the keys to the effectiveness of this approach. There are many reasons for a company to choose team retailing because this strategy:

- Provides strategic advantage.
- Meets the needs of demanding, global customers more effectively.
- Establishes long-term, quality relationships with clients.
- Provides competitive advantage through a different kind of customer interface strategy.
- Increases flexibility in terms of how customers are served.
- Increases sales and potential profitability.
- Forces resource allocation decisions at the field level.
- Generates better input to new product and services development.

Team retailing effectiveness also depends on selection and training. Training to back up the appropriate selection is so important, in fact, that IBM sends their team leaders to a customized executive development program on how to be an effective key account executive. It's a tough, rigorous program and in addition, IBM requires recertification every few years. Also at the organizational level, in order for a team retailing approach to be successful, the supplier organization must have a strong market orientation as part of its organizational culture.

INTERNET RETAILING

Shopping behavior of consumers has shifted with the increasing applications of Internet forcing retailers to redefine their roles to secure their place in the Internet age. E-commerce supported with secured Internet is making an impact, providing the benefits of capturing a global audience 24 hours a day, seven days a week. Lately, the longer working day is driving customers away from queues at shopping malls and turning them to the convenience of the Internet. Unlike non-electronic retailing, Internet retailing does not have well-developed typologies that identify underlying factors for differentiating on-line buyers. There are three primary types of Internet retailers including intermediaries, product focused retailers, and retailers in micro-segment. The strongest differentiator of sites may be the size and scope of the product assortment [47]. In travel and tourism industries, retail services stores are losing business to Internet competitors. The future losers will be the toy stores that do not have a strong online presence. The advantage of online shopping is that a shopper can check to see if the product is in stock, determine its suitability, and make a purchase, all without standing in a queue or fighting for a parking space. Shopping patterns are changing as customers seek added value from their shopping experience, including dramatic cost savings. Customers wishing to send a bouquet to Europe can either visit an *Interflora* accredited florist and make the purchase over the counter or visit the *Interflora* web site. Internet retailing can be distinguished from conventional practices for the following reasons:

- Easy to place an order
- Long product-line and choices
- Competitive prices
- Faster service and delivery
- Comprehensive information about what is being offered
- No sales pressure
- Convenient payment procedures.

Large manufacturers of consumer goods recognize the added benefit of the Internet, especially the one-to-one relationships it offers. Some large manufacturers have used the Internet to introduce customized shopping options, thus becoming retailers themselves and providing yet another challenge to the traditional store owner. Shoppers can choose the hair, eye and dress color of the doll they purchase by visiting the Barbie web site of Mattel Company where shoppers may feel the difference in service that the traditional stores cannot offer. Customization services attract collectors or shoppers prepared to wait for their purchases to be created and delivered. Consumers through Internet shopping often hesitate in revealing their demographics and purchasing patterns, including date of birth, address, average spending, product preferences, and hobbies due to security reasons. Web-based businesses largely use this information as a platform to create an interactive loyalty program and database marketing. Although consumers can research high-price items such as cars and real estate via the Internet, and analyze information on the attributes of offerings, the deal is still more effectively done face to face as the confidence of buyers is boosted in personal negotiation. A retailer provides necessary personal contact that the Internet cannot offer. However, a successful retail store must build confidence over the virtual shopping and add value to its consumers´ shopping experience by giving them that 'something extra' to ensure their continued patronage. In contrast, certain industries such as music have won a significant percentage of the market away from retail outlets. There will always be a place for retailers who serve impulse and recreational purchasers. The conventional retail stores need to re-invent store ambience as their online competitors often compete offline and online. While some Internet-only retailers survive and prosper, for most, multichannel retailing provides a sustainable and attractive blend of new and existing retail formats. The complementary roles of stores and Internet are frequently noted, yet the renewed significance of catalogues is often ignored. A study revealed that four major components in channel choice emerge consistently, which include risk reduction, product value, ease of shopping, and experience [48].

Consumers process information in different ways, and determine their cognitive styles. Some companies have begun testing Web sites dynamically adapt to the cognitive styles of visitors to enable them receive information in a style comfortable for them. A Web site could offer additional technical details about a product to a user who has an analytic cognitive style, while a visitor who has a holistic cognitive style would receive information with decreased complexity. Such enhanced information platforms have the potential to increase the effectiveness of Web communications and influence buyer behavior. A recent study involving a Web site, from which consumers could purchase broadband services, found that, the company could increase potential purchases by 20 percent, when the Web site is designed to match individual consumers' cognitive styles. While there are many dimensions of cognitive style that could be used, it is observed that analytic or holistic, impulsive or deliberative, visual or verbal, and leader or follower may be stimulating designs of the web site to attract on-line retailing [49].

CONSUMER ACCEPTANCE OF INTERNET CHANNEL

The real breakthrough in e-commerce came with the increased commercial use of the WWW and electronic mail. The use of WWW as a new channel of distribution is still limited but continuously growing [50]. Electronic commerce has the capabilities to function as a new type of non-store retailing, in which the Internet establishes the direct link between the consumer and the retailer or manufacturer bypassing the conventional stores. Even though the non-store retailing has been in practice for quite some time, it has represented only 2.5% of total retail sales in Europe during early 90's. Besides, there are some prominent reasons for the corporate sector business to consider the Internet as a direct marketing channel. The distinguished advantages that World Wide Web offers to the consumer as a channel of distribution are as given below [51]:

- 24 hours and 7 days a week availability of retailing
- Instant gratification because of instant information availability
- The interactivity remains with consumer which means the consumer controls the flow of information unlike the conventional retail stores.

However, there are some disadvantages in the non-store retailing in terms of perceived higher risk than the conventional retail stores and also in terms of security issues on all transactions. Large companies have announced joint efforts and task forces working on security issue and they believe that the cryptography designs in near future would help in solving the security problems to some extent [52]. There are few measures which offer distinguished risk relief to the consumers for all web based transactions in terms of providing money-back guarantee, offering well-known brands and selling at reduced prices as compared to the on store retailing [53]. The retailing functions performed by the WWW may be classified into three broad groups:

- Non-store information channel,
- Non-store reservation channel with a traditional mail or courier service delivery, and
- Non-store purchasing and physical delivery channel

The Internet is perceived to increase market efficiency, and thus reduce price dispersion. Internet retailers of homogeneous products may choose to actively differentiate their stores and offerings in order to avoid competing just on price. It has been observed in a study that high traffic is associated with lower market prices. However, Internet retailers can mitigate the tendency to compete on price by relying on external linkages from other websites. Furthermore, Internet retailers may use economies of scale and participation with *shopbots* in conjunction with charging lower prices. Surprisingly, Internet retailer use of economies of scope is associated with higher prices. This indicates that firms may be in a position to avoid price competition by selling a wider product variety [54]. The basic practice of retailing has undergone remarkable, fundamental changes with the rise of non-store retailing comprising retailing by person to person, catalogues, telephone, and the Internet. Coupled with consumers' increased willingness to buy via these alternative channels, the traditional brick-and-mortar stores are not necessarily a requirement in today's retailing environment. Given the dramatic changes affecting retailing in general, the question of what makes a successful retail establishment is relevant for practitioners and researchers [55].

A growing middle class and growth in retail sales have provided high potential to buy products cash in China. Yet outside of the main cities in the vast expanse of rural China, with about 750 million population, the reliance on cash makes it difficult for consumers to spend and for retailers to sell. China has just 530 point-of-sales terminals and automatic teller machines (ATM) per million people, far below the 10,000 per million found in the United States. Accordingly, cash is used in 83 percent of all payment transactions in China, compared with just 21 percent in the United States. With most of these terminals and ATMs in China's cities, practically all rural transactions are cash based. This offers an opportunity in China to wean rural consumers off their reliance on cash to add more ATMs and POS terminals. However, it is estimated that such an effort would cost at least $2 billion and add just 130 terminals and ATMs per million people. Installing equipment and extending the telecommunications network in remote areas would also take a long time. There are alternative mobile payments solutions gaining exposure around the world, but these are not well suited to the needs of rural China. Seoul and Tokyo have both introduced a system that allows a transaction to be completed using a mobile phone with a special built-in chip and an in-shop non-contact reader. However, the need to install the reader and to use special and expensive mobile handsets renders the solution inadequate for rural China. A short messaging service (SMS) based payment system is, cost effective, versatile and ubiquitous. Users simply send an SMS message specifying the mobile-phone number of the payee and the amount to transfer, along with a personal identification number. Within seconds, the payee receives both a confirmation message by SMS and the money in the designated account. The payer receives a confirmation message [56].

The information channel, which is largely used by the consumers, is being served currently by various powerful search engines on the web. In addition to this channel the ordering/re-ordering channel is also functioning for reservation of goods and services. The internet cannot provide physical delivery for goods other than software and the channels have to depend on the air or surface supply chain links. This involves cumbersome procedures to be handled like export-import duties, local laws and security of payment and delivery of goods and services. The consumer acceptance of the goods and services ordered/reserved via internet is more crucial at this juncture. The web marketing does not help consumers in making buying decision through *touch, feel and pick* process. However, the internet for virtual shopping is one of the preferred channels for the consumers of the same region or state or country [57]. The web retailers should appoint the franchisee in various regions to handle the delivery of goods and establish link between the web retailers and remote consumers.

Information communication technology (ICT) and consumer shopping behavior are two of the most dynamic motivators of the emerging global economy. Both retailing and ICT increasingly provide strategic opportunities and powerful tools for enhancing customer value, the market expansion, and the development of brand equity in the competitive marketplace. Applications of Internet have intensified towards sharing information among players in the marketplace, attract consumers on e-commerce, and online markets. The Internet is becoming more sophisticated across various routes to market. Online transactions are ushering not only in products sales but also in the area of services marketing, in order to improve the commercial practices and augment customer value. In view of changing shopping dynamics in the global marketplace, it is observed that reaching a new information threshold of universal, ubiquitous communication has brought the virtual shopping into a new stage of interactivity, accelerating both wired and wireless management and marketing [58]. The online method has been proved ideal for the tourism industry in terms of enabling information retrieval and electronic transactions. The success of certain online-tourism businesses, such as Expedia, Travelocity, and Orbitz, along with an increasing prevalence of online shopping for tourism products and services are the good examples of e-commerce. The web retailers in association with the franchisees should develop interactive home system for consumers and interactive business system for the business-to-business transactions. However, the electronic retailing channel provides the consumers a scope for wide selection, screening of products out of long catalogue, reliability on brands presented on the net and comparison of price and product features as per the need and utility.

TECHNOLOGY SHIFTS IN RETAILING

History of retailing dates back with the history of technology in society. A bird's eye view at the evolution of retailing reveals that technology has played a role as the primary enabler of change. As technology grows sophisticated, the consumer's expectations also swell exponentially. In fact the convergence of a few key technologies is enabling that change. Smart cards payment technology has driven new revolution in retailing as this technology has not only helped in increasing the quick buying decisions of consumers but also attracted large mass of potential customers into retail gamut. Smart cards have offered a wide variety of applications that could revolutionize payment transactions, reduce costs, and spur online purchasing. Despite the benefits these electronic purses offer, a number of issues inhibit their widespread use, especially in open systems. The smart cards can store various types of encrypted information as well as cash balances and digital signatures. A secret key can be used to secure e-commerce transactions as well as protect the card contents. These keys are vulnerable to attack, however, and the stored-value feature is attractive to international money launderers. Despite some risk factors the smart cards are globally accepted by the retailers [59].

There has been a significant change in retail trading over the years. Modernization, systematization, and consolidation are the catch phrases and keys to understanding retail. The present age is of rocket science retailing which is a blend of the traditional forecasting systems with the prowess of information technology. It fuses data and instinct with computer models to create a high-tech forecasting system supported by a flexible supply chain. The need is to evaluate not what the retailer sold but what it could not sell and what it could have sold, had the inventory been available. Merchandise decisions have become more complex and the penalties for errors even steeper. To reduce the fallouts and to increase the customer satisfaction, merchandise planning has become all the more important. A new set of software tools and sophisticated techniques have emerged, which promise to revolutionize the entire merchandising chain, from buying to stocking to pricing. The latest techniques used for efficient inventory management include vendor managed inventory, forecasting techniques, inventory classification, computer assisted ordering, distribution centers and direct store delivery [60].

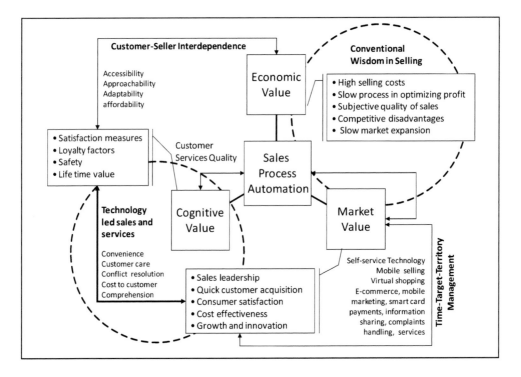

Figure 10.2. Customer Value and Sales Automation Convergence Paradigm.

The impact of technology on various functions of retailing has been increasing. As the number of channels for a retailer increases, managing the dynamics of customer behavior in the rapidly emerging multi-channel environment becomes complex. Building and retaining a long-term association with customers requires that relationship management applications should be able to accommodate the various channels. Multi-channel customers are the most valuable customers and hence multi-channel integration would improve customer loyalty and retention [61]. Besides self-service retail stores and grocery stores, the technology has enormously supported the buying process of consumers for capital goods like automobiles. The purchase of a car is a complex process when compared with other retail experiences. Despite the range of purchase channels available and the increased level of accessible information, the majority of customers still choose to buy a car through a traditional dealer network. However, since the end of twentieth century the computer assisted buying process has been well received by customers [62]. The convergence of sales automation process and delivering customer value is important for achieving success in sales performance. The Figure 10.2 exhibits the factors driving customer value following the sales force automation in a firm.

Sale automation process induces three types of values among customers that include economic value, market value, and cognitive value as illustrated in Figure. Conventional selling practices are not cost effectives and quality of sales in traditional sales approach is subjective as it largely depends on the personality of a salesperson, unlike in automated sales process. E-shopping is influenced by time and attractiveness of virtual sales offers and effectiveness of customer relations. These factors vary widely in reference to consumer segments and markets attractiveness and induce compulsive buying behavior among customers, which is judged by the satisfaction in spending and perceiving pleasure of buying, occasionally exercising choice and passing time in knowing new products, services, technologies and understanding fellow customers [63]. Retailers using a "store as the brand" strategy invest in creating a specific, unique shopping experience for their target customers and encourage leisure and group buying behavior where delivery of customer satisfaction seems to be an effective source of differentiation. Change-of-season sales are most frequently introduced with attractive sales promotions in reference to price discounts or two for one price basis and linked with objectives of moving a volume of stock. Retail promotional sales also include general sales, and these are linked with other promotional objectives and activities such as increasing profit and inventory management. A company's information technology (IT) structure and its brand architecture are intended to minimize transaction costs both within the organization and between the organization and its customers. Business-to-consumer (B-to-C) e-commerce fundamentally alters the structure of those transaction costs relevant to the IT structure and the brand architecture. Manufacturing companies typically implement contemporary changes in the IT structure and the brand architecture to support consumers in retail buying as B-to-C e-commerce is highly important to them and that these changes result in a stronger integration within and between the IT structure and the brand architecture [64].

The rapid development of electronic commerce has seen emerging electronic service retailers attracting the interest of, and gaining the patronage of, both service providers and customers. However, there is consensus that the e-commerce industry in general has not been able to cope with all the challenges of, and to realize the true potential of, the technology-based marketplace. Some research studies argue that although the Internet marketplace possesses unique characteristics, there are certain traditional values that remain central to business success in all markets [65]. In an increasingly competitive market there is a keen interest among retailers to understand as much as possible about consumer behavior. Advances in technology have presented retail marketers with many new research tools with which to monitor such behavior. Video monitoring technology offers an objective and accurate research tool for retailers to keep a watch on consumer behavior. However it is a wrong notion for those managers, who believe that filling their websites with a broad array of information, diverts attention from their company's core offerings. A new global study, however, has revealed that such information increases customer attractiveness and using consumers' desire for providing better value is the strongest predictor of superior shareholder value for e-commerce companies [66].

It has been recognized that enhancing the role of technology in a service organization would serve to reduce costs and improve service reliability. The new information technology is becoming an important factor in the future development of financial servicesindustry, especially banking industry. However, it is argued that there remains an important role for customized relationships in the delivery of any service proposition [67]. A large number of customers use the Internet as a medium of business (electronic-commerce). Association of self-service technologies with customers indicates that six attributes common to the diffusion model including perceived convenience and financial benefits, risk, previous use of the telephone for a similar purpose, self-efficacy, and Internet use, play a significant role in the performance of retail banking operations. It can be stated that banks adopting information technology based capital-intensive techniques are more efficient as both cost and profit frontier gain competitive advantage in the financial markets [68].

Information and internet-based technologies have fostered new supply chain initiatives in food retailing, which have significantly contributed towards enhancing the efficiency of store format, membership in a chain, unionization, and adoption of variety of information diffusion tools. The major breakthrough has emerged in supply chain management in the fields of data sharing, decision sharing, and technologies that support product assortment, pricing, and merchandising decisions [69]. Among many innovative technological processes, radio frequency identification (RFID) is gaining popularity among retail firms for managing the inventory and just-in-time supplies. RFID tag has the potential to significantly reduce costs in retailing. Although this technology is still not cost effective for all manufacturers, it may soon become mainstreamed as its advantages outweigh its initial investment. With many manufacturers and vendors becoming early adopters, the cost barrier for RFID will quickly be eliminated. It is unlikely, however, that RFID tags will replace bar codes in the near future because of the start-up cost to retailers and suppliers [70]. RFID tags enhance the operational efficiency as the retailers can store much more information about products than bar codes can, and unlike bar codes they don't have to be seen by a scanner to be recorded. In fact, RFID tags can signal their presence to scanners a few yards away even when obscured by packaging, so the contents of a closed container can be quickly scanned and recorded. It is observed in many research studies that logistics and information technology strategies are developed and implemented in a parallel way by both manufacturers and multiple retailers. Applying improved technology suggests that multinational firms possess greater operational efficiency at both secondary and in-store distribution operations as compared to firms using conventional practices, which is largely attributed to their integration of logistics and information technology operations [71].

CONSUMER BEHAVIOR ON TECHNOLOGY SHIFTS

There is no defined critical path of success for domestic and international firms in retailing consumer goods across different marketplaces in the world. Exceptionally, couple of companies like Fuji, Kodak and Coca-cola that have spread their business in over 100 countries developed gradually. The characteristics of the global market place are diverse and international marketing approaches are different. The companies need to adopt a strong rationale for grouping the countries into segments. The multinational and the global corporation are different as the former operates in a number of countries and carries adjustment in the production and marketing practices in each country at a highly relative costs. [72] The global corporation operates with the stanch loyalty at relatively low costs with standardization. Coca-cola and Pepsi-Cola companies have standardized their products globally according to the regional and ethnic preferences of consumers. The most effective world competitors integrate quality and trust attributes into their cost structure. Such companies compete on the basis of appropriate value of price, quality, trust and delivery systems. These values are considered by the companies in reference to the product design, function and changing consumer preferences like fashion. The multi-national retailing firms develop strong information base about the business environment in a marketplace and put their efforts on adapting to the given consumer preferences and retail environment, and set gradual penetration process in the marketplace. On the contrary, the global retailing firms attempt to acquire customer for the economic reasons of lower prices by standardizing its marketing operations. For example Wal-Mart has emerged with its Every Day Low Prices (ELDP) policy, as one of the most favored retail store by a large number of customers in developed and developing countries who are sensitive to the price. The global retailing companies treat the consumer segments as composed of a many independent and customized preferences that lead to loyalty formation over a long association with the firms and their brands. There are four major drivers that propel consumers towards configuring their buying behavior which include market, benefit based, cost, and technology. Of these, the market driver is most significant in developing consumer behavior.

The *market driver* spots the needs of common customers, global customers, global channels in reference to products that serve existing demand of consumers, products that bridge the gap between latent demand and availability of products, and develop consumer education for those products about which consumers do not have any awareness. The common customers needs become a compelling factor for the retailers when customers of different profiles appear with the similar or identical needs within the same product category. Emergence of international retailers and increased accessibility to the virtual and brick-and-mortar retail channels has created homogenous groups of customers. However, some markets that typically deal with the culture bound products like food and beverages, apparel, and entertainment strongly resist the shift towards globalization and remain multi-domestic, serving to the different customer preferences and differentiated products across the countries. On the contrary the global customers need the same products or services across the marketplaces like the case of Kodak films or Hilton Hotels. The global distribution and logistics companies offer seamless transport, storage and delivery services to support smooth operations of retailers. The retailing companies can expand internationally provided the channel infrastructure is met with the distribution needs of the company to serve customers and enhance their shopping value. Hence integrated networks thrive to bring the multinational companies close to the global distributors and retail channels in order to generate systems effect on consumers and hold them loyal to the store. Convergence marketing is applied to the same marketing ideas on brand names, packaging, advertising and other components of marketing-mix in different marketplaces to serve varied consumer segments. Nike's campaign anchoring the basketball champion Michael Jordan pulled-up the brand in many countries. This is how the good ideas of multinationals get leveraged to win the consumer confidence world over.

> The Coca-Cola bottling system grew up with roots deeply planted in local communities. This heritage served the Company well today as people seek brands that honor local identity and the distinctiveness of local markets. The Coca-Cola Company believes that employees and brands are the two core assets on which this global enterprise is built. Each plays a vital role in the Company's promise to benefit and refresh everyone it touches. The company has rationalized the value chain, implementing uniform practices and procedures across distribution network, and identified additional opportunities to streamline the operations through countrywide network.Firms with growth aspirations have several ways of reaching their goals. Large companies have long sensed the potential value of investing in external start-ups, but more often than not, they fail to get it right. The managerial dimension influences the operational efficiency of internal fit of the corporate ventures. The success of a company is associated with high levels of commitment, competitive skills and dynamics in functional management of the venture.Managers define clear sales tasks to ensure significant results from the salespeople and evaluate timely performance feedback. A firm may focus on administering sales activities in four major processes including territory-target based deployment of sales force, efficiency in managing client account, working with improved information systems, and minimizing field level task conflicts. Hence, limiting the number of accounts for each salesperson may work out as a good option to improve the efficiency of sales activities in the market. Field sales managers may be encouraged to adapt management by objectives, performance appraisal, and monthly reviews to encourage their salespeople to do their work effectively. However, some companies fail to coordinate salespeople's efforts with other customer-facing teams and while operating at cross-purposes, these functions conflict over roles and resources jeopardizing the relationship with customers. To correct such misalignments, it is necessary to match sales management practices with strategic priorities of the firm and allow effective coordination among customer-facing teams to enable seamless service for customers [73].

The *benefit based driver* rides on consumer decision making towards products and services in reference to perceived benefits and shared benefits. Consumers hold perceived benefit upon using the product by way of owning (purchasing) the product or services while the shared benefit is derived by the consumers when its use value is felt upon conceiving the word of mouth information. Thus, companies manufacturing consumer goods and retailers continuously look for matching their product and retailing strategies appropriately with consumer preferences. However, the emergence of strong global competitors has served to develop the market infrastructure for the local companies and also helped in transfer of technological skills enabling the domestic company to explore the scope of expansion. Consumers evaluate benefit based buying decisions in reference to perceived use value, value for money, social implications, opportunity cost of buying, and advantages of the product or services in satisfying relative needs [74]. In emerging markets like India and China, where large segment of consumers introspects their buying decisions based on direct and indirect benefits, global manufacturing and retailing firms focus their business strategies on the basis of quantum of emerging middle class in these countries, mix of local and global leadership need to be fostered to explore and exploit business opportunities, and accepting on local partners [75].

The *cost drivers* that commonly affect consumer decisions include product search cost, final price and services costs like maintenance, insurance, and security. These factors are largely based on the scale of retailing that involve the inventory management and service provider functions either of the manufacturing firm or of an outsourced agency. When a new automobile plant is set-up, it aims at designing, manufacturing or assembling and delivering a particular model by penetrating into the neighboring markets to gain the advantages of economies of scale. The high market share multi-domestic companies derive gains from spreading their production activities across multiple product lines or diversified business lines to achieve advantage through the scope of economies. The manufacturing and marketing activities of Proctor and Gamble, Unilever, Colgate-Palmolive may illustrate this global attribute that is explained by the cost drivers that appear on part of the customers.

The lowering of trade barriers made globalization of markets and production a theoretical possibility, and technological change has made it a tangible reality. Since the end of World War II, the world has seen major advances in communications, information processing, and transportation technology including, most recently, the explosive emergence of the Internet and World Wide Web. The *technology drivers* play a significant role in global business. Global expansion of the multinational companies has been highly stimulated by the technological advancements in designing, manufacturing and marketing of consumer and industrial products. Services have also been improved by many technological breakthroughs. The internet revolution has triggered e-commerce as an open access channel and a strong driving force for the global business in the consumer and industry segments. Improved transport and communication now makes it possible to be in continuous contact with producers anywhere in the world. This makes it easier for companies to split production of a single good over any distance. Storage and preservation techniques have revolutionized the food industry for example, so that the idea of seasonal vegetables is no longer relevant today as anything can be exported all year round from anywhere.

In addition, the IT revolution has made the movement of investment capital around the globe an almost immediate process ensuring that financing opportunities across the developed and developing world have both expanded and become more flexible. However, non-economic drivers of global integration, from travel to telephone traffic maintained their forward momentum, making the world more integrated at the end of 2002 than ever before. Technological upgrading, in the form of introduction of new machinery and improvement of technological capabilities, provides a firm with the means to be successful in competition. In the process of introducing better technologies, new lower-cost methods become available, which allow a firm to increase labor productivity, i.e., the efficiency with which it converts resources into value. Firms adopt these newer methods of production if they are more profitable than the older ones. The ability of a firm to take advantage of technical progress is also enhanced if the firm improves its entrepreneurial and technological capabilities through two strategies, namely learning and adaptation, and innovation. The latter is a process of searching for, finding, developing, imitating, adapting, and adopting new products, new processes, and new organizational arrangements. Because rivals do not stand still, the firm's capacity to develop these capabilities, as well as its ability to compete, depends on the firm's maintaining a steady pace of innovation [76]. Containerization has revolutionized the transportation business, significantly lowering the costs of shipping goods over long distances. Before the advent of containerization, moving goods from one mode of transport to another was very labor intensive, lengthy, and costly. It could take days to unload a ship and reload goods onto trucks and trains.

Technological changes are the main impetus behind new market opportunities. The extent of such changes may be explained from super technologies to the appropriate and intermediate technologies. The strategic choices have wide ranging ripple effects through the organization that determine the key success factors and growth performance. Some companies would be making right strategic choices by improving the implementation process of competitive advantages. These companies are guided by the shared strategic vision and are driven by the responsive attitude towards the market requirements. They emphasize the continuous strive to satisfy the customers. A strategic vision in managing markets may be understood as the guiding theme that explains the nature of business and the future projections thereof. These projections or business intentions depend on the collective analysis of the environment that determines the need for new developments or diversifications. The vision should be commissioned on a concrete understanding of the business and the ability to foresee the impact of market forces on the growth of business. The vision motivates the organization for collaborative business planning and implementation.

Strategic thrust has a significant magnitude and direction in sailing the business though the turbulent situation. The factors associated with the competitive advantage and business investments uphold the strategic thrust to achieve the business objectives though the positive channel efforts. The competitive advantage may be assessed in reference to the superior customer value and the lowest delivered cost. Such combination of the strategies may be termed as competitive superiority that explains cost effective delivery strategy to enhance the customer value. An overall edge is gained by performing most of the activities at a cost lower than competitors. This would enable the company to optimize its cost of delivery of the new products and simultaneously enhance the value of customer value to up-hold the strategic thrust of the company.

SELECT READINGS

Kirkpatrick, D. L. (2001), Reorganizing the Sales Force, Managing Change Effectively, Butterworth-Heinemann, Boston, MA.

Schmitt B H. and Alex S (1997), *Marketing Aesthetics: The Strategic Management of Brands, Identity and Image*. New York: The Free Press.

Thompson, K.T. (2005), *Sales automation done right*, Ardexus Sales Ways, Toronto, Canada.

REFERENCES

[1]   Moriarty, R. T., and Swartz, G. (1989), Automation to boost sales and marketing, *Harvard Business Review*, 67 (1), 100-108.

[2]   Durkin, M. (2007), Understanding registration influences for electronic banking, *The International Review of Retail, Distribution and Consumer Research*, 17(3), 219-231.

[3]   Bush, A.J., Moore, J. B.,and Rocco, R. (2005), Understanding sales force automation outcomes: A managerial perspective, *Industrial Marketing Management*, 34 (4), 369-377.

[4]   Rajagopal (2006), Managing effective customer services through call centre: A case of Roche diagnostics Mexic° (B), Discussion case, EGADE Business School, ITESM, Mexico City.

[5]   Clark, P., Rocco, R. A. and Bush, A. J. (2007), Sales force automation systems and sales force productivity: Critical issues and research agenda, *Journal of Relationship Marketing*, 6(2), 67-87.

[6]   Jones, D. L. (2004), Hospitality industry sales force automation: Organizational and individual levels of adoption and the implications on performance, productivity and profitability, *Journal of Hospitality & Leisure Marketing*, 11(2), 173-185.

[7]   Such behavior was noticed during informal discussion with the sales promoters.

[8]   Rajagopal (2010), *ING Commercial America: The virtual sales office*, Discussion Case, Center for International Cases, ITESM, Mexico.

[9]   Johnson, D.S. and Bharadwaj, S. (2005), Digitization of selling activity and sales force performance: An empirical investigation, *Journal of the Academy of Marketing Science*, 33 (1), 3-18.

[10] Kanaracus, C. (2008), Gartner: CRM market up 23 percent in '07, *IDG News Service* http://www.pcworld.com/businesscenter/article/148010/gartner_crm_market_up_23_pe rcent_in_07.html, also see Wailgum, T. (2008), The future state of the CRM market, CIO.com   http://www.pcworld.com/article/148135-2/the_future_state_of_the_crm_ market.html

[11] Schminke, M., Ambrose, M.L., and Neubaum, D.O. (2005), The effect of leader moral development on ethical climate an employee attitudes, *Organizational Behavior and Human Decision Processes*97 (2), 135-151.

[12] Rapp, A., Agnihotri, R., and Forbes, L. (2008), The sales force technology-performance chain: The role of adaptive selling and effort, *Journal of Personal Selling and Sales Management,* 28(4), 335-350.

[13] Wang, D. and Kumbhakar, S. C. (2009), Strategic groups and heterogeneous technologies: an application to the US banking industry, *Macroeconomics and Finance in Emerging Market Economies*, 2(1), 31-57.

[14] Sullivan, Richard J and Wang Zhu (2005), *Internet Banking: An Exploration in Technology Diffusion and Impact*, Federal Reserve Bank of Kansas City, Payments System Research Working Paper # PSR-WP-05-05, 1-45.

[15] Cho, S. D., and Chang, D. R. (2008), Salesperson's innovation resistance and job satisfaction in intra-organizational diffusion of sales force automation technologies: The case of south Korea, *Industrial Marketing Management,* 37(7), 841-847.

[16] Pays, P. A. and de Comarmond, F. (1996), An intermediation and payment system technology, *Computer Networks and ISDN Systems*, 28 (7), 1197-1206.

[17] Hanninen, S., and Sandberg, B. (2006), Consumer learning roadmap: a necessary tool for new products, *International Journal of Knowledge and Learning*, 2(3), 298-307.

[18] de Blasio, G. (2008), Urban-Rural Differences in Internet Usage, e-Commerce, and e-Banking: Evidence from Italy, *Growth and Change*, 39 (2), 341-367.

[19] Aldlaigan, A., and Buttle, F. (2005), Beyond satisfaction: customer attachment to retail banks, *The International Journal of Bank Marketing*, 23 (4), 349-359.

[20] Donner, J. and Tellez, C. (2008), Mobile banking and economic development: linking adoption, impact, and use, *Asian Journal of Communication*,18(4), 318-332.

[21] Yousafzai, S., Pallister, J. and Foxall, G. (2009), Multi-dimensional role of trust in Internet banking adoption, *Service Industries Journal*, 29(5), 591-605.

[22] Wang Zhu (2005), *Technology Innovation and Market Turbulence: A Dot com Example*, Federal Reserve Bank of Kansas City, Payments System Research Working Paper # PSR-WP-05-02, 1-49.

[23] Ahn, H. J. (2010), Evaluating customer aid functions of online stores with agent-based models of customer behavior and evolution strategy, *Information Sciences,* 180(9), 1555-1570.

[24] Linden, G., Smith, B., and York, J. (2003), Amazon.com recommendations: item-to-item collaborative filtering, *IEEE Internet Computing*7, 76–80.

[25] Varnali, K., and Toker, A. (2010), Mobile marketing research: The-state-of-the-art. *International Journal of Information Management,* 30(2), 144-151.

[26] Rajagopal and Rajagopal, A. (2010), Impact of customer centric technologies in retail banking, *International Journal of Business Competition and Growth*, 1 (2), *In Press*.

[27] Bunker, D., Kautz, K. H., and Nguyen, Thanh, A. L. (2007), Role of value compatibility in IT adoption, *Journal of Information Technology*, 22 (1-2), 69-78.

[28] Hu, R., Yang, D., and Qi, R. (2009), *A transaction management mechanism for building trust in mobile commerce*, International Conference on Management of e-Commerce and e-Government, ICMeCG 2009, 408-411.

[29] Anitsal, I., and Flint, D. J. (2006),Exploring customers' perceptions in creating and delivering value technology-based self-service as an illustration, *Services Marketing Quarterly*, 27 (1), 57-72.

[30] Consoli, D. (2008), Systems of innovation and industry evolution: The case of retail banking in the UK, *Industry & Innovation*, 15(6), 579-600.

[31] Bhaskar, R. (2008). Information technology systems deliver competitiveness for ABC parcel services. Journal of Cases on Information Technology, 10(3), 1-9.

[32] de Young, R. (2001), *The financial progress of pure-play internet banks*, BIS Papers, No. 7, November.

[33] Moreau, R. (2005), *Vending in emerging countries*, Euromonitor International Online, July 26 http://www.euromonitor.com/Vending_in_emerging_countries.

[34] Baek, K. J., and Segev, A. (2005), A web services-enabled marketplace architecture for negotiation process management, *Decision Support Systems*, 40(1), 71-87.

[35] Curran, J. M., and Meuter, M. L. (2005), Self-service technology adoption: comparing three technologies, *Journal of Services Marketing*, 19 (2), 103-113.

[36] Ryder, S. G., Ross, K. G., and Musacchio, J. T. (2008), Optimal services policies under learning effects, *International Journal of Services and Operations Management*, 4 (6), 631-651.

[37] Rajagopal (2006[a]), Measuring Customer Value Gaps: An Empirical Analysis in the Mexican Retail Market, *Economic Issue*, Vol.10, No. 1, 19-40.

[38] Boujena, O., Johnston, W. J., and Merunka, D. R. (2009), The benefits of sales force automation: A customer's perspective, *Journal of Personal Selling and Sales Management*, 29(2), 137-150.

[39] Hunt, S. D., and Morgan, R. M. (1995), The comparative advantage theory of competition, *Journal of Marketing*, 59 (2), 1-15.

[40] Ghandforoush, P., Sen, T. K., Tegarden, D. P., and Ramaswamy, R. (2010), Designing systems using business components: A case study in call centre automation, *International Journal of Electronic Customer Relationship Management*,4(2), 161-179.

[41] Gupta, S., Lehmann, D. R., and Stuart, J. A. (2004),Valuing customers. *Journal of Marketing Research*, 41 (1), 7-18.

[42] Salesforce.com (2009), Annual Report, 2009. For details on CRM support tools of the company see corporate web site http://www.salesforce.com

[43] Hwang, Y., and Kim, D. J. (2007), Customer self-service systems: The effects of perceived Web quality with service contents on enjoyment, anxiety, and e-trust, *Decision Support Systems*,43 (3) 746-760.

[44] Bughin, J., Erbenich, C., and Shenkan, A. (2007), How companies are marketing online: A McKinsey Global Survey, *McKinsey Quarterly*, September.

[45] Claret, J., Mauger, P., and Roegner, E. V. (2006), Managing a marketing and sales transformation, *McKinsey Quarterly*, August.

[46] Kumra, G. (2007), Leading change: An interview with the managing director of Tata Motors, *McKinsey Quarterly*, January.

[47] Mottner, S., Thelen, S., and Karande, K. (2003), A typology of internet retailing: An exploratory study, *Journal of Marketing Channels*, 10(1), 3-23.

[48] Mcgoldrick, P. and Collins, N. (2007), Multichannel retailing: profiling the multichannel shopper, The International Review of Retail, Distribution and Consumer Research, 17(2), 139-158.

[49] Urban, G. L., Hauser, J. R.,Liberali, G., Braun, M., and Sultan, F. (2009), Morph the Web To Build Empathy, Trust and Sales, *Sloan Management Review*, 50 (4), 43-51.

[50] Mondahl, M. (1995), Forecasts of the core economy, Forrester Research Report, http://www.forester.com,September 1995. Also see Hambercht and Quist http://www.hambercht.com

[51] Hoffman, D. L. and Novak, T. P. (1996), Marketing in hypermedia computers, *Journal of Marketing*, 60 (3), 50-68.

[52] McCarthy, V. (1997), Web Security: How Much is Enough ?*Datamation,* 43 (1), January, 112-117.

[53] Van den, P. D. and Leuis Joseph: Mail Order verses Retail Store Buying- The Role of Perceived Risk and Risk Reduction Strategies, International Review of Retail,Distribution and Consumer Research, 4, October 1996, 351-371.

[54] Bailey, J. P., Faraj, S., and Yao, Y. (2007), The Road More Travelled: Web Traffic and Price Competition in Internet Retailing, *Electronic Markets*,17(1), 56-67.

[55] Crittenden, V. L., and Wilson, E. J. (2002), Success factors in non-store retailing: exploring the Great Merchants Framework, *Journal of Strategic Marketing*, 10(4), 255-272.

[56] Jan, B., Chris,Ip., and Anna, Y. (2007), Developing New Rural Payment System in China, *South China Morning Post* on March 31.

[57] Rajagopal (1999),:*Virtual Shopping: Strengths and Weaknesses*, Working Paper, Administrative Staff College of India, 5-18.

[58] Chen,C. (2006), Identifying significant factors influencing consumer trust in an online travel site, *Information Technology & Tourism*, 8 (3/4), 197–214.

[59] Kearns, G. S., and Loy, S. (2003), Global acceptance of stored-value smart cards: analysis of inhibitors and facilitators, *International Journal of Services Technology and Management*, 3 (4), 417-428.

[60] Kumar, B., and Banga, G. (2007), Merchandise planning: An indispensable component of retailing, *The Icfaian Journal of Management Research*, 6 (11), 7-19.

[61] Ganesh, J. (2004), Managing customer preferences in a multi-channel environment using Web services,*International Journal of Retail & Distribution Management*, 32 (3), 140-146.

[62] Reed, G., Story, V., and Saker, J. (2004), Information technology: changing the face of automotive retailing? *International Journal of Retail & Distribution Management*, 32 (1), 19-32.

[63] Watkins, M.and Bond, C. (2007), Ways of experiencing leisure, *Leisure Sciences*, 29 (3), 287-307.

[64] Treiblmaier, H., and Strebinger, A. (2008), The effect of e-commerce on the integration of IT structure and brand architecture, *Information Systems Journal*, 18 (5), 479-498.

[65] La, K. V., and Kandampully, J. (2002), Electronic retailing and distribution of services: cyber intermediaries that serve customers and service providers, *Managing Service Quality*, 12 (2), 100-116.

[66] Eisingerich, A. B., and Kretschmer, T. (2008), In E-commerce more is more, *Harvard Business Review*, 86 (3), 20-21.

[67] Durkin, M. (2007), Understanding registration influences for electronic banking, *The International Review of Retail, Distribution and Consumer Research*, 17 (3), 219-231.

[68] Casolaro, L. and Gobbi, G. (2007), Information Technology and Productivity Changes in the Banking Industry, *Economic Notes*, 36 (1), 43-76.

[69] Timothy, P. and Robert, K. (2007), Evaluating food retailing efficiency: the role of information technology, *Journal of Productivity Analysis*, 27 (2), 101-113.

[70] Pate, S. S., Blaylock, K., and Southward, L. (2007), RFID: An interface survey between students and retailers, *The Marketing Review*, 7 (3), 273-281.

[71] Bourlakis, M., and Bourlakis, C. (2006),Integrating logistics and information technology strategies for sustainable competitive advantage, *Journal of Enterprise Information Management*, 19 (4), 389-402.

[72] Theodore, L. (1998),*The globalization of markets* in International and Global Marketing-Concept and Cases, (Ed.) Meloan W Taylor and Graham L John, Irwin McGraw Hill, Boston, 13-23, 1998.

[73] For details see Rajagopal (2008), Organizational Buying and Sales Administration in the Retail Sector, Journal of Retail and Leisure Property, 7 (1), 3-20.

[74] Thomas, C. F. (2000*), Kodak Vs. Fuji – The Battle for Global Market Share,* Institute of Global Business Strategy, Lubin School of Business, Pace University, New York, 1-23.

[75] Prahalad, C. K. and Lieberthal, K. (2003), End of corporate imperialism, Harvard Business Review, 81 (8), 109-17.

[76] Asian Development Bank (2003),*Drivers of Change. Globalization, Technology and Competition*, Section III, Competitiveness in Developing Asia, Asian Development Outlook, 2003.

AUTHOR BIOGRAPHY

Dr. Rajagopal is Professor of Marketing at EGADE Business School of Monterrey Institute of Technology and Higher Education (ITESM) in Mexico City Campus and is Fellow of the Royal Society for Encouragement of Arts, Manufacture and Commerce, London. He is also Fellow of the Institute of Operations Management and Chartered Management Institute, United Kingdom. Dr. Rajagopalis listed with biography in various international directories including Who's Who in the World and Outstanding Intellectual of the 21st Century since 2008.He has wide specialization on the marketing related topics that include competitor analysis, marketing strategy, consumer behavior, selling systems, international marketing, services marketing, and new product development. He teaches in undergraduate, graduate, doctoral programs, and executive MBA programs. Dr. Rajagopal holds graduate and doctoral degrees in Economics and Marketing respectively from Ravishankar University in India. He has to his credit 35 books on marketing and rural development themes and over 400 research contributions that include published research papers in national and international refereed journals. He has imparted training to senior executives and has conducted 56 management development programs. His research contributions have been recognized by the Government of Mexico and he has been awarded the status of National Researcher (SNI-level II) since 2004.

INDEX

T